COLLISIONAL PROCESSES IN THE SOLAR SYSTEM

ASTROPHYSICS AND SPACE SCIENCE LIBRARY

VOLUME 261

COLLISIONAL PROCESSES IN THE SOLAR SYSTEM

Edited by

MIKHAIL YA. MAROV

Keldysh Institute of Applied Mathematics,
Moscow, Russia

and

HANS RICKMAN

Astronomical Observatory,
Uppsala, Sweden

SPRINGER-SCIENCE+BUSINESS MEDIA, B.V.

Library of Congress Cataloging-in-Publication Data

Collisional processes in the solar system / edited by Mikhail Ya. Marov and Hans Rickman.
    p. cm. -- (Astrophysics and space science library ; v. 261)
    Includes index.
    ISBN 978-94-010-3832-4   ISBN 978-94-010-0712-2 (eBook)
    DOI 10.1007/978-94-010-0712-2
    1. Collisions (Astrophysics) 2. Solar system. I. Marov, Mikhail IAkovlevich. II.
Rickman, Hans. III. Series.

QB466.C64 C65 2001
523.2--dc21

                                                        2001029529

ISBN 978-94-010-3832-4

Printed on acid-free paper

TABLE OF CONTENTS

PREFACE

The exploration of our Solar System is rapidly growing in importance as a scientific discipline. During the last decades, great progress has been achieved as the result of space missions to planets and small bodies – asteroids and comets – and improved remote-sensing methods, as well as due to refined techniques of laboratory measurements and a rapid progress in theoretical studies, involving the development of various astrophysical and geophysical models. These models are based, in particular, on the approach of comparative planetology becoming a powerful tool in revealing evolutionary processes which have been shaping the planets since their origin. Comets and asteroids, being identified as remnants of planetary formation, serve as a clue to the reconstruction of Solar System history because they encapsulated the primordial material from which the planets were built up. At the same time, these interplanetary carriers of original matter and messengers from the past, being triggered by dynamical processes well outside our neighboring space, were responsible for numerous catastrophic events when impacting on the planets and thus causing dramatic changes of their natural conditions.

In the crossroads of astronomy and geophysics, recent years have seen a growing understanding of the importance of collisional processes throughout the history of the Solar System and, therefore, the necessity to get more insight into the problem of interactions of planets and small bodies. This importance is clearly manifested by the observed cratering records on planetary surfaces and such dramatic events as the explosions of comet D/Shoemaker-Levy 9 fragments in Jupiter's atmosphere in 1994, that of the Tunguska object over Siberia in 1908, and the Chicxulub event dating back to the end of the Cretaceous. The significance of impact processes in planetary evolution is demonstrated not only by the obvious fact that comets and asteroids scar the planetary surfaces, but also by the progressively collected evidence that they dramatically influenced the evolution of the atmosphere and hydrosphere of the Earth and atmospheres of other planets. Indeed, deposition of cometary-like volatiles through impact processes is considered to be an important contributor to the origin of secondary atmospheres on the inner planets. These impacts, at later times, probably have repeatedly

vii

interfered with the Earth's biological evolution and may once have been essential for even starting this evolution.

Currently part of the Near-Earth Objects (NEO) population, closely approaching the Earth's orbit, represents a potential threat to our civilization. At the same time, the asteroidal and cometary populations have been continuously shaped by planetary gravity, resulting in a preferential flux of meteorites originating in resonant zones of the main belt, and, further out in the Solar System, the formation of dynamical structure in the recently discovered Edgeworth–Kuiper belt populated by trans-Neptunian small bodies. In addition, comets are known to be continually tossed around by close encounters with the planets, and an understanding of their dynamics is necessary in order to assess the cometary inventory of the Solar System and its slow "evaporation" into interstellar space. Indeed, comets provide the link of our Solar System with the Galaxy in several respects, *e.g.*, through tidal perturbation of the Oort cloud by the Galactic mass distribution and thus modulation of comet flux thrown inward the Solar System, and by probing pristine material of the presolar cloud delivered by the comets. The various processes accompanying their interactions with the planets offer an opportunity to understand better both the nature of these primitive small bodies and their role in planetary cosmogony and Galactic evolution, thereby bridging the tiny place of our habitat to the entire Galaxy.

The explosive growth of knowledge in the field of space exploration and, in particular, the interdisciplinary character of the subject of collisions in the Solar System, put the scientific interests of planetary and Galactic astronomers, as well as specialists in geophysics, dynamics, and cosmochemistry, in a common focus. This is why the subject was included in the program of the XXIII General Assembly of the International Astronomical Union in Kyoto in 1997, where a Joint Discussion (JD 6) entitled "Interactions between planets and small bodies" was held. It offered an invaluable opportunity to gather a wide community for free and open-minded discussions. The topic and program of the meeting brought together astronomers dealing with both observations and theory and allowed to give an outlook to the formation and evolution of planetary systems in general, being also relevant to the problem of Galactic evolution. Short versions of some and an overview of the other presented papers and posters were published in 1998 in the IAU Proceedings (*Highlights in Astronomy*, **11A**, pp. 219–276, ed. J. Andersen, assisted by M. Marov and H. Rickman). Simultaneously, we supported the suggestion by Prof. W.B. Burton to publish a book focused on the collisional processes in the Solar System as a volume of the Astrophysics & Space Science Library (ASSL) book series. It allowed us to encourage the JD participants to publish their full size and upgraded papers and, additionally, to broaden up the team of contributors to the volume. We considered the idea to issue a book focused on the intriguing topic of interactions of Solar System bodies and their implications as a challenge and feasible to implement with appreciation of the colleagues we have invited. World recognized experts in the field are in the list of authors, which makes the book a comprehensive edition representing state-of-the-art knowledge of the collisional aspect of Solar System evolution, and allows to consider it as a valuable reference book.

As the work on the book approached its final phases, we received the sad news of the passing away of one of the authors, *Paolo Farinella*. His outstanding achievements and warm personality are described in a memorial added to his chapter by his co-author, D.R. Davis. In fact, his career paralleled the explosive growth of the field covered by this book while spanning most of its essential aspects. May his bright memory continue to inspire the young ones who enter the field in ever increasing numbers.

The overall content of the volume comprises 21 papers divided into five sections, which cover essentially all problems of primary importance and interest. These are: Impactors and Cratering in the Solar System; Comets and Asteroids: Observations and Models; Accretion and Formation Processes; Kuiper Belt; and Near-Earth Asteroids. The basic approach to the treatment of the problems involved is coherently conjoint with experimental evidence including such that became available only in recent years. Obvious highlights include the explosion in the rate of discoveries of trans-Neptunian objects in the Edgeworth-Kuiper belt and the beginning of physical studies of these objects, significantly increasing our understanding of the outer parts of the Solar System; the advent of comets Hyakutake and Hale-Bopp leading to a dramatic improvement in our knowledge of the physics and chemistry of comets; better understanding of the processes that have affected the asteroid population in the main belt and their injection into inner planet crossing orbits, including identification of new routes from the main belt to Earth by way of Mars-crossing orbits, with advancing capability to explain the observed number of large NEAs and Mars crossers, and the dynamical evolution and the collision probabilities for the Aten and Apollo groups; detection of water in the stratospheres of all the giant planets which implies infall of material to these planets associated with cometary encounters; and a great acceleration of the discoveries of NEOs larger than 1 km diameter due to the increasing capability of automated wide-field cameras and search software, allowing us to move toward a posture in which any future impact (including near-term threat to the Earth) can be accurately and cost-efficiently predicted to meet the objectives of the Spaceguard Survey.

In-depth study of collisional processes opens new horizons in the understanding of Solar System evolution. Most important, the concept of an intrinsically related and strongly interacting system of constituent bodies means that not only transport of material of different composition, which played a crucial role in the evolution of planets, but even collisions of catastrophic scale should be regarded as rather routine events throughout the Solar System history, in contrast with the basic idea of the Solar System as an ordered rather than chaotic entity. Indeed, the concept of chaotic planetary formation is increasingly supported by recent discoveries of unusual planetary-mass companions around other stars and also by new comprehensive computer modelling of the evolution of protoplanets. Both gravitational scattering and catastrophic collisions are identified as key constituents of chaotic processes responsible for the putative extrasolar planetary systems and, in an isolated case, for the unique configuration of the Solar System.

This new volume of the ASSL book series reflects significant advances in the study of a new field of Solar System exploration that has now reached a fascinating stage and continuously attracts growing attention and interest of specialists in astronomy, space science, geophysics and many other related fields. Our objective was to stimulate interaction among scientists of different disciplines in this particular field in order to integrate various approaches, promote a steady convergence of some controversial ideas towards an acceptable concept, and coordinate future studies. The topics covered by the volume are in symbiotic relation and the subject continues to develop at a remarkable pace. There is no doubt that coming years will demonstrate tremendous progress of such a multidisciplinary approach to fertilize a general understanding of Solar System origin and evolution.

The book would not have eventuated had it not been for the agreement of our colleagues to submit their papers in support of the project. Their efforts, colloborative spirit, and patience in due course of back and forth communication, reviewing and subsequent revision of the manuscripts are highly appreciated. We in no way attempted to modify the authors' original concepts or to use this book to promote our own ideas, with an obvious exemption for the two papers we have co-authored. Essentially every paper is based on its original version with the author's corrections in response of the reviewing remarks and with applying only minor editing.

We acknowledge Prof. Burton's initiative to undertake this effort. Our special thanks are due to Kluwer Academic Publishers for generous support in the book production.

M. MAROV   H. RICKMAN

SIZE-FREQUENCY DISTRIBUTIONS OF PLANETARY IMPACT CRATERS AND ASTEROIDS

B.A. IVANOV
Institute for Dynamics of Geospheres,
Russian Academy of Sciences
Leninsky Prospect 38/6, Moscow, Russia 117939

AND

G. NEUKUM AND R. WAGNER
DLR Institute for Planetary Exploration
Rudower Chaussee 5, Berlin, Germany

Abstract. The size-frequency distributions (SFD) for projectiles which formed craters on terrestrial planets and asteroids Gaspra, Ida, and Mathilde are compared using modern cratering scaling laws. The result shows the relative stability of these distributions during the past 3.7 Gy (Orientale basin and younger formations). The derived projectile size-frequency distribution is compared with the size-frequency distribution of main-belt asteroids. The recent Spacewatch data demonstrate the spectacular similarity of the size distribution of asteroids with diameters larger than 1 km and the population of crater-forming projectiles derived from the cratering data. Consequently one can suppose that the efficiency of the new projectile delivery to planetary crossing orbits does not depend on asteroid size. The migration of large main belt bodies to Mars-crossing orbits or to resonances seems to play an important role in the generation of planet-crossing impactors.

1. Introduction

The continuing slow evolution of the Solar System results in the exchange of matter between small bodies (asteroids and comets) and planets and their satellites. Some number of small bodies may be created as ejecta at large-scale planetary impacts. The chaotic evolution of small body orbits

1

M. Ya. Marov and H. Rickman (eds.), Collisional Processes in the Solar System, 1–34.
© 2001 *Kluwer Academic Publishers. Printed in the Netherlands.*

due to various resonances creates a flux of planet-crossing bodies. Planet-crossers can strike a planet forming a crater or a meteor. Impact craters and meteors yield footprints of Solar System evolution. These footprints give additional data to the traditional astronomical observations of small bodies.

Astronomical observations (and meteor records) give a snapshot of the small body population. The stability of the observed picture needs additional modeling of celestial mechanics problems to study the stability of orbits and possible evolutionary paths. Planetary cratering records show a picture integrated through the whole geologic lifetime of the studied surfaces. We need to compare surfaces of various ages on different planets to reveal spatial and temporal variations of the crater-forming projectile flux.

To compare crater distributions on different planets in a rigorous way one needs to take into account several parameters such as gravity, atmosphere, crustal strength, density and structure of these bodies, as well as differences in the projectile flux, size-frequency distribution and impact velocity spectrum. This problem is under investigation since a long time. Comprehensive discussions may be found in several excellent papers (Basaltic Volcanism, 1981; Hartmann, 1977; Strom and Neukum, 1988).

The goals of the present paper are:

- to estimate the size-frequency distribution (SFD) for the projectiles that produced the observed lunar crater population;
- to compare the impact crater SFD on terrestrial planets;
- to compare the crater-forming body ("projectile") SFD with the asteroid SFD for the Main Belt and planet-crossing asteroids.

2. General approach

2.1. SIZE-FREQUENCY DISTRIBUTION

The general approach to comparing the crater SFD with the projectile SFD has been formulated by Neukum and Ivanov (1994), who used the approach of one ("average") impact velocity and one ("average") impact angle. Here we attempt to extend this approach solving the inverse problem, transforming the lunar crater SFD into the corresponding projectile SFD. The model-derived projectile SFD is used to construct the size-frequency distributions for impact craters on other planetary bodies.

Suppose we have a dated lunar surface which has been exposed during a given time. Below we will use 1 Gy (10^9 years) as time unit. During this time a number of impact craters have been formed on the surface. To describe the size-frequency distribution of these craters we use the standard crater SFD derived by Neukum (1983). The basic form of this SFD is the cumulative number of craters, N, per km^2 with diameters larger than a given value D. For the time period of 1 Gy, $N(D)$ may be expressed (Neukum, 1983) as:

$$\log_{10}(N) = \sum_{j=0}^{11} a_j \times [\log_{10}(D)]^j \qquad (1)$$

where D is in km, N is the number of craters per km^2 per Gy, and the coefficients a_j are given in Table 1. "Old" and "new" coefficients are discussed below in Sect. 5. A similar equation is used here to present the projectile SFD derived below in Sect. 4. Coefficients for this projectile SFD are also listed in Table 1. In the projectile SFD column the first coefficient $a_0 = 0$ is taken for simplicity. This coefficient determines the absolute number of projectiles. The absolute value of a_0 for projectiles may be found by fitting to observational data (see Fig. 15 below).

TABLE 1. Coefficients in Eq. (1)

Coefficient	"Old" $N(D)$ (1983)	"New" $N(D)$ (1999)	$R(D)$ for projectiles
a_0	-3.0768	-3.0876	0
a_1	-3.6269	-3.557528	$+1.375458$
a_2	$+0.4366$	$+0.781027$	$+1.272521 \times 10^{-1}$
a_3	$+0.7935$	$+1.021521$	-1.282166
a_4	$+0.0865$	-0.156012	-3.074558×10^{-1}
a_5	-0.2649	-0.444058	$+4.149280 \times 10^{-1}$
a_6	-0.0664	$+0.019977$	$+1.910668 \times 10^{-1}$
a_7	$+0.0379$	$+0.086850$	-4.260980×10^{-2}
a_8	$+0.0106$	-0.005874	-3.976305×10^{-2}
a_9	-0.0022	-0.006809	-3.180179×10^{-3}
a_{10}	-5.18×10^{-4}	$+8.25 \times 10^{-4}$	$+2.799369 \times 10^{-3}$
a_{11}	$+3.97 \times 10^{-5}$	$+5.54 \times 10^{-4}$	$+6.892223 \times 10^{-4}$
a_{12}	$-$	$-$	$+2.614385 \times 10^{-6}$
a_{13}	$-$	$-$	-1.416178×10^{-5}
a_{14}	$-$	$-$	-1.191124×10^{-6}

Eq. (1) is valid for D from 0.01 km to 300 km. Other representations of the size-frequency distribution will be used below: (1) the differential distribution: dN/dD; (2) the R-distribution: $R = D^3 \times dN/dD$. The differential distribution is practically the number of craters δN in the diameter range from D to $D + \delta D$. As we have an analytical representation of $N(D)$, we can simply differentiate Eq. (1):

$$\frac{dN}{dD} = \frac{N}{D} \times \sum_{j=1}^{11} a_j \times [\log_{10}(D)]^{j-1} \tag{2}$$

where N is expressed by Eq. (1). Note: mathematically the derivative dN/dD is negative, since $N(D)$ is a decreasing function. However, for practical purposes, dN/dD should be the number of craters per unit interval of diameter, so it should have a positive value. Below we use the convention that dN/dD is positive and equal to the absolute value of that given by Eq. (2). The function $N(D)$ is close to a power-law D^m where m is in the range from -4 to -1.5. Consequently dN/dD is also close to a power-law with the index $(m-1)$. It is not very convenient for graphical presentation, as N and dN/dD vary over many orders of magnitude for the range of crater diameters of interest. To avoid this inconvenience one may use the so-called R-distribution, which is simply dN/dD multiplied by D^3:

$$R(D) = D^3 \times \frac{dN}{dD} \tag{3}$$

If dN/dD is proportional to D^{-3}, then $R(D)$ is a constant; if dN/dD is proportional to D^{-2}, then $R(D)$ is proportional to D; if dN/dD is proportional to D^{-4}, then $R(D)$ is proportional to D^{-1}, etc.

When one processes real discrete data on numbers of craters, he needs to use an agreement on the R-plot definition (see Arvidson *et al.*, 1978). Let us use bin boundaries of the crater counts such that D_i and D_{i+1} have a constant ratio. There is a standard agreement to take $D_{i+1}/D_i = 2^{1/2}$. A discretized form of the R-representation is

$$R(D_{av}) = \frac{2^{3/4}}{2^{1/2}} D_i^2 [N(D_{i+1}) - N(D_{i-1})] \tag{4}$$

where $D_{av} = (D_i D_{i-1})^{1/2}$. Note that the R value ascribed to the average bin diameter D_{av} is expressed through the left bin diameter D_i.

The same equations may be used to describe the size-frequency distribution of projectiles. Below we express projectile dimensions via the projectile diameter D_P. One may use Eqs. (3–4) for projectiles, changing D into D_P.

2.2. RESTORATION OF THE SIZE-FREQUENCY DISTRIBUTION FOR PROJECTILES

Let us assume that there is a population of projectiles in space distributed in size according to some law $N(D_P)$, where N is the number of projectiles with diameters larger than a given value, D_P. These projectiles strike the planetary surface and form impact craters, with a size-frequency relation $N(D)$.

To connect $N(D)$ and $N(D_P)$ one needs to know the relationship $D(D_P)$. In many previously published models $N(D)$ and $N(D_P)$ have been assumed to be some power-law dependences of the crater diameter, D, and on the projectile diameter, D_P. As the standard lunar cratering curve deviates from a simple power law, the model should take this fact into account. Below we demonstrate a general formulation of the problem of restoration of the projectile population from the crater size-frequency distribution.

Let the crater diameter, D, be connected with the projectile and impact parameters as

$$D = D(D_P, v, \alpha)$$

where D_P is the projectile diameter, v is the impact velocity and α is the impact angle (vertical impact corresponds to $\alpha = 90°$).

Let the projectiles have some velocity distribution, and thus the probability to have an impact velocity in the range from v to $v + dv$ is a function $f_v(v)dv$. The function $f_v(v)$ is normalized by

$$\int_{v_{min}}^{v_{max}} f_v(v)\, dv = 1$$

The minimum velocity of impact is equal to the escape velocity of the planetary body, v_{esc}. The maximum velocity depends on the orbital parameters of the target body and projectiles.

Let the projectiles have some distribution of the impact angle. The share of impacts in the range of angles from α to $\alpha + d\alpha$ is a function $f_\alpha(\alpha)d\alpha$ with the same normalization as for velocity

$$\int_{\alpha_{min}}^{\alpha_{max}} f_\alpha(\alpha)\, d\alpha = 1$$

The maximum impact angle is naturally 90° (vertical impact). The minimum value of the impact angle may be determined as the angle below which, at $\alpha < \alpha_{min}$, craters become elongated. From the restricted amount of experimental data (Gault and Wedekind, 1978) 15° may be used as an estimate of α_{min}.

Any crater of a given diameter D may be formed by impact of projectiles of various diameters, depending on the impact velocity and impact angle. So we cannot connect any crater of a diameter D with a single value of the projectile diameter D_P. This is possible only in the model of "one velocity – one impact angle" investigated by Neukum and Ivanov (1994). Beyond that model, to connect size-frequency distributions of craters and projectiles, we need to sum the number of all projectiles with all possible velocities and angles of impact, which form a crater of the same diameter D. On differential form this condition is expressed as

$$\frac{dN}{dD} = \int\limits_{v_{min}}^{v_{max}} dv \int\limits_{\alpha_{min}}^{\alpha_{max}} d\alpha \left[\frac{dN}{dD_P} \frac{dD_P}{dD} \right] f_v(v) f_\alpha(\alpha) \qquad (5)$$

where both values in square brackets should be calculated for the projectile diameter $D_P(D, v, \alpha)$, for which a crater of diameter D is formed upon impact with the velocity v at the angle α.

The integral equation (5) allows to calculate the projectile size-frequency distribution for known crater size-frequency distribution, $D_P(D, v, \alpha)$, and velocity and angle distributions of the impacts. For known projectile SFD and impact velocity spectra, Eq. (5) produces a model impact crater SFD (a production function) for a given planetary surface.

3. Scaling relations for impact cratering

According to the experimental and theoretical results for a given planetary body, the total range of crater diameters may be divided into several specific ranges, where different target parameters control the relationship between projectile mass and velocity and final crater dimensions. For smaller craters one may suppose the strength regime of cratering, where the final crater dimensions are governed by projectile parameters and the target material strength. Larger craters are formed in the gravity regime, where the energy needed to uplift target material in the planetary gravity field is larger than the energy dissipated due to plastic work. The largest craters experience structural modification after the transient cavity formation; modification processes give rise to craters with central peaks, multiring basins and other forms of complex craters and basins. Boundary diameters for the above listed regimes of cratering vary from planet to planet, and the ratio, for example, of crater rim diameter to projectile diameter is not a constant even for the same projectile velocity.

3.1. VERTICAL IMPACT – SIMPLE CRATERS

During the last 20 years large progress in the understanding of cratering scaling relations has been achieved. Reviews of the main ideas of scaling have been done recently by Melosh (1989), Schmidt and Housen (1987) and Holsapple (1993).

Here we use the most thoroughly elaborated approach of Schmidt and Housen (1987), which is partially confirmed by their experimental data from high-G centrifuge experiments. The approach is based on the concept of a coupling parameter (CP) which is a combination of projectile parameters and target density. The coupling parameter allows to compare the cratering efficiency for various impact conditions. For the formal convenience we introduce CP here as a dimensional parameter to show later the proper velocity scaling.

Let us use the coupling parameter, CP, in the form

$$CP = D_P(\delta/\rho)^\nu[v\sin(\alpha)]^\mu \qquad (6)$$

where D_P is projectile diameter, δ is projectile density, ρ is target density, v is impact velocity, ν and μ are experimentally derived exponents, and α is the impact angle (the effect of oblique impact is briefly discussed below in Sect. 3.3).

For strength craters Schmidt and Housen (1987) supposed a scaling relation of the form

$$\frac{\rho V}{m} = A\left(\frac{\rho}{\delta}\right)^{1-3\nu}\left(\frac{Y}{\rho v^2}\right)^{-3\mu/2} \qquad (7)$$

where m is the projectile mass, V is crater volume, Y is the effective target material strength, and A is a coefficient of proportionality. One can see in the Eq. (7) that the value of $\left(\frac{Y}{\rho}\right)^{1/2}$ plays the role of a velocity scale.

Here we rewrite this equation collecting all parameters of the projectile into a single value of the coupling parameter:

$$\frac{V}{(CP)^3} = A_V\left(\frac{Y}{\rho}\right)^{-3\mu/2} \qquad (7a)$$

where A_V is a coefficient of proportionality. Supposing geometric similarity of all craters, the crater radius, R, is proportional to the cube root of the volume, $V^{1/3}$. For such similar craters Eq. (7) may be rewritten in the form

$$\frac{R}{CP} = A_2\left(\frac{Y}{\rho}\right)^{-\mu/2} \qquad (8)$$

For gravity craters the scaling relationship by Schmidt and Housen (1987) is

$$\frac{\rho V}{m} = B_1\left(\frac{\rho}{\delta}\right)^{\frac{2+\mu-6\nu}{2+\mu}}\left(\frac{gD_P}{v^2}\right)^{\frac{-3\mu}{2+\mu}} \qquad (9)$$

where g is surface gravity, and B_1 is a numerical constant.

In contrast to the case of the strength crater the value of a scale velocity here is $(gD_P)^{1/2}$. It means that the impact velocity (tens of kms^{-1}) is compared with the free fall velocity over vertical distance of D_P in the gravity field. The alternative way is to chose as the scale of velocity the value $v_{scl} = \left(V^{1/3}g\right)^{1/2}$. This is the velocity of the free fall from the height comparable with some linear crater size.

Making the same rearrangement as before, one may derive:

$$\frac{V}{(CP)^3} = B_V \left(Vg^3\right)^{-\mu/2} \tag{9a}$$

Thus for geometrically similar craters

$$\frac{R}{CP} = B_2 \left(Rg\right)^{-\mu/2} \tag{10}$$

Eqs. (8) and (10) have the same functional form; all projectile related parameters are collected in a single coupling parameter; the right sides of these equations depend only on the target parameters.

For a given effective strength of the target and planetary gravity one may define a boundary crater radius, R_{sg}, where the R/CP ratio will be the same for strength and gravity presentations:

$$\frac{R_{sg}}{CP} = B_2 \left(R_{sg}g\right)^{-\mu/2} = A_2 \left(\frac{Y}{\rho}\right)^{\mu/2} \tag{11}$$

For a given Y/ρ and g, Eq. (11) defines an effective boundary between the gravity and strength regimes of cratering. Not far from R_{sg} strength and gravity give comparable inputs to the control of final crater dimensions, so one needs to combine Eqs. (8) and (10) to make the strength-gravity transition in a smooth way. The simplest procedure is to add both members with the exponent $-\mu/2$:

$$\frac{R}{CP} = \frac{B_2}{\left(A_3\frac{Y}{\rho} + Rg\right)^{\mu/2}} \tag{12}$$

where A_3 and B_2 are "evolved" constants from Eqs. (8) and (10). A similar approach has been proposed earlier by Holsapple and Schmidt (1979). The strength member $A_3(Y/\rho)$ may be rewritten in the form

$$A_3\frac{Y}{\rho} = R_{sg}g \tag{13}$$

With this definition Eq. (12) may be rewritten in the form

$$\frac{R}{CP} = \frac{B_R}{[g\left(R_{sg} + R\right)]^{\mu/2}} \tag{14}$$

or, for crater diameters,

$$\frac{D}{CP} = \frac{B_D}{[g\,(D_{sg} + D)]^{\mu/2}} \tag{14a}$$

or, for crater volumes,

$$\frac{V^{1/3}}{CP} = \frac{B_V}{\left[g\left(V_{sg}^{1/3} + V^{1/3}\right)\right]^{\mu/2}} \tag{14b}$$

In Eqs. (14, 14a, 14b) the values

$$[g\,(R_{sg} + R)]^{1/2}$$

$$[g\,(D_{sg} + D)]^{1/2}$$

$$\left[g\left(V_{sg}^{1/3} + V^{1/3}\right)\right]^{1/2}$$

play the role of velocity scales, which balance the dimensional form of the CP parameter introduced before in Eq. (6). The comparison of the scaling equations (14a,b) with experimental data is presented by Neukum and Ivanov (1994).

Figure 1 illustrates the geometrical meaning of Eq. (14) showing the gentle transition from strength to gravity regimes of cratering. The influence of gravity begins to be noticeable at crater diameters larger than $0.25D_{sg}$; the influence of strength may be important up to $D = 4D_{sg}$. The transition from simple to complex craters, also shown on Fig. 1, will be discussed later.

Numerical values of the strength-gravity boundary diameter D_{sg} are still poorly known. Data from centrifuge experiments by Holsapple and Schmidt (1979) presented in the form of Eq. (7a) give for the alluvium $D_{sg} = 6\pm2$ m and for the model oil-based clay $D_{sg} = 100 \pm 50$ m. Cratering data for shallow buried high explosive (HE) detonations in alluvium at the depth of burst (DOB) equal to one charge radius show $D_{sg} = 10\pm5$ m. HE explosions at the surface of clay soils produced craters, whose scaled volumes increased with the scale of the event. The strength-gravity transition here may be approximately estimated in the diameter range of 50 to 100 m for typical terrestrial soils.

Data for hard rocks are very limited but indicate a decrease of the effective strength as the scale of the events increases (Schmidt, 1980). The estimated value of D_{sg} in the range of 100 ± 50 m seems to be reasonable as a first approximation on Earth. Numerical simulation by Bryan *et al.* (1979) shows that gravity played an important role during formation of

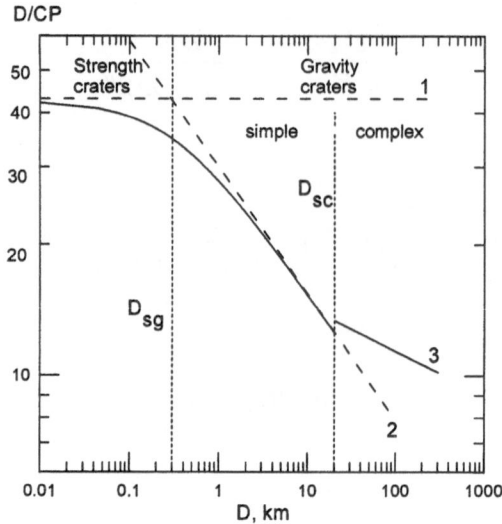

Figure 1. The ratio of the crater diameter, D, to the coupling parameter, CP (Eq. 6) for various regimes of impact cratering.

Meteor Crater, Arizona ($D = 1.2$ km). So the value of D_{sg} seems to be well below 1 km.

On other terrestrial planets the strength-gravity transition may be preliminarily fixed from morphometry. Housen *et al.* (1983) have shown that all simple gravity craters have geometrically similar rim profiles and maximum rim heights, h, linearly proportional to the crater diameters. (The same conclusion was published in Russian earlier by Ivanov, 1979). Based on the linear relationship $h(D)$ one may conclude that the value of D_{sg} may be as small as 300 m on the Moon (lowest diameter used by Pike, 1977) and 600 m on Mars (lowest diameter used by Pike and Davis, 1984). The same range for the lunar D_{sg} value (about 300 m) was estimated by Moore *et al.* (1974) from the radial extent of continuous ejecta (see also discussion in Croft, 1985). Scaled to the terrestrial gravity (Eq. 13), the above-mentioned values correspond to D_{sg} values of ~ 50 m which seem to be too small for rocky surfaces. It is hard to evaluate the accuracy of estimates based on linear $h(D)$ relations taking into account the gradual character of the strength-to-gravity transition and the wide scatter of rim height measurements both on the Moon and on Mars. At the same time dynamic weakening of rock during crater formation may lead to relatively small effective values of strength in comparison with the strength of single rock fragments. The possible dense fracturing due to meteor bombardment (on the Moon) and near surface weathering (on Mars) may result in small apparent values of target strength, which gradually increase with depth. If so, our model with a single value of effective strength would be too simple.

An upper limit of the effective strength may be deduced from modeling of the crater collapse. Melosh (1979) estimates the apparent strength (cohesion) of planetary crusts as around 3 MPa (30 bar) for all terrestrial planets. Note that some perspective collapse models treat the strength not as a target material constant but as a complex function of a crater size, gravity and other parameters (Melosh and Ivanov, 1999). The model used here is the simplest end-member model, which needs to be elaborated much more thoroughly in the future.

Here we use the lunar value of $D_{sg} = 300$ m and the gravity scaling to other planets (Eq. 13).

3.2. MODIFIED (COLLAPSED) CRATERS

Above some critical diameter, D_*, impact craters on all planetary bodies have so-called complex morphology (central peaks, concentric rings, etc.). Formation of complex craters seems to be connected with the modification of a transient cavity in the gravity field of a planet. The processes affecting the shape of the transient cavity after it reaches the maximum volume are generally referred to as modification processes, and the complex set of such processes may be referred to as the crater modification stage. In specific media like ductile metals, plasticene, or wet clay, the modification stage occurs as a stress relaxation which does not change dramatically the final shape of the transient cavity. Craters in these materials are close to hemispherical. Thus the main morphometric parameter, the ratio of crater depth to crater diameter, is close to 0.5.

In media like sand or fragmented rocks there are two mechanisms that modify the transient cavity shape. The first mechanism is dry friction leading to growth of the transient cavity after the moment when the cavity depth reached its final value. The second mechanism is the slumping of the steep transient cavity walls under the action of gravity. These two mechanisms result in a similar crater shape close to a paraboloid of revolution with the depth to diameter ratio in the range from 0.2 to 0.25. These craters are named "simple craters". Both the aforementioned mechanisms can act together, but typically the first one acts mostly for small-scale impact and explosion laboratory craters 0.1 to 1 m in diameter. The second mechanism acts for large natural craters with diameters above 100 m approximately. Wall slumping is the main modification process for simple craters. The result of slumping is a breccia lens that overlies the true floor of the crater, which is the remnant of the transient cavity floor. For example, the well known Meteor Crater, Arizona, with a diameter of 1.2 km has a depth of the true floor (called true depth) of approximately 300 m (true depth to diameter ratio of 1/3). The true floor is covered by a breccia lens with a thickness of 150 m. The resulting apparent depth of the crater is $300 - 150 = 150$ m below the original target surface. The crater depth measured from the uplifted rim crest to the visible (apparent) crater floor is finally 0.19, expressed as depth/diameter ratio (Roddy, 1978).

The largest simple impact crater found on Earth is the crater Brent (Canada) with a diameter of 4 km. On the Moon all craters with diameters less than ~ 15 km are simple ones with a constant ratio of the rim to visible floor depth to the rim crest diameter of 0.2 (*e.g.*, Pike, 1977).

Larger impact craters begin to be complex craters. Complex craters have smaller depth-to-diameter ratios than simple craters. The geological study of terrestrial complex impact craters shows the uplift of sub-crater layers above the pre-impact level (see an introductory set of papers in Roddy *et al.*, 1977). At the center of the crater this uplift creates a central mound or central peak. A ring depression (or circular trough) surrounds the central mound. The ring depression is filled with fragmented material (allogenic breccia) and impact melt. On other planets one can see only the summit of the central mound above the flat level of a relatively flat breccia/melt surface. The central mound manifests the main modification mechanism for complex crater formation: the uplift of a transient cavity bottom (some researchers use the term rebound). The bottom uplift is accompanied by subsidence of the crater rim. In gross the process may be called a transient crater collapse.

In craters of larger diameters the central mound may have a central depression on the top. The rim of this central depression forms an inner crater rim. Such craters are called double ring craters.

The largest craters studied on other planets have a complex structure with three or more rings. These craters are called multi-ring basins.

The systematic change of crater morphology (simple bowl-shaped $\rightarrow$ double ringed $\rightarrow$ multi-ringed) has been found on all planetary bodies with a solid crust. For terrestrial planets (Mercury, Venus, Earth, the Moon and Mars) the critical diameters of these morphological transitions are approximately in inverse proportion to the surface gravity (Pike, 1980). Hence one may suppose that gravity is a main driving force of transient crater collapse. Investigations of transient cavity collapse in a gravity field shows that the process may be modeled with traditional rock mechanics if one ascribes very specific mechanical properties to the rock in the vicinity of a crater: the effective strength of rock needed is around 30 bar (Melosh, 1977); the effective angle of internal friction is below 5° (McKinnon, 1978). Rock media with such properties may be referred to as "temporarily fluidized". The nature of this "fluidization" is as yet poorly understood (see the review of hypotheses by Melosh, 1977). Melosh (1979, 1982) suggests an acoustic (vibration) nature of the fluidization. This model now seems to be the best approach to the problem (Ivanov and Melosh, 1999).

An open question is how to implement Melosh's model (or other possible models) into hydrocodes for numerical simulation of dynamic crater collapse. Some first attempts were presented by Ivanov and Kostuchenko (1997).

The gravity-aided collapse of a transient cavity makes the diameter of the final modified crater larger than for a hypothetical simple crater formed by the same impact in the absence of the gravity collapse. Before appropriate progress in the understanding of crater collapse will be achieved, one can use semi-empirical models to describe the complex crater formation. Currently one can use Croft's model (Croft, 1985; see also Chapman and McKinnon, 1986) to find the transient cavity diameter, D_t, for an observed crater with a rim diameter D:

$$D_t = D_*^{0.15} D^{0.85} \tag{15}$$

for $D > D_*$. The value of the critical diameter D_*, which defines the critical crater diameter where the collapse begins, depends on the target material strength and gravity. On Earth the value of D_* is around 4 km for crystalline rocks. For other terrestrial planets D_* varies approximately in inverse proportion to the surface gravity (Pike, 1980).

Using (15) to estimate D_t, and using (14a) to find the value of CP and (6) for relating to an assumed impact velocity and impact angle, one can estimate the projectile diameter needed to form a crater with a given final diameter D. Numerical modeling of the impact events allows to calculate the impact melt volume. Comparison of observed and calculated volumes of impact melt for large terrestrial craters serves as the best known way to check the approach discussed above. A reasonably good fit between theory and observations gives confidence in the validity of the scaling laws presented here (Ivanov, 1981; Pierazzo et al., 1997). For gravity craters a simpler expression may be derived:

$$D_t = 1.16(\rho/\delta)^{1/3} D_P^{0.78} (v \sin \alpha)^{0.43} g^{-0.22} \tag{16}$$

(see, for example, Pierazzo et al., 1997).

A general review of the simple-to-complex transition has been published by Pike (1980). Most of these data reflects boundary diameters between ranges of different morphology styles of impact craters. Now it is hard to say what is the best value of D_* to use in Eq. (15) for estimating the crater diameter increase due to modification on a given planetary body. As a first approximation for terrestrial planets one may use the inverse proportionality of D_{sg} and D_* to gravity g (Pike, 1980).

3.3. OBLIQUE IMPACT

The main experimental results for the cratering efficiency of oblique impacts have been published by Gault and Wedekind (1978). For small-scale laboratory craters they derived two trends of the cratering efficiency of an oblique impact with the angle α (measured from the horizontal surface) in terms of the ratio of the crater volume formed by oblique impact, $V(\alpha)$, to the crater volume, $V(90°)$, formed by the vertical impact of the same projectile with the same velocity. According to these experiments, for granite

$$\frac{V(\alpha)}{V(\alpha = 90°)} = (\sin \alpha)^2 \tag{17}$$

and for a non-cohesive sand

$$\frac{V(\alpha)}{V(\alpha = 90°)} = \sin \alpha \tag{18}$$

According to the coupling parameter approach the angular dependence of the cratering efficiency may be added directly to the expression for CP (Eq. 6). Following the analysis of the problem by Melosh (1989), here we use the simplest assumption that the cratering efficiency of an oblique impact depends on the vertical component of the impact velocity, $v \sin \alpha$.

Following Gilbert (1893) and Shoemaker (1962), we assume the angular distribution of planetary impacts in the form

$$f_\alpha(\alpha) = dN/d\alpha = \sin 2\alpha \tag{19}$$

This distribution has a maximum at $\alpha = 45°$.

One should mention here that Eq. (19) describes the angular distribution of impacts of projectiles of a given diameter. As the crater size decreases when the angle of impact decreases, larger (and less numerous) projectiles are needed to create a crater of a given diameter. Consequently, the angular distribution of impacts for craters of a fixed diameter differs from Eq. (19). For a projectile SFD in the form of a power law

$$N(D_P) = A \times D_P^{-n}$$

with a constant exponent, the average angle of impact for a specific crater diameter varies from 58° for $n = 4$ to 52° for $n = 1.5$. This effect results in a smaller fraction of craters created by oblique impacts among craters of the same diameter in comparison with Eq. (19).

4. Distribution of impact velocity

One may list three more or less reliable sources for establishing such a distribution:

1. Measured velocities of small meteors. The best collection known to the authors has been published in Russian by McCrosky et al. (1979). This catalog of bolides includes several hundred events. The question is how independent are these data and what is the input of meteor showers with close orbits and impact velocities.

2. Astronomical catalogs of Earth-crossers (Shoemaker, 1977; Rabinowitz et al., 1994). These data need to be recalculated from orbital parameters to impact probabilities and velocities. Shoemaker did this with Öpik's theory (see, for example, Shoemaker and Wolfe, 1982). A larger compendium of observational data may be found in the "astorb.dat" file updated at Lowell Observatory (http://naic.edu/ nolan/astorb.html). This file contains orbital osculating elements for $> 30,000$ small bodies including a relatively large number of planet-crossing asteroids: 18 for Mercury, 99 for Venus, 218 for Earth/Moon, and 450 for Mars. These data seem to have a large observational bias; however we use them to make some statistically meaningful estimates.

3. A theoretical model by Rabinowitz (1993) who tried to estimate the total ("debiassed") population of Earth crossers from observational data. The aforementioned Öpik model may be used to convert the Rabinowitz theoretical prediction into impact velocity and probability data.

The astronomical approach has a specific disadvantage: the lifetime of observed orbits is much smaller than the collisional time scale. Within their lifetime orbits may evolve significantly (see, for example, Milani et al., 1989). So we are forced to apply the idea of some kind of equilibrium: the exchange of different "cells" of the phase space of planet crossers (semimajor axis a, eccentricity e and orbit inclination i) is in a steady state: if a body with current parameters a, e, i changes the orbit due to some perturbation or resonant interaction, statistically, another body will change its orbital parameters into the same a, e, i.

Table 2 summarizes the impact velocity distributions based on data listed above for the Earth/Moon system, average velocities of impact and the average gravitational focusing factor

$$g_{enh} = 1 + (v_{esc}/U_{inf})^2$$

of the projectile flux enhancement due to the gravitational attraction to the target body (here v_{esc} and U_{inf} are the escape velocity and the average velocity of projectiles "at infinity"). The average velocities are similar for

TABLE 2. Comparison of impact velocity frequencies on Earth and on the Moon binned in 2 km/s intervals

v_{imp} (km s^{-1})	McCrosky et al. (1979)	Shoemaker (1977)	Estimates based on Rabinowitz (1993)	astorb.dat (1998)
11	0.017	0	0.048	0
13	0.091	0.138	0.130	0.2506
15	0.122	0.272	0.145	0.2498
17	0.138	0.244	0.142	0.1910
19	0.107	0.112	0.130	0.1095
21	0.11	0.087	0.0819	0.0671
23	0.093	0.045	0.0582	0.0515
25	0.074	0.027	0.0692	0.0094
27	0.037	0.03	0.0575	0.0307
29	0.064	0.012	0.0293	0.0125
31	0.06	0.024	0.0292	0.0208
33	0.032	0	0.0353	0.0034
35	0.032	0.012	0.0166	0.0030
37	0.013	0	0.0120	0.0003
39	0.001	0	0.00969	0.0000
41	0	0	0.00447	0.0004
43	0	0	0.00022	0
The average impact velocity on Earth:				
$\{v_{imp}\}$ (km s^{-1})	21.49	17.95	20.18	17.3
g_{enh}	1.53	1.79	1.66	1.7
Same ("at infinity") projectile flux on the Moon:				
$\{v_{imp}\}$ (km s^{-1})	20.17	15.48	19.05	14.7
g_{enh}	1.02	1.04	1.03	1.06

all sets of data, giving somewhat larger velocities for observed meteors and the debiased population of Earth crossers.

For other planets we can use only the observed projectile population, and most estimates here are done with the "astorb.dat" data file. Figure 2 shows impact velocity spectra for terrestrial planets used in this paper for first-order estimates. Such spectra binned into 2 km s^{-1} intervals (Table 2 and Fig. 2) were used to find the projectile SFD from the lunar data and impact crater SFD on Mercury, Venus and Mars.

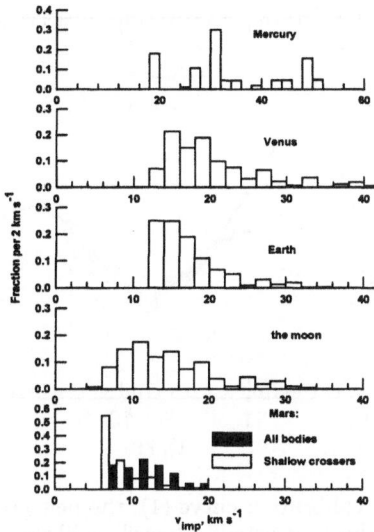

Figure 2. The impact velocity spectra for terrestrial planets used in this paper for first-order estimates.

5. The projectile size-frequency distribution

Combining the estimates discussed above, one can solve numerically Eq. (5) and find the projectile SFD, dN/dD_P, for a given impact crater SFD, dN/dD. To simplify the problem in this paper, we test the end-member hypothesis of a purely asteroidal projectile flux onto the terrestrial planets. Having such an estimate, we can later return to the problem of the cometary fraction in the observed crater population. For the same reason we assume the same projectile density ($2.7 \ \mathrm{Mg\,m^{-3}}$) for all estimates.

Eq. (5) was solved numerically for the analytical input of the lunar impact crater SFD in the form of Eq. (2). To facilitate the planetary estimates, the solution is approximated in the same form as Eqs. (1) and (2) with a polynomial of 14th degree valid for projectile diameters from 1 m to 100 km. The estimates for the largest projectile sizes should be used with caution, as the source crater SFD in this range is based on the largest lunar basins, all of which are very old and do not appear in younger crater populations.

Polynomial coefficients for the R-plot are listed in Table 1. The estimated SFD is used below to produce a model ("lunar analogue") for Mercury, Venus, Earth, and Mars. The projectile SFD is also compared with the recent data on the main belt SFD. In Sect. 7 the projectile SFD will be shown graphically.

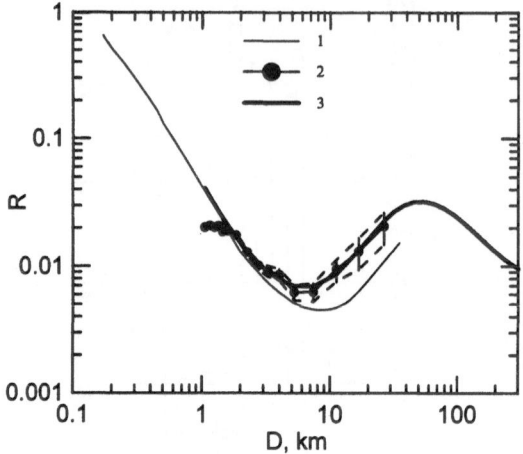

Figure 3. The "old" lunar calibration curve (1), the new crater count for the Orientale basin with error bars, and the "corrected" ("new") calibration curve (3). For $D < 2$ km and $D > 20$ km the "new curve" has the same shape as the "old" one.

6. Inter-planet comparison

In this section we compare the "theoretical" planetary SFD, derived above from the lunar data, with published measurements for selected areas on Mercury, Venus, Earth, and Mars. The most convenient areas to illustrate the projectile ("production") SFD are surfaces once renovated by a minor resurfacing after the main "renovation" event.

The Moon. The new count of 6700 lunar impact craters in the area of 7.8×10^5 km^2 renovated by the Orientale impact allows to re-estimate the size-frequency distribution (Ivanov *et al.*, 1999). The curve may be approximated by a 11th degree polynomial for 100 m $< D <$ 200 km (Fig. 3). "New" coefficients in Eq. (1) are listed in Table 1 in comparison with "old" coefficients from Neukum (1983).

Figure 4 illustrates the new Orientale count and the crater SFD fit to this data and to the typical highland count near the Tsiolkovsky area, farside, published by Ronka *et al.* (1981). Most of the craters were formed > 3 Gy ago, so the question remains how the projectile population has evolved since the Orientale formation time. The problem of the stability in time of the production SFD for lunar craters is under discussion since many years (*e.g.*, Strom, 1977; Hartmann, 1984; Strom and Neukum, 1988; Neukum and Ivanov, 1994). Here we show only selected examples of lunar SFD distributions of various ages to illustrate the main features of the non-power-law SFD. Figures 5 and 6 illustrate published crater counts for Mare Imbrium and Mare Crisium in comparison with the proposed standard distribution.

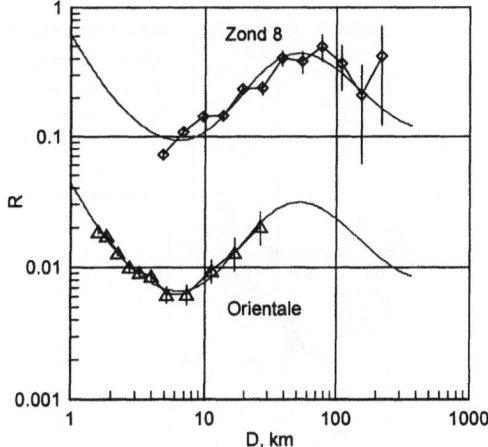

Figure 4. The SFD analytical fit to the lunar mare and highland crater counts.

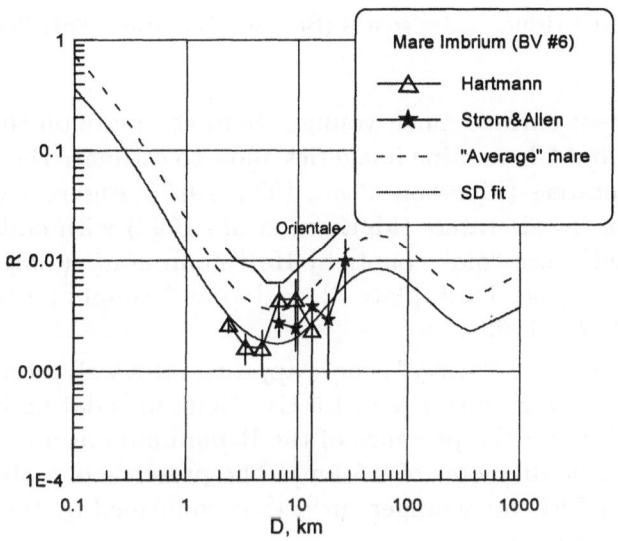

Figure 5. Mare Imbrium counts (Basaltic Volcanism, 1981).

While the visible misfit of some counts should be discussed separately (see, for example, the discussion of a possible resurfacing by Hartmann, 1995), here we point out the presence of a definite R-minimum in the R-plots. Therefore, the general deviation of the mare crater SFD from a simple power law is well illustrated by individual crater counts. It seems that the suggested "flattening" of the "average mare" crater SFD is the result of the averaging of individual crater counts for separate areas. It is difficult

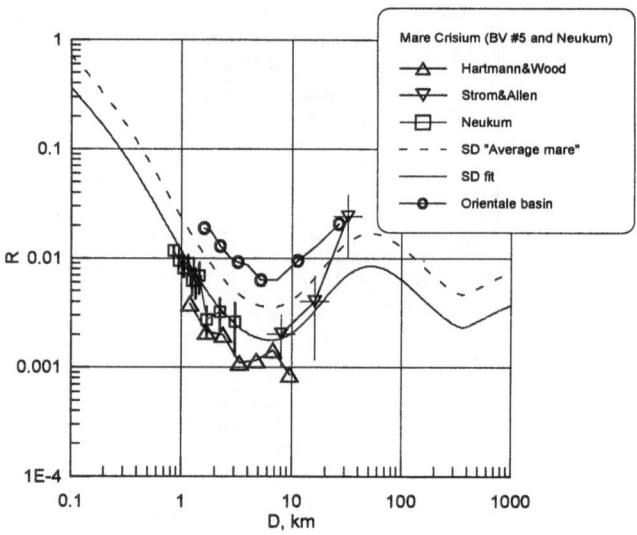

Figure 6. Mare Crisium crater counts (Basaltic Volcanism, 1981; Neukum, 1980).

to find a coherent surface much younger than the mare on the Moon. The recent Galileo and Clementine imageries allow to estimate the SFD of rayed (Copernican) craters (McEwen *et al.*, 1993, 1997). Figure 7 compares the SFD for farside rayed craters (McEwen *et al.*, 1997) with crater counts for Copernicus itself – see "old" counts by Hartmann *et al.* (1981; see Basaltic Volcanism, 1981, page 1112, plate 3), and "new" counts by Neukum cited by McEwen *et al.* (1993).

The non-power-law ("wavy") curve approximates well the size-frequency distribution for craters with age of 1.5 Gy (Neukum's dating in McEwen *et al.*, 1993), confirming the presence of the R-minimum at a crater diameter of 6 km (projectile diameter of 0.5 km). The presence of a steep branch of the lunar crater SFD for younger surfaces is confirmed by the crater count for Aristarchus (Fig. 8).

Mercury. The most convenient areas to illustrate the projectile ("production") SFD are surfaces once renovated by a minor resurfacing after the main "renovation" event. The mare surface in the Caloris basin is one of few suitable areas on Mercury. Figure 9 compares direct measurements (Basaltic Volcanism, 1981) and calculated SFD (the "lunar analog"). The good coincidence of these data shows a definite similarity of projectile SFDs on the Moon and Mercury in the projectile diameter range from 1 to ~ 100 km with a steep part for smaller craters and the "R-minimum" for craters with $D \sim 8$ km. However, the age of the Caloris basin is comparable to the age of the Orientale basin. For this reason it would be very instructive to make

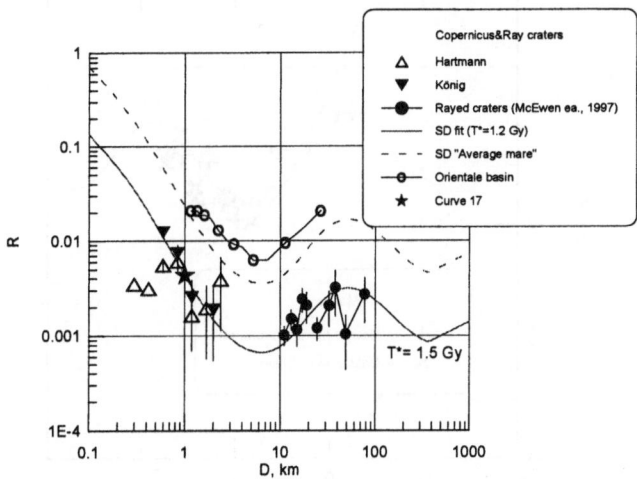

Figure 7. The SFD for farside rayed craters (McEwen *et al.*, 1997) with crater counts for Copernicus itself.

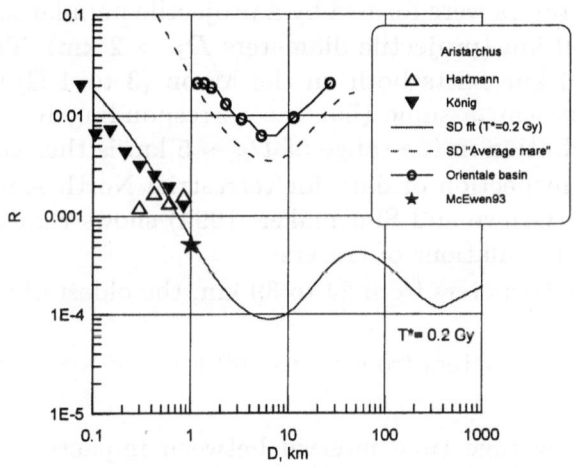

Figure 8. The crater count for Aristarchus.

a similar comparison for younger surfaces.

Venus. Magellan data allow to compare the lunar data averaged over the last 3 Gy with a planetary surface of $\sim$ 0.5 Gy age. The presence of the atmosphere may be taken into account using a model of projectile

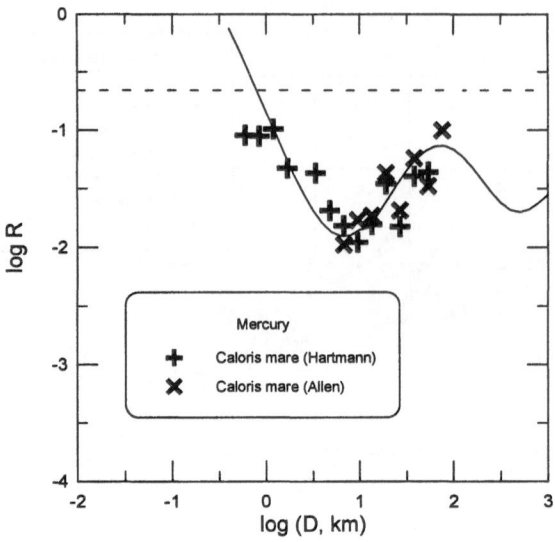

Figure 9.   The crater count for the Caloris mare basin on Mercury.

atmospheric passage. A model and model results for Venus and Titan were presented by Ivanov *et al.* (1997). The resulting comparison (Fig. 10) shows that Venusian craters were formed by a projectile population with a similar SFD for $D > 10$ km (projectile diameters $D_P > 2$ km). The R-maximum at $D \sim 50 - 70$ km exists both on the Moon (3 to 4 Gy) and on Venus (~ 0.5 Gy). One can assume that the corresponding R-maximum in the projectile distribution in the range of $D_P \sim 5$ km is thus stable in time.

Earth. The inspection of data for terrestrial North American and European cratons (Grieve and Shoemaker, 1994) shows that it is possible to distinguish two populations of craters:
– 8 craters with diameters from 24 to 39 km, the oldest of which is 115 My old, and
– 8 craters with diameters from 55 to 100 km and ages ranging from 214 to 370 My.

Adding the average time interval between impacts to the oldest age in each set, one can estimate the time of accumulation as ~ 135 My and 380 My, respectively, for the younger and older populations. For a proper balance between crater diameter bin width and the number of craters per bin, only two bins for each age sub-population may be used to represent the crater production rate.

We assume that craters smaller than ~ 20 km in the younger set and smaller than ~ 45 km in the older set are depleted by erosion processes. Figure 11 shows the incremental plot for these data sets in comparison

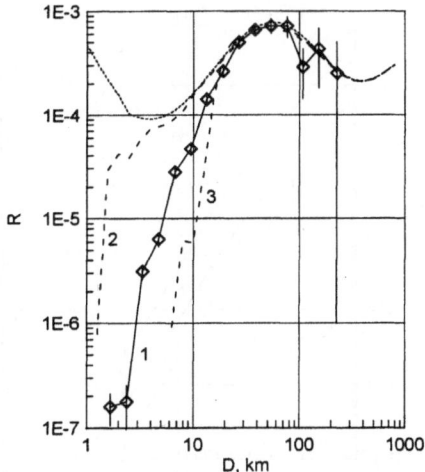

Figure 10. The R-plot for the size-frequency distribution of Venusian craters (1) in comparison with the lunar curve recalculated for the Venusian conditions with the Schmidt-Holsapple scaling law and the Croft model (Eq. 15) with $w=0.85$ (2) and $w=0.92$ (3).

with their "lunar analogues" calculated for assumed values of accumulation time and a constant flux (see a discussion of the possible flux variability by Grieve and Shoemaker, 1994, and McEwen *et al.*, 1997). As an illustrative estimate, the "lunar analogue" for the Sudbury age is also plotted here to show that, statistically, Sudbury may be the largest impact crater formed in the area of 17.6×10^6 km^2 – the total area of measured cratons.

Figure 12 shows the R-plot of terrestrial data recalculated to the reference age of 1 Gy assuming a constant crater production rate. This assumption does not contradict the lunar crater SFD recalculated to the terrestrial conditions.

The scarcity of statistically reliable terrestrial data prevents any opportunity to check the shape of the SFD curve; however, the density of well dated terrestrial craters witness in favor of minor flux variations during the last 0.5 Gy.

Small projectiles cannot form impact craters, being decelerated and fragmented in the atmosphere. However, these bodies give rise to meteor and bolide phenomena. During the last decade satellite monitoring has allowed to estimate the energy of several tens of bright bolides, giving an opportunity to count the flux of bodies (Nemtchinov *et al.*, 1997). Assuming that all bolide projectiles have the average NEA velocity and probability of impact, one can estimate the modern population of small NEAs. Below we compare these data with the lunar-derived projectile population (see Fig. 15).

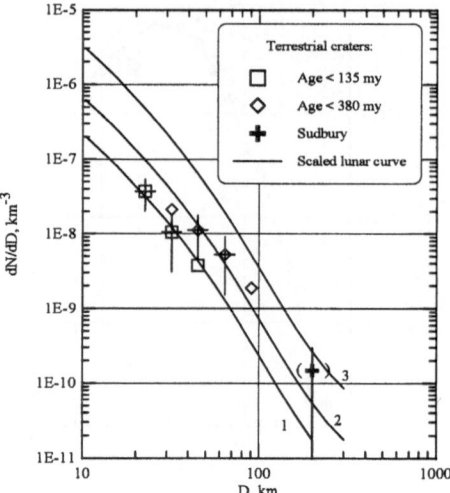

Figure 11.   The differential SFD for terrestrial craters in comparison with data for cratons (North American + European).

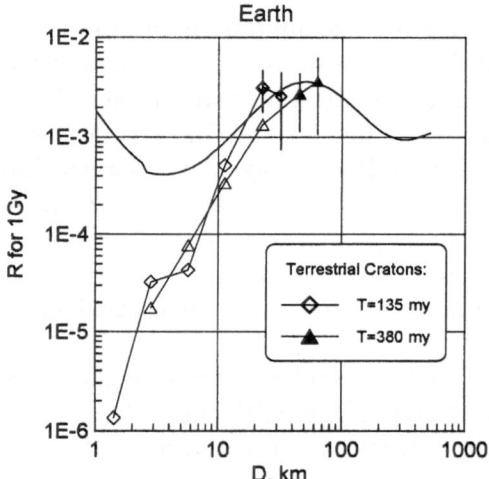

Figure 12.   The same data as in Fig. 11 in a R-plot. Data are divided by 0.135 and 0.380 to put them at the 1 Gy position for checking the flux variation.

Mars. The new data from Mars Global Surveyor (Hartmann *et al.*, 1999a,b) show the same trends as for other terrestrial planets (Figs. 13, 14). The small craters on Mars have a large rate of obliteration, so one should add some model of crater "equilibrium" to analyze the data. How-

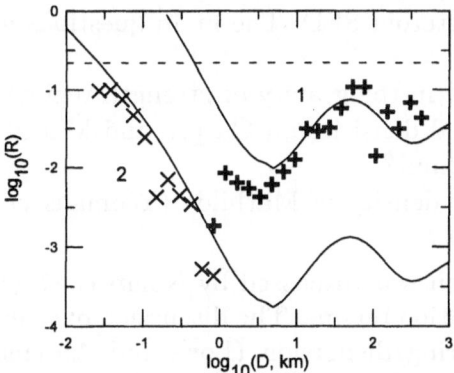

Figure 13. Mars: the crater SFD for heavily cratered terrain (1) and a relatively young volcanic caldera floor (Hartmann *et al.*, 1999a).

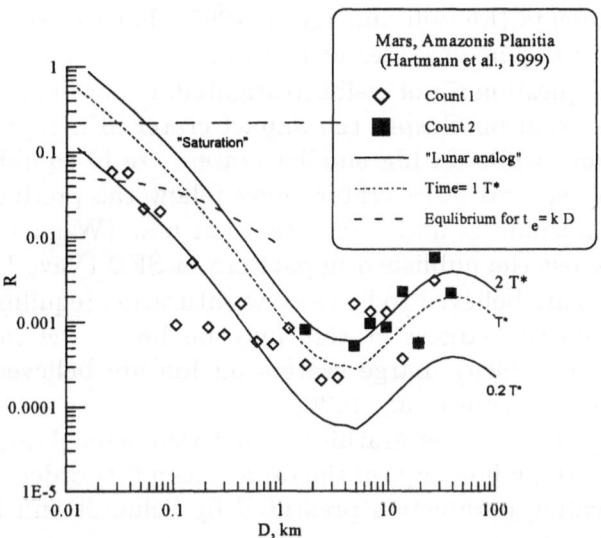

Figure 14. Crater counts for Amasonia Planitia on Mars. T_* marks a reference position of the SFD curve, calculated assuming the same production flux as on the Moon.

ever, the presence of the same R-maxima and R-minima in the SFD is clear.

Craters on asteroids. The Galileo and NEAR spacecraft returned images of three asteroids: Gaspra, Ida, and Mathilde (Belton *et al.*, 1992, 1994; Chapman *et al.*, 1996a,b; Veverka *et al.*, 1997). All three bodies are covered by impact craters. An up to date discussion of asteroid impact cratering is given by Bottke and Greenberg (this volume). Assuming an average velocity of impact of 5.5 km/s in the Main Belt, it is possible to estimate the small (projectile) asteroid SFD. The main questions for the projectile size estimates are:

– are craters formed in the gravity or strength regimes?

– are the crater areal densities on Gaspra and Mathilde close to the saturation/equilibrium limit?

– how does the low density of Mathilde (Yeomans et al., 1997) effect the cratering scaling law?

The first question was discussed by Nolan et al. (1996) in favor of the gravity crater formation regime. The discussion of strength/gravity regimes for asteroid shattering/dispersion (Love and Ahrens, 1995; Melosh and Ryan, 1997) confirms that the dispersal of asteroids due to catastrophic collision may be controlled by self-gravity for bodies as small as 0.5 to 1 km in diameter.

Consequently, all studied asteroids may be constructed of reaccreted shattered fragments (Melosh and Ryan, 1997). For this reason we assume here the gravity regime of crater formation.

The second question is not well investigated. Following Chapman (1995) we assume here that on Gaspra the impact crater SFD represents the production function, while on Ida smaller craters are in equilibrium (saturation), and only several large craters may follow the production function. The case of Mathilde is under investigation now (Wagner and Neukum, 1999). Here we use the published impact crater SFD (Veverka et al., 1997).

Ida's craters are believed to be close to saturation (equilibrium) at small diameters, while the largest craters may be below the saturation limit (Chapman et al., 1996a). Large craters on Ida are believed to be in the gravity regime (Asphaug et al., 1996).

The scaling of craters on Mathilde is not well defined due to the unusually low density (high porosity) of the target. As a first order approximation, here we use scaling parameters presented by Schmidt and Housen (1987) for the loosest soil.

With the aforementioned assumptions a model projectile distribution for all imaged asteroids may be constructed (Fig. 15). For comparison the lunar-derived projectile SFD is also shown (dashed line in Fig. 15). The model results for craters on asteroids seem to demonstrate the presence of the R-minimum of the projectile SFD curve approximately in the same range of projectile diameters as for NEAs.

7. Size-frequency distribution of asteroids in the Main Belt

Earth-based astronomical observations and the satellite infrared survey (IRAS) have revealed all Main Belt asteroids with diameters larger than about 40 km (van Houten et al., 1970; Gradie et al., 1989; Cellino et al.,

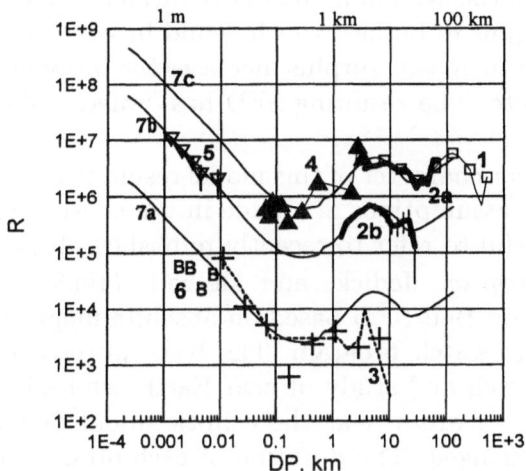

Figure 15. The R-plot for observed+PLS asteroids (1) in comparison with the Space-watch data (Jedicke and Metcalfe, 1998) for all Main Belt (2a) and inner belt (2b) asteroids, the derived projectile population for Mathilde (4) and Gaspra (5), the estimated NEA population (Rabinowitz *et al.*, 1994) and preliminary data on bolides observed from satellites (Nemtchinov *et al.*, 1997). The NEA estimates derived from cratering on terrestrial planets is shown for comparison with NEA data (7a). The same curve is raised to fit the data for the inner belt (7b) and the Main Belt (7c).

1991). For smaller diameters one usually supposes a power-law SFD: the differential distribution (number dN of asteroids in a range of diameters from D to $D + dD$)

$$dN/dD \propto D^{-k} \tag{20}$$

where the value of k may vary from 2.95 (the so-called PLS-slope, after the Palomar-Leiden Survey – see van Houten *et al.*, 1970) up to 3.5 (the so-called Dohnanyi slope – see Dohnanyi, 1969). $k = 3.5$ is a typical value for a self-similar cascade of fragments. Davis *et al.* (1994) used a geometrical average of two possible power-law distributions – the PLS distribution and estimations by Cellino *et al.* (1991), to analyze the IRAS data.

Deviations from a simple power-law SFD of impact craters, considered above, suggest that the asteroid SFD also deviates from a simple power law at diameters smaller than the limit of completeness of detection. For large bodies (~ 100 km in diameter) the non-power-law SFD is usually discussed as an intrinsic feature of the initial distribution of small bodies before the Main Belt accumulation (Davis *et al.*, 1985). A possible mechanism for such deviations is based on results of modeling of impact evolution in the Main Belt – some models give a "wavy" SFD (Campo Bagatin *et al.*, 1994a,b).

In these models the fragmentation cascade is depleted in the smallest fragments, which leave the system by non-gravitational forces. Thus at a larger size there is a surplus of bodies, which would be destroyed by these "missing" impactors. In turn, this surplus increases the probability of destruction of larger bodies, etc. The resulting SFD has "waves" of deviations from a simple power law.

Despite the clear character of this model result, it is necessary to keep in mind that several assumptions are made in the construction of the model. Therefore it is useful to refer to recently published observational results.

Spacewatch program. Jedicke and Metcalfe (1998) have published an analysis of the Main Belt SFD based on absolute magnitudes measured as a part of the Spacewatch program. The basic purpose of the program is the automated search and study of near-Earth asteroids (NEA); however, during the survey of areas near the ecliptic about 60,000 Main Belt asteroids were also imaged. The duration of each observation was too small to determine the orbit, and statistical methods were used to estimate the distribution of orbital parameters from that of the magnitudes. The application of standard methods of the observation imcompletness modeling have allowed to estimate the distribution of absolute magnitudes in the Main Belt. The SFD is estimated from magnitudes using relative numbers of asteroids of various types and corresponding typical albedos in the Main Belt (Gradie *et al.*, 1989). The total sensitivity permits SFD estimates for asteroid diameters above 2 km. For comparison, the more direct IRAS SFD estimates are given for diameters exceeding ~ 20 km (Cellino *et al.*, 1991). However, it is necessary to note that estimations by Jedicke and Metcalfe (1998) for bodies with absolute magnitude $H < 11.5$ exceed the number of actually observed asteroids by a factor of about 100.

In Fig. 15 data from direct observations (following Davis *et al.*, 1989) and estimations by Jedicke and Metcalfe (1998) are shown. The SFD of projectiles, received above from the SFD for impact craters, is used to approximate the data. Comparing these data one can note the following:

1. The direct astronomical data show a relative R-minimum at asteroid diameters $D \sim 30-40$ km; the depth of this minimum may vary for different semimajor axes a.

2. Estimations for the diameter range < 10 km demonstrate the presence of a second R-maximum at $D \sim 4-5$ km. This maximum is well visible in the inner and mid Main Belt ($2.0 < a < 2.6$ AU) and may be assumed for the exterior zone, being on the limit of detectability.

3. The position of the second R-maximum coincides with a similar maximum for the size distribution of cratering projectiles.

4. In the range of diameters from 2 to 20 km the SFD shapes for projectiles and inner belt asteroids look similar within the limits of accuracy

of available data.

Let us note once again that R-maxima and R-minima of the SFDs correspond to deviations from a simple power law with an index equal 3.

The bulk size-frequency distribution of asteroids of various types certainly gives only a general picture. The SFD for asteroids of different types can differ essentially from each other (Gradie *et al.*, 1989). However, for more detailed discussion the accumulation of much more observational data is necessary.

The SFD deviations from a power law for impact craters, considered above, suggests that the SFD for asteroids in the Main Belt – the main source of projectiles for cratering on terrestrial planets – also deviates from a simple power law at diameters smaller than the limit of completeness of detection. Recent modeling of asteroid collisional evolution gives a new clue to the origin of such deviations.

The numerical modeling of asteroid collisions by Love and Ahrens (1996) demonstrated the strength to gravity transition in the catastrophic events at unexpectedly low asteroid diameters ∼ hundreds of meters. The estimates by Melosh and Ryan (1997) also show the importance of self-gravity for the reaccumulation of fragments after the catastrophic disruption into a "rubble pile".

The revised impact strength model was used by Durda *et al.* (1998) in a standard numerical model of collisional evolution. The modeling demonstrated a very simple idea:

– Just above the transitional target diameter, larger critical projectiles ("killers") are required;

– The number of "killers" is less than for the range of pure strength confinement; hence at larger diameters a surplus of bodies exists; this surplus increases the probability of destruction of larger bodies, etc.

– Finally some equilibrium is established where the SFD of asteroids deviates in a wavy manner from a simple power law.

In Fig. 16 the modeling results by Durda *et al.* are compared to the projectile SFD, described above. In the range of diameters less than 100 m the discrepancy is large, demanding an additional discussion beyond the scope of the present paper. However, the idea of the transition from strength to gravity confinement of asteroids at catastrophic collisions allows to reproduce a wavy deviation from the power law. Moreover, the positions of R-maxima and R-minima correspond to observational data. An important result for the interpretation of cratering on terrestrial planets is that according to the model the specific SFD shape is reached for less than 1 Gy and remains approximately stable during more than 4 Gy. Without considering the depletion due to ejection into NEA orbits, the number of bodies

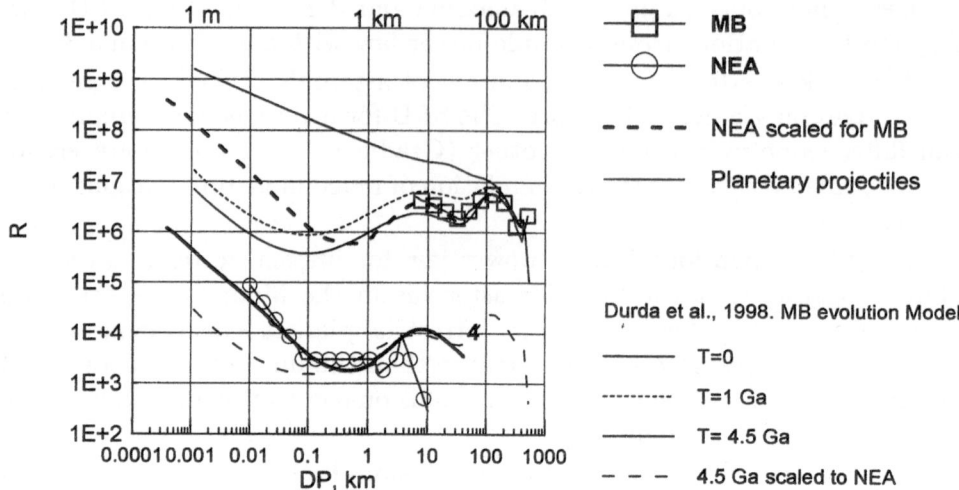

Figure 16. The results of modeling by Durda *et al.* (1998) in comparison to the projectile SFD.

in the Main Belt according to the model by Durda *et al.* decreased by less than three times during the last 3.5 Gy.

8. Conclusions and discussion

1. The application of the cratering scaling laws allows to estimate the size-frequency distribution (SFD) of projectiles from the measured SFD of lunar impact craters.

2. The estimated SFD of projectiles has a complex form with wavy deviations from a simple power law.

3. The estimated SFD of projectiles allows to reproduce the impact crater SFD on all terrestrial planets. One can assume that the majority of impact craters were formed by one population of projectiles, and the shape of the projectile SFD did not change dramatically for the last 4 Gy.

4. The SFD of the crater-forming bodies within the limits of accuracy of available data is similar to the SFD of asteroids in the Main Belt. Within the same limits of accuracy the contribution of comets to the crater formation is relatively insignificant, provided that the cometary SFD does not replicate the wavy SFD for asteroids. Asteroids, ejected from the Main Belt into NEAs, present the main source of crater-forming objects on the terrestrial planets.

5. Similarities of SFDs of asteroids with planet-crossing orbits and asteroids of the Main Belt speak in favor of the chaotic migration of bodies

in the Main Belt as the main mechanism of delivery of asteroids and their fragments into resonant phase areas, where they are rather quickly injected into planet-crossing orbits. The migration of bodies is caused by the gravitational action of large planets and does not depend on their size. In contrast the "direct impact" delivery of fragments into resonances should result in some sorting, since in general smaller fragments of a catastrophic disruption have larger velocities and, consequently, larger chances to reach a resonant phase area. Hence the direct delivery seems to result in a NEA enrichment in small bodies and, hence, in a steeper SFD of NEAs in comparison with the Main Belt asteroids (Rabinowitz, 1997). Probably this mechanism is partially responsible for the divergence of the estimated SFD for NEAs and the model SFD for the Main Belt, shown in Fig. 8. Anyway, the hypothesis about the main role of chaotic migration looks reasonable at least for bodies lager than ~ 1 km.

6. If the NEA and Main Belt SFD's are similar at diameters > 1 km, the cratering rate should be proportional to the number of bodies of the appropriate size in the Main Belt. The models of collisional evolution (*e.g.*, Durda *et al.*, 1998) result in an about three-fold reduction of the number of asteroids for the last 3.5 Gy. Consequently the cratering rate on the terrestrial planets should not have changed by more than three times during the same interval. This conclusion seems to correspond to a conclusion, made by many planetologists: for the last 3 Gy, the cratering rate may be treated as a constant within a factor of two.

Thus the general picture of cratering on the terrestrial planets begins to look more similar and self-consistent. The questions for future study include the cometary contribution in impact crater formation, and the possible difference of SFD's of NEAs and the Main Belt at small (< 1 km) sizes.

References

Arvidson, R., Boyce, J., Chapman, C., Cintala, M., Fulchignoni, M, Moore, H., Neukum, G., Schultz, P., Soderblom, L., Strom., R., Woronov, A. and Young, R. (1978) Standard techniques for presentation and analysis of crater size-frequency data, *Icarus* **37**, 467–474.

Asphaug, E., Moore, J.M., Morrison, D., Benz, W., Nolan, M.C. and Sullivan, R.J. (1996) Mechanical and geological effects of impact cratering on Ida, *Icarus* **120**, 158–184.

Basaltic Volcanism Study Project (1981) Chapter 8 "Chronology of planetary volcanism by comparative studies of planetary cratering", in *Basaltic Volcanism on the Terrestrial Planets*, Pergamon Press, N.Y., pp. 1049–1127.

Belton, M.J.S. and 19 co-authors (1994) First images of asteroid 243 Ida, *Science* **265**, 1543.

Belton, M.J.S. and 9 co-authors (1992) Galileo encounter with 951 Gaspra – First pictures of an asteroid, *Science* **257**, 1647–1652.

Bryan, J.B., Burton, D.E., Cunningham, M.E. and Lettis, L.A. (1978) A two-dimensional computer simulation of hypervelocity impact cratering: Some preliminary results for Meteor Crater, Arizona, *Proc. Lunar Planet. Sci. Conf. 9th*, pp. 3931–3964.

Campo Bagatin, A., Cellino A., Davis, D.R., Farinella, P. and Paolicchi, P. (1994a) Wavy size distribution for collisional systems with a small-size cutoff, *Planet. Space Sci.* **42**, 1049–1092.

Campo Bagatin, A., Farinella, P. and Petit, J.-M. (1994b) Fragment ejection velocities and the collisional evolution of asteroids, *Planet. Space Sci.* **42**, 1099–1107.

Cellino, A., Zappalà, V. and Farinella, P. (1991) The asteroid size distribution from IRAS data, *Mon. Not. R. Astr. Soc.* **253**, 561–574.

Chapman, C.R. (1995) Galileo observations of Gaspra, Ida, and Dactyl: Implications for meteoritics, *Meteoritics* **30**, 496.

Chapman, C.R. and McKinnon, W.B. (1986) Cratering of planetary satellites, in *Satellites*, (J.A. Burns and M.S. Matthews eds.) Univ. Arizona Press, Tucson, pp. 492–580.

Chapman, C., and 7 co-authors (1996a) Cratering on Ida, *Icarus* **120**, 77–86.

Chapman, C., Veverka. J. , Belton, M., Neukum, G. and Morrison, D. (1996) Cratering on Gaspra, *Icarus* **120**, 231–245.

Croft, S.K. (1985) The scaling of complex craters, *Proceedings of 15th Lunar Planet. Sci. Conf., J. Geophys. Res.* **90**, C828–C842.

Davis, D.R., Ryan, E.V. and Farinella, P. (1994) Asteroid collisional evolution: results from current scaling algorithm, *Planet. Space. Sci.* **43**, 599–610.

Davis, D., Weidenshilling, S.J., Farinella, P., Paolicchi, P. and Binzel, R.P. (1989) Asteroid collisional history: Effects on sizes and spins, in *Asteroids II*, (R. Binzel, T. Gehrels, and M.S. Matthews eds.), Univ. Arizona Press, Tucson, pp. 805–826.

Davis, D.R., Chapman, C.R., Weidenschilling, S.J. and Greenberg, R. (1985) Collisional history of asteroids: Evidence from Vesta and the Hirayama families, *Icarus* **62**, 30–35.

Dohnanyi, J.W. (1969) Collisional model of asteroids and their debris, *J. Geophys. Res.* **74**, 2531–2554.

Durda, D., Greenberg, R. and Jedicke, R. (1998) Collisional models and scaling laws: A new interpretation of the shape of the Main-belt asteroid distribution, *Icarus* **135**, 431–440.

Gault, D.E. and Wedekind, J.A. (1978) Experimental studies of oblique impact, in *Proc. Lunar Planet. Set. Conf. 9th*, Pergamon Press, N.Y., pp. 3843–3875.

Gradie, J.C., Chapman, C.R. and Tedesco, E.W. (1989) Distribution of taxonomic classes and the compositional structure of the asteroid belt, in *Asteroids II*, (Binzel R.P., Gehrels T., and Matthews, M.S. eds.), Univ. Arizona Press, Tucson, pp. 316–335.

Grieve, R.A.F. and Shoemaker E.M. (1994) The record of the past impacts on Earth, in *Hazards due to Comets and Asteroids*, (T. Gehrels, Ed.) Univ. Arizona Press, Tucson, pp. 417–462.

Hartmann, W.K. (1977) Relative crater production rates on planets, *Icarus* **31**, 260–276.

Hartmann, W.K. (1984) Does crater "saturation equilibrium" occur in the Solar System?, *Icarus* **60**, 56–74.

Hartmann, W.K. (1995) Planetary cratering I: Lunar highlands and tests of hypotheses on crater populations, *Meteoritics* **30**, 451–467.

Hartmann, W.K., Berman, D., Esquerdo, G.A. and McEwen, A. (1999a) Recent Martian volcanism: New evidence from Mars Global Surveyor (abstract), *LPSC XXX*, CD-ROM edition, No. 1270.

Hartmann, W.K., Malin, M.M., McEwen, A., Carr, M., Soderblom, L., Thomas, P., Danielson, E., James, P. and Veverka, J. (1999b) Evidence for recent volcanism on Mars from crater counts, *Nature* **397**, 586–589.

Holsapple, K.A. (1993) The scaling of impact processes in planetary sciences, *Ann. Rev. Earth.Planet. Sci.* **21**, 333–373.

Holsapple, K.A. and Schmidt, R.M. (1979) A material-strength model for apparent crater volume *Proc. Lunar Planet. Sci. Conf. 10th*, pp. 2757–2777.

Housen, K.R., Schmidt, R.M. and Holsapple, K.A. (1983) Crater ejecta scaling laws: Fundamental forms based on dimensional analysis, *J. Geophys. Res.* **88**, 2485–2499.

Ivanov, B.A. (1979) Simple model of cratering, *Meteoritika* no. 38, pp. 68–85, in Russian.

Ivanov, B.A. (1981) Cratering mechanics, in *Advances in Science and Technology of*

VINITI, Ser. Mechanics of Deformable Solids **14**, VINITI Press, Moscow, pp. 60–128, in Russian – see also English translation: NASA Tech. Memorandum 88477/N87-15662, 1986.

Ivanov, B.A. and Kostuchenko, V.N (1997) Block oscillation model for impact crater collapse, *Lunar and Planetary Science Conference 28th*, CD-ROM, abstract no. 1655.

Ivanov, B.A., Neukum, G. and Wagner, R. (1999) Impact craters, NEA, and main belt asteroids: Size-frequency Distribution, *Lunar and Planetary Science Conference 30*, CD-ROM edition, abstract no. 1583.

Ivanov, B.A., Basilevsky, A.T. and Neukum, G. (1997) Atmospheric entry of large meteoroids: Implication to Titan, *Planet. Space Sci.* **45**, 993–1007.

Jedicke, R. and Metcalfe, T.S. (1998) The orbital absolute magnitude distributions of Main Belt asteroids, *Icarus* **131**, 245–260.

Love, S. and Ahrens, T.J. (1996) Catastrophic impacts on gravity dominated asteroids, *Icarus* **124**, 141–155.

McCrosky, R., Chao, K. and Posen, A. (1979). Data on bolides of Prairie Network, *Meteoritika* no. 37, Nauka Press, Moscow, pp. 44-59 (in Russian).

McEwen, A.S., Gaddis, L.R., Neukum, G., Hoffman, H., Pieters, C.M. and Head, J.W. (1993) Galileo observations of post-Imbrium lunar craters during the first Earth-Moon flyby, *J. Geophys. Res.* **98** no. E9, 17,207–17,231.

McEwen, A.S., Moore, J.M. and Shoemaker, E.M. (1997) The Phanerozoic impact cratering rate: Evidence from the farside of the Moon, *J. Geophys. Res.* **102**, 9231–9242.

McKinnon, W.B. (1978) An investigation into the role of plastic failure in crater modification, in *Proc. Lunar Planet. Sci. Conf. 9th*, Pergamon Press, NY, pp. 3965–3973.

Melosh, H.J. (1977) Crater modification by gravity: A mechanical analysis of slumping, in *Impact and Explosion Cratering*, Pergamon Press, NY, pp. 1245–1260.

Melosh, H.J. (1979) Acoustic fluidization: a new geologic process?, *J. Geophys. Res.* **84**, 7513–7520.

Melosh, H.J. (1982) A schematic model of crater modification by gravity, *J. Geophys. Res.* **87**, 371–380.

Melosh, H.J.(1989) *Impact Cratering: A Geologic Process*, Oxford University Press, N.Y. & Clarendon Press, Oxford, 245 pp.

Melosh, H.J. and Ivanov, B.A. (1999) Impact crater collapse, *Annu. Rev. Earth Planet. Sci.* **27**, 385-415

Melosh, H.J. and Ryan, E.V. (1997) Note: Asteroids shattered but not dispersed, *Icarus* **129**, 562–564.

Milani, A., Carpino, M., Hahn, G. and Nobili, A.M. (1989) Dynamics of planet-crossing asteroids: Classes of orbital behavior, *Icarus* **78**, 212–269.

Nemtchinov, I.V., Svetsov, V.V., Kosarev, I.B., Golub', A.P., Popova, O.P., Shuvalov, V.V., Spalding, R.E., Jacobs, C. and Tagliaferri, E. (1997) Assessement of kinetic energy of meteoroids detected by satellite-based light sensors, *Icarus* **130**, 259–274.

Neukum, G. (1983) *Meteoritenbombardement and Datierung Planetarer Oberflächen*, Habilitation dissertation for faculty membership, Univ. of Munich, 186 pp.

Neukum, G., and Ivanov, B.A. (1994) Crater size distribution and impact probabilities on Earth from lunar, terrestrial-planet, and asteroid cratering data, in *Hazards due to Comets and Asteroids*, (T. Gehrels, Ed.), Univ. Arizona Press, Tucson, pp. 359–416.

Nolan, M.C., Asphaug E., Melosh, H.J. and Greenberg, R. (1996) Impact craters on asteroids: Does gravity or strength control their size?, *Icarus* **124**, 359–371.

Pierazzo, E., Vickery, A.M. and Melosh, H.J. (1997) A reevaluation of impact melt production, *Icarus* **127**, 408–423.

Pike, R. (1977) Size-dependence in the shape of fresh impact craters on the moon, in *Impact and Explosion Cratering*, (Eds. Roddy D.J., Pepin R.O., and Merrill R.B.), Pergamon Press, N.Y., pp. 489–510.

Pike, R.J. (1980) Control of crater morphology by gravity and target type: Mars, Earth, moon, in *Proc. Lunar. Planet Sci. Conf 11th*, Pergamon Press, N.Y., pp. 2159–2189.

Pike, R.J. and Davis, P.A. (1984) Toward a topographic model of Martian craters from photoclinometry (abstract), *Lunar and Planetary Science XV*, pp. 645–646.

Rabinowitz, D. (1993) The size-distribution of the Earth-approaching asteroids, *Astrophys. J.* **407**, 412–427.

Rabinowitz, D., Bowell, E., Shoemaker, E. and Muinonen, K. (1994) The population of Earth-crossing asteroids, in *Hazards due to Comets and Asteroids*, (Ed. T. Gehrels), University of Arizona Press, Tucson, pp. 285–312.

Rabinowitz, D.L. (1997) Are main-belt asteroids a sufficient source for the Earth-approaching asteroids? Part II. Predicted vs observed size distribution, *Icarus* **130**, 287–295.

Roddy, D.J. (1978) Pre-impact geologic conditions, physical properties, energy calculations, meteorite and initial crater dimensions and orientation of joints, faults, and walls at Meteor Crater, Arizona, *Proc. Lunar. Sci. Conf. 9th*, Pergamon Press, NY, pp. 3891–3930.

Roddy, D.J., Pepin, R.O. and Merrill, R.B., eds. (1977) *Impact and Explosion cratering*. Pergamon Press, N.Y. 1301 pp.

Ronca, L.B., Basilevsky, A.T., Kryuchkov, V.P. and Ivanov, B.A. (1981). Lunar craters evolution and meteoroidal flux in pre-mare and post-mare times, *The Moon and the Planets* **245**, 209–229.

Schmidt, R.M. and Housen, K.R. (1987) Some Recent Advances in the Scaling of Impact and Explosion Cratering, *Int. J. Impact Engng.* **5**, 543–560.

Schmidt, R.M. (1980) Meteor Crater: Energy of formation-implications of centrifuge scaling, in Proc. *Lunar Planet. Sci. Conf. 11th*, pp. 2099–2128.

Shoemaker, E.M. and Wolfe, R. (1982) Cratering time scales for the Galilean satellites, in *Satellites of Jupiter*, (Ed. D. Morrison), Univ. of Arizona Press, Tucson, pp. 277–339.

Shoemaker, E.M. (1977) Astronomically observable crater-forming projectiles, in *Impact and Explosion Cratering*, (Eds. D.J. Roddy, R.O. Pepin, and R.B. Merrill), Pergamon Press, New York, pp. 639–656.

Strom, R. (1977) Origin and relative age of lunar and mercurian inter-crater plains, *Phys. Earth Planet. Interiors* **15**, 156–172

Strom, R.G. and Neukum, G. (1988) The cratering record on Mercury and the origin of impacting objects, in *Mercury*, (Vilas, F., Chapman, C.R. and Matthews, M.S., eds.), Univ. of Arizona Press, Tucson, pp. 336–373.

van Houten, C.J., van Houten–Groeneveld, I., Herget, P. and Gehrels, T. (1970) The Palomar–Leiden survey of faint minor planets, *Astron. Astrophys. Suppl.* **2**, 339–448.

Veverka, J. and 16 co-authors (1997) NEAR's flyby of 253 Mathilde: Images of a C asteroid, *Science* **278**, 2109–2114.

Wagner, R. and Neukum, G. (1999) Impact crater count on Mathilde, *Abstracts presented to the EGS XXIV General Assembly, The Hague, 1999*, 185.

Yeomans, D.K. and 12 co-authors (1997) Estimating the mass of asteroid 253 Mathilde from tracing data during the NEAR flyby, *Science* **278**, 2106–2109.

THE METEOROIDAL INFLUX TO THE EARTH

ZDENĚK CEPLECHA
Astronomical Institute of the Academy of Sciences,
25165 Ondřejov Observatory, Czech Republic

Abstract. The influx of different meteoroid populations for the mass range from 2×10^{-8} kg to 2×10^3 kg was derived independently from double- and multi-station photographic and television observations with mass, velocity and other data available for individual meteoroids. The total influx is simply the sum of all partial fluxes of different meteoroid populations. Characteristics of individual meteoroid populations are also given and summarized in Table 3. These results are combined with fluxes of interplanetary particles and bodies from other sources of information (as given in Table 2) in order to derive cumulative and incremental mass influx to the Earth in the entire mass range from 10^{-21} kg to 10^{15} kg.

1. Introduction

The meteor phenomenon originates from meteoroids that are large enough to produce light during their collisions with the Earth's atmosphere. We can make use of this atmospheric phenomenon for investigation of the influx of meteoroids to the Earth. We are limited to a size range from about 0.01 mm (the smallest size still producing light depends mostly on velocity) to about 10 m, the largest meteoroids recorded during the atmospheric trajectory.

Part of the meteoroid population is clearly linked to comets as shower meteors derived from cometary meteoroid streams. Another part is linked to asteroids, and we can study some of these bodies in laboratories as meteorites. Meteoroids not linked to any specific shower (meteoroid stream) or comet or asteroid are called sporadic, and they form the most important influx to the Earth. Meteoroids belonging to a specific stream may bring more mass than sporadic meteoroids to the Earth only during very short time intervals, mostly during so called meteor storms. From the point of view of an average mass influx to the Earth, meteor showers (even storms)

M. Ya. Marov and H. Rickman (eds.), Collisional Processes in the Solar System, 35–50.

have little significance, and moreover they are limited mostly to millimeter and centimeter sizes. This chapter deals only with sporadic meteors.

The ability to penetrate into the atmosphere depends strongly on the meteoroid velocity. Especially the mass loss due to severe *ablation* causes a practical upper velocity limit of about 30 $km\,s^{-1}$ for the occurence of a meteorite fall. Well documented meteorite falls (double- or multi-station records of atmospheric trajectories) were used for calibration of relative masses derived for the majority of well documented meteors (and artificial meteors) from their brightness and velocity as function of time during their atmospheric trajectories (McCrosky *et al.*, 1971; Halliday *et al.*, 1981, 1984; Ceplecha 1961, 1997; Ayers *et al.*, 1970).

2. Observational Methods

Different observational methods are capable of getting various data on meteoroids and their interactions with the atmosphere. As an ideal we would like to have a "complete" set of data on an individual meteor, i. e. geometrical data (position of the trajectory in the atmosphere), dynamical data (height and distance along the trajectory as function of time; velocity; deceleration; "dynamical" mass), photometric data (integrated light intensity in the complete pass-band as function of time, "photometric" mass), spectral data (intensity radiated in individual spectral lines and molecular bands as function of time), ionizational data (density of ions and free electrons as function of time), and orbital data (elements of the Keplerian orbit at encounter with the Earth). If a meteorite fall were recovered after such a complete recording, we would have an ideal case for testing our theories of meteoroid interaction with the atmosphere and thus also for calibrating the scale of masses which the meteoroids had before the collisions. This has never been achieved in full complexity, but we have enough records with precise recording techniques, namely double- and multi-station direct and spectral photographic recording of meteors, to use them for determining the influx at different meteoroid masses.

The fluxes derived from visual observations (counting of meteors of different brightness) have little significance for deriving meteoroid influx, because of the intrinsic combination of mass and velocity effects on the luminosity. Meteoroids observed by counting radar echoes (duration and distance, without data on mass and velocity for individual meteoroids) were mostly used in the past for deriving the meteoroid influx, and we should consider such fluxes with enough precaution. Again the duration of the radar echo depends on a combination of velocity and mass in a way which cannot be simply decoded from the counts. Thus we want to use multi-station radar observations with velocity, mass and other data available for individual meteoroids. There are not many such observations published.

The meteoroid fluxes presented in this chapter for the mass range from 2×10^{-8} kg to 2×10^3 kg are based on statistical studies of data derived for individual meteoroids recorded by three different, double- or multi-station photographic systems, and by low-light-level double-station television systems. The three photographic systems are: Super-Schmidt cameras recording meteoroids somewhat brighter than observed by a visual observer with masses from 5×10^{-7} to 10^{-3} kg; classical small camera systems recording meteoroids from about zero stellar magnitude with masses from 10^{-4} to 0.5 kg; fireball networks recording bolides from 0.1 kg to 2×10^3 kg. The lower mass limits are given by the sensitivity of the system, while the upper limits are given by the time the systems were used for recording meteors. The television systems yielded data on meteoroids from 2×10^{-8} to 5×10^{-6} kg.

3. Meteoroid Populations

Different meteoroid populations were first recognized independently by Jacchia (1958) and Ceplecha (1958). The differences in beginning heights proved to be the most important tool for recognition of different meteoroid populations among Super-Schmidt and small-camera meteors when and if the right dependence on velocity was considered (Ceplecha, 1967, 1968, 1988). This was also expanded to faint TV meteors (Hawkes *et al.*, 1984; Jones *et al.*, 1985). Two main discrete levels of meteor beginning heights were found. They are separated by 10 km difference. The lower level was denoted A, the higher was denoted C (Table 3). The C–group of meteoroids was recognized to contain two populations of orbits: one with ecliptically-concentrated short-period orbits was denoted C1 and the other with random orbital inclinations of long-period orbits was denoted C2. Classical meteor showers with known parent comets are of type C1 or C2; thus the cometary origin of meteoroids of the whole C-group is quite convincing.

In addition, two smaller groups were found among Super-Schmidt, small-camera and TV meteors. The intermediate group B has typical orbits with small perihelion distances and aphelion close to Jupiter. The other smaller group was originally referred to as a group "above C" accounting for very high beginning heights. Today the notation D is used for this group. The Draconid meteor shower is a typical member of this group. It contains extremely friable meteoroids with the lowest known densities of all types of solid cometary material coming to the Earth (Ceplecha, 1968).

Details on the four populations of big meteoroids (fireballs, bolides) can be found elsewhere (Ceplecha and McCrosky, 1976; Ceplecha, 1977, 1983, 1985, 1994; Sekanina, 1983; Wetherill and ReVelle, 1981a, b). Four inde-

Table 1. Survey of meteoroid populations among photographic and television meteors. a is the semimajor axis, e is the eccentricity, i is the inclination, ρ_M is the meteoroid bulk density in $[1000 \times kg\ m^{-3}]$, σ is the ablation coefficient in $[s^2\ km^{-2}]$, % obs is the relative observed number of meteors in percent, ast.met. means "asteroidal meteors"

mass range / group	Television cameras b) from 2×10^{-8} to 5×10^{-6} kg % obs	a	e	i	Super-Schmidt b) from 5×10^{-7} to 10^{-3} kg % obs	a	e	i	Small cameras b) from 10^{-4} to 0.5 kg % obs	a	e	i	Fireball networks b) from 0.1 to 2×10^3 kg group	% obs	a	e	i	ρ_M	σ	assumed composition / parent bodies
ast. met.	<1	0.7 [a]	0.39 [a]	18° [a]	1	2.4	0.64	15°	5	2.5	0.64	10°	irons	3	1.8	0.5	6°	7.8	.07	iron meteorites / asteroids
A	27	1.6	0.55	14°	50	2.3	0.61	1°	39	2.5	0.64	4°	I	29	2.4	0.68	6°	3.7	.014	ordinary chondrites / asteroids
B	2	2.1	0.95	29°	3	2.4	0.92	5°	5	2.5	0.90	6°	II	33	2.3	0.61	5°	2.0	.042	carbonaceous chondrites / comets, asteroids
c) C1	21	1.7	0.63	16°	7	2.2	0.80	6°	11	2.5	0.80	5°	c) IIIA	11	2.4	0.82	4°	1.0	.08	dense cometary material / inner parts of comets
C2	18	≈∞	0.99	random	32	≈∞	0.99	random	21	≈∞	0.99	random	IIIAi	11	≈∞	0.99	random	.75	.10	regular cometary material / short period comets
c) C3	28	1.3	0.60	random	6	1.9	0.72	random	9	2.1	0.77	random	c) C3	4	2.7	0.67	random	.75	.10	regular cometary material / long period comets
D	3	2.6	0.66	18°	1	3.3	0.70	25°	10	3.1	0.77	10°	IIIB	9	3.0	0.70	13°	.27	.21	soft cometary material / short period comets of Giacobini-Zinner type

a) only one meteor No. 8111104060 recognized as "asteroidal"; its elements are given.
b) total mass range: individual groups differ due to different distribution of velocities.
c) C3 corrected for random i (instead of i > 35°) by adding the correspondiong part of C1 (IIIA) to C3.

pendent methods gave results differing in details, but the existence of four groups of fireballs according to various ablation abilities of their bodies, as found originally by Ceplecha and McCrosky (1976), has been confirmed. A brief summary of the fireball groups follows. Group I fireballs are of the smallest ablation ability and of the greatest bulk density. Among them are the *Příbram, Lost City* and *Innisfree* meteorite falls. Group II fireballs belong to meteoroids of somewhat lower density and greater ablation ability than the group I meteoroids. It is proposed that the group II fireballs belong to carbonaceous bodies, which mostly disintegrate in the atmosphere and only the most compact members of their population can reach the ground as carbonaceous chondrites of CI and CM types. This meteoric material may be both of asteroidal and cometary origin. Wetherill and ReVelle (1981b) prefer rather the cometary source. The average ablation coefficient of the type I fireball is 0.014 s^2/km^2, while the same value for the type II fireball is 0.042 s^2/km^2 (based on individual determinations of total ablation coefficient from the most precise photographic records of fireballs).

Group IIIA contains bodies with a high ablation ability and small bulk density less than 1000 $kg\,m^{-3}$. The cometary origin of these meteoroids is evident, because cometary shower meteors belong to this group. Also the two systems of orbits, i. e. the short-period ecliptically-concentrated system and the long-period randomly-inclined system (IIIAi) exist among fireballs of group IIIA, in complete analogy to cometary orbits. In addition a third system or subgroup among these IIIA fireballs, denoted C3, has short-period but randomly-inclined orbits. This C3 system forms only 4% of all fireballs, but becomes the most important for meteoroids with masses below 10^{-6} kg (Hawkes *et al.*, 1984; Jones *et al.*, 1985). The mass-selective ejection-velocities from long-period comets is suspected to be the main reason for existence of such a system of orbits of cometary meteoroids.

The fourth group of fireballs was denoted IIIB. Bodies of this group have the highest known ablation ability and the smallest bulk density of several hundreds of $kg\,m^{-3}$. Fireballs of the Draconid meteor shower belong to this group. The cometary origin is also evident. It is rather surprising that this group contains relatively more bodies among fireballs than among fainter meteors. Evidently more cometary material comes to the Earth in meter and ten meter size ranges than estimated before photographic observations on fireballs became available.

Table 3 contains a survey of all meteoroid populations as delineated among bodies observed by different techniques. The whole range covers masses from 2×10^{-8} kg to 2×10^3 kg. Table 3 contains the fraction of bodies of the given group among all observed meteor bodies. "Characteristic" orbits are also given for each group, but these values are maxima (and sometimes medians) of very broad irregular statistical distributions.

TABLE 2. Sources of observational flux-data on Earth-crossing interplanetary bodies

Mass range (kg)		method	author
from	to		
10^{-21}	10^{-9}	lunar microcraters space probes	Grün et al. (1985)
10^{-16}	6×10^{-10}	space probes	Naumann (1966) Hoffman et al. (1975a,b) Grün & Zook (1980)
3×10^{-9}	3×10^{-6}	radar meteors	Elford et al. (1964) Verniani & Hawkins (1965)
10^{-6}	10^{4}	photographic and television meteors	Ceplecha (1988, 1996, 1997)
8×10^{5}	10^{15}	asteroid discoveries by Spacewatch	Rabinowitz (1993, 1994, 1996) Rabinowitz et al. (1993)
10^{12}	10^{15}	photographic Earth-crossing asteroids	Shoemaker et al. (1990)

It is not possible to separate these populations using the orbital elements alone. The average value of the bulk density of the bodies, ρ_m, and the average value of the ablation coefficient σ, are given for each group. The proposed probable "composition and structure" of the groups, as given in the last column of Table 3, is established on much observational evidence (Ceplecha et al., 1998, p. 421). Parent bodies of group I are evidently asteroids, the most important of them having orbits probably close to the 3:1 Kirkwood gap (Wetherill, 1985) delivering meteorites to the Earth also through the ν_6 resonance and with an overwhelming help of the Yarkovsky effect (Vokrouhlický and Farinella, 1998; Farinella and Vokrouhlický, 1999).

4. Meteoroid Influx onto the Earth

In the past many authors estimated the total interplanetary matter arriving on the Earth by extrapolating results from a limited size-range assuming a power law size distribution with a constant exponent. Namely, visual observations of meteors and also the observations by radar occupy very narrow size-ranges and such an extrapolation cannot yield a good estimation of the total influx. In this chapter we attempt to estimate the total influx of bodies onto the Earth from 10^{-21} kg to 10^{15} kg by combining results of different methods and authors (Table 2) following the same procedures as described in more details by Ceplecha (1992).

TABLE 3. The relation of meteoroid masses, m_o, used in Ceplecha's previous flux papers (1988, 1992) to the new meteoroid masses, m, used in this chapter (after the new calibration by the Lost City mass) (Ceplecha, 1996).

log m_o kg	log m kg	log m_o kg	log m kg	log m_o kg	log m kg	log m_o kg	log m kg	log m_o kg	log m kg
−3.00	−3.00	−1.60	−1.72	−0.20	−0.52	1.20	0.78	2.60	1.98
−2.80	−2.80	−1.40	−1.55	0.00	−0.34	1.40	0.96	2.80	2.14
−2.60	−2.60	−1.20	−1.39	0.20	−0.16	1.60	1.14	3.00	2.31
−2.40	−2.41	−1.00	−1.22	0.40	0.03	1.80	1.32	3.20	2.49
−2.20	−2.22	−0.80	−1.05	0.60	0.22	2.00	1.49	3.40	2.67
−2.00	−2.05	−0.60	−0.88	0.80	0.41	2.20	1.65	3.60	2.86
−1.80	−1.88	−0.40	−0.70	1.00	0.59	2.40	1.82	3.80	3.06

There are two main changes according to the absolute calibration of fluxes as used in Ceplecha (1992). A recent dynamical solution of the Lost City fireball and meteorite fall (Ceplecha, 1996) (photographic records re-measured and better theoretical model of atmospheric motion and ablation including gross-fragmentation used) yielded new and more reliable luminous efficiencies. They enabled more reliable values of masses to be derived for individual meteoroids in the mass range of hundreds of kilograms. This change was rather large: the newly determined luminous efficiencies for hundred kilogram meteoroids proved to be about one order of magnitude larger than experimental values obtained for bodies in the mass-range of grams (Ceplecha et al., 1996). Luminous efficiencies for these smaller masses were derived from rocket experiments with firing iron bodies of gram masses into the atmosphere (Ayers et al., 1970). Thus the absolute mass scale is now well defined at two ranges of masses, at grams and at hundred kilograms. The mass scale on the flux curve (Ceplecha, 1992) needs to be corrected accordingly (Ceplecha, 1996). The relation of meteoroid masses used in Ceplecha's previous flux papers (1988, 1992), m_o, to meteoroid masses used in this work, m, after corrections to the new values of the Lost City mass (Ceplecha, 1996), are given in Table 3.

The second change is due to more bodies discovered by the Spacewatch Telescope, which made their statistics more reliable. Also the assumption of a peculiar orbital distribution of these bodies, which was suspected as a possible explanation for the very large fluxes derived, proved to be of less significance than expected. The new fluxes were determined taking into consideration the unusual orbital distribution. Approximately three times lower fluxes of 10-meter-sized bodies than in Rabinowitz' (1993) original report were newly derived (Rabinowitz, 1994, 1996).

TABLE 4. Cumulative numbers N (all bodies with masses larger than the given mass m) of interplanetary bodies coming to the entire Earth's surface per year.

log m kg	log N	log m kg	log N	log m kg	log N	log m kg	log N	log m kg	log N	log m kg	log N
15.0	-7.71	9.0	-3.18	3.0	2.85	-3.0	7.10	-9.0	14.85	-15.0	17.77
14.8	-7.49	8.8	-2.97	2.8	2.99	-3.2	7.38	-9.2	15.06	-15.2	17.84
14.6	-7.29	8.6	-2.77	2.6	3.13	-3.4	7.71	-9.4	15.26	-15.4	17.92
14.4	-7.09	8.4	-2.55	2.4	3.27	-3.6	8.05	-9.6	15.43	-15.6	17.99
14.2	-6.90	8.2	-2.33	2.2	3.40	-3.8	8.30	-9.8	15.59	-15.8	18.07
14.0	-6.72	8.0	-2.09	2.0	3.54	-4.0	8.46	-10.0	15.74	-16.0	18.15
13.8	-6.55	7.8	-1.85	1.8	3.67	-4.2	8.59	-10.2	15.87	-16.2	18.23
13.6	-6.38	7.6	-1.61	1.6	3.80	-4.4	8.70	-10.4	16.00	-16.4	18.32
13.4	-6.23	7.4	-1.36	1.4	3.93	-4.6	8.78	-10.6	16.11	-16.6	18.42
13.2	-6.08	7.2	-1.10	1.2	4.06	-4.8	8.86	-10.8	16.22	-16.8	18.52
13.0	-5.94	7.0	-0.85	1.0	4.18	-5.0	8.95	-11.0	16.31	-17.0	18.63
12.8	-5.81	6.8	-0.60	0.8	4.31	-5.2	9.05	-11.2	16.40	-17.2	18.74
12.6	-5.69	6.6	-0.36	0.6	4.43	-5.4	9.16	-11.4	16.48	-17.4	18.86
12.4	-5.57	6.4	-0.12	0.4	4.55	-5.6	9.30	-11.6	16.56	-17.6	18.99
12.2	-5.46	6.2	0.12	0.2	4.67	-5.8	9.46	-11.8	16.63	-17.8	19.12
12.0	-5.35	6.0	0.34	0.0	4.79	-6.0	9.65	-12.0	16.70	-18.0	19.26
11.8	-5.24	5.8	0.56	-0.2	4.91	-6.2	9.89	-12.2	16.77	-18.2	19.40
11.6	-5.14	5.6	0.76	-0.4	5.04	-6.4	10.16	-12.4	16.84	-18.4	19.54
11.4	-5.03	5.4	0.96	-0.6	5.17	-6.6	10.47	-12.6	16.91	-18.6	19.69
11.2	-4.92	5.2	1.15	-0.8	5.30	-6.8	10.80	-12.8	16.98	-18.8	19.84
11.0	-4.79	5.0	1.33	-1.0	5.43	-7.0	11.16	-13.0	17.05	-19.0	20.00
10.8	-4.67	4.8	1.50	-1.2	5.57	-7.2	11.55	-13.2	17.12	-19.2	20.15
10.6	-4.53	4.6	1.66	-1.4	5.71	-7.4	11.96	-13.4	17.19	-19.4	20.31
10.4	-4.39	4.4	1.82	-1.6	5.86	-7.6	12.38	-13.6	17.26	-19.6	20.47
10.2	-4.24	4.2	1.98	-1.8	6.01	-7.8	12.82	-13.8	17.33	-19.8	20.63
10.0	-4.08	4.0	2.13	-2.0	6.17	-8.0	13.25	-14.0	17.40	-20.0	20.79
9.8	-3.91	3.8	2.28	-2.2	6.33	-8.2	13.65	-14.2	17.47	-20.2	20.95
9.6	-3.74	3.6	2.42	-2.4	6.50	-8.4	14.00	-14.4	17.54	-20.4	21.12
9.4	-3.56	3.4	2.56	-2.6	6.68	-8.6	14.32	-14.6	17.62	-20.6	21.28
9.2	-3.37	3.2	2.71	-2.8	6.88	-8.8	14.60	-14.8	17.69	-20.8	21.45

The method used for deriving the total meteoroid influx in the meteor mass range from 2×10^{-8} kg to 2×10^{3} kg (Ceplecha, 1992) is based on independent handling of each meteoroid population of Table 3 separately. The total influx is then just the sum of partial influxes of individual populations (groups). This procedure takes into account the different parameters of individual populations, which would be mixed together in case of taking all meteoroids as one homogeneous statistical sample. Because the interaction of meteoroids with the atmosphere enables us to make a classification of each meteoroid, the fluxes in the meteor mass range are very reliable. The variety of meteoric bodies coming to the Earth is enormous: if it is expressed in ratio of the air density at the terminal point of two meteoroids of the most differing composition and structure, but with the same velocity, the same mass and the same slope of the trajectory, then this ratio is 1:1000.

The resulting cumulative number fluxes of interplanetary bodies onto the Earth derived from data cited in Table 2 are given in Table 4. Fluxes across this vast range of masses cannot be expressed by a simple power law, which suggests several interlaced populations of different origin, as was already demonstrated for meteoroids entering the atmosphere as sporadic meteors with masses from 2×10^{-8} to 2×10^3 kg. The absolute precision of the influx (including the precision of all the calibrations) varies somewhat with mass and absolute numbers observed by different techniques. At small masses, at $\log m = -5$ the standard deviation in $\log N$ after taking into account the calibration to absolute values is $\epsilon_{\log N} = \pm 0.5$; at the first center of calibration of meteoroid masses, at $\log m = -3$, $\epsilon_{\log N} = \pm 0.3$; at $\log m = 0$, $\epsilon_{\log N} = \pm 0.3$; at $\log m = 2$, $\epsilon_{\log N} = \pm 0.5$; at $\log m = 3$, $\epsilon_{\log N} = \pm 0.6$; between $\log m = 3$ and $\log m = 6$, $\epsilon_{\log N} < \pm 0.7$; at $\log m = 6$, $\epsilon_{\log N} = \pm 0.3$; the value $\epsilon_{\log N} = \pm 0.3$ is then typical for most of the rest of masses up to about $\log m = 13$, while at $\log m = 15$ the error quickly increases to about $\epsilon_{\log N} = \pm 0.8$, because of scarcity of observed bodies and the large difference between results from Spacewatch and from classical photographic work.

Figure 1 presents the cumulative numbers in the mass range of 10^{-6} kg to 10^{15} kg, i.e. fluxes for meteoroids and large interplanetary bodies. Symbols in Figure 1 have meanings as follows. The thick line: resulting fluxes of all bodies. The thin line from $\log m = -6$ to $\log m = 3$ denoted 'I': fluxes of the type I meteoroids (stony bodies, mostly ordinary chondrites; average bulk density of 3700 $\mathrm{kg\,m^{-3}}$; for larger masses they are meteorite dropping). The short thin line from $\log m = 4.5$ to $\log m = 6.5$ denoted 's': fluxes from recording sounds of superbolides (the dots are individual values of cumulative numbers defining the thin short line (ReVelle, 1997)). The arrow at $\log m = 5.4$ denotes those satellite-observed superbolides, which were interpreted as belonging to type I (stony bodies); because separation of type I and type II superbolides was not possible (Ceplecha et al., 1997), only an upper limit can be given (the base of the arrow) and the arrow symbol means: 'less than the given value'. Short-dashed curve: the cumulative flux curve of all bodies as published in Ceplecha (1992) before calibration with the new Lost City mass. Long-dashed curve: the cumulative flux curve of all bodies as published in Ceplecha (1996) (without calibration for the

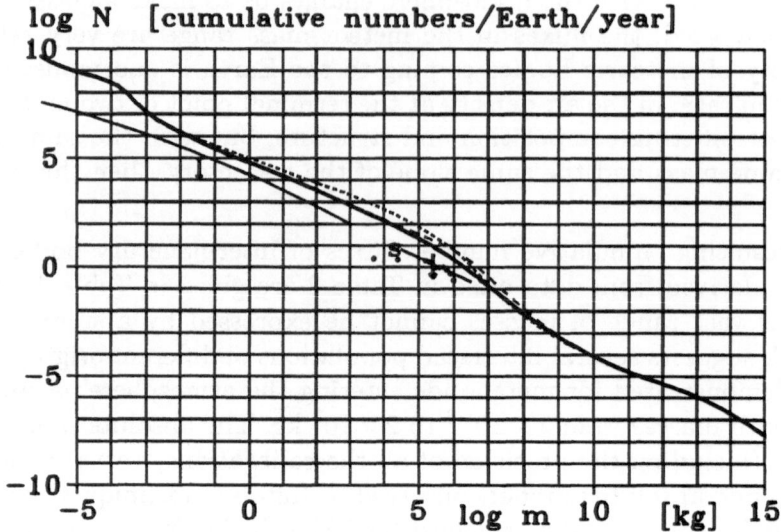

log N [cumulative numbers/Earth/year]

log m [kg]

Figure 1. Common logarithm of cumulative numbers N (all bodies with masses larger than the given mass m) of interplanetary bodies coming to the entire Earth's surface per year are plotted against common logarithm of the mass m. Explanation of lines and symbols in the text.

new Spacewatch statistics in Rabinowitz (1994)).

Percent numbers of different meteoroid types as functions of mass are given in Ceplecha (1988), but the correction to the new mass scale has to be derived from Table 3 of this chapter. It is also important that meteoroids of the type I population penetrate to the lowest heights in comparison to the other populations (on assumption of the same velocity, the same mass and the same slope of the trajectory) and so they belong to the strongest observed and recognized meteoroid component. The terminal heights of type I meteoroids are by 11 km lower on average than terminal heights of type II meteoroids (Ceplecha, 1988, 1994), by 25 km lower than type IIIA, and by 37 km lower than type IIIB. Even very large bodies of type IIIB, much larger than 10^3 kg, terminate at extremely high altitudes due to enormous fragmentation of highly uncohesive cometary material as e.g. the Šumava bolide (Borovička and Spurný, 1996).

Type II meteoroids deserve special attention. Their average bulk density is 2000 kg/m^{-3} and they are easily breakable at pressures lower than 0.6 MPa (6 bars), i.e. at pressures ten to hundred times less than the strength of carbonaceous chondrites (Ceplecha *et al.*, 1993). This population was also suspected to be composed of at least two unseparated components, an asteroidal and a cometary component (Wetherill and ReVelle,

1981b): the recovered carbonaceous chondrites represent only a part of the strongest bodies of this type, those which at atmospheric penetration are unusually survivable and form end-members of the type II group of bolides. The total component of strong bodies is guessed at some 10% by Wetherill and ReVelle (1981b). Thus most of the type II fireballs seem to be composed of weak cometary material and thus only masses larger than 10^7 kg can penetrate to lower atmospheric heights (Ceplecha, 1992). The thick line in Figure 1 representing total meteoroid influx to the Earth, nearly represents also the weak cometary component in the mass range from 10^{-6} kg to 10^3 kg (values by 10% below the curve are inside the thickness of the line).

The acoustic data are based on AFTAC infrasound data from deep penetrating superbolides (altogether 10 events). They were taken from ReVelle (1997) and are represented by a short thin line in Figure 1 in the mass range from 3×10^4 kg to 3×10^6 kg; individual points representing cumulative numbers are also given. Acoustic signals can originate only from very deep penetrating superbolides, i.e. only from superbolides of type I in the mass range 10^4 to 10^7 kg. We can thus interpret the flux curve of superbolides with recorded sounds as a flux curve of type I superbolides.

We arrived at a rather consistent interpretation of all available flux data in the meteor mass range. The flux at the largest masses is mostly asteroidal (by "asteroidal" I mean rather the structure of bodies in the sense of "strong stones"). Going to smaller masses the flux starts to change to include more and more weaker cometary bodies beginning from 10^8 kg to smaller masses. Going to even smaller masses weak cometary bodies become overwhelming. This way one actually argues indirectly that a majority of the smallest Spacewatch objects and superbolides are weak bodies similar to those belonging to bolides of types II, IIIA, and IIIB (with insignificant admixture of the strong component of type II bolides). Thus bodies "in the gap" of observational data with masses between the largest observed bolides and the smallest observed Spacewatch objects are perhaps mostly of weak cometary structure and composition.

The arrow in Figure 1 at $\log m = 5.4$ belongs to satellite observations of unseparated type I and II superbolides calibrated by the total flux at this mass. The position of the arrow is in good agreement with the acoustic data and with the proposed cometary structure and composition of bodies (superbolide objects) with masses between the largest observed bolides and the smallest observed Spacewatch objects.

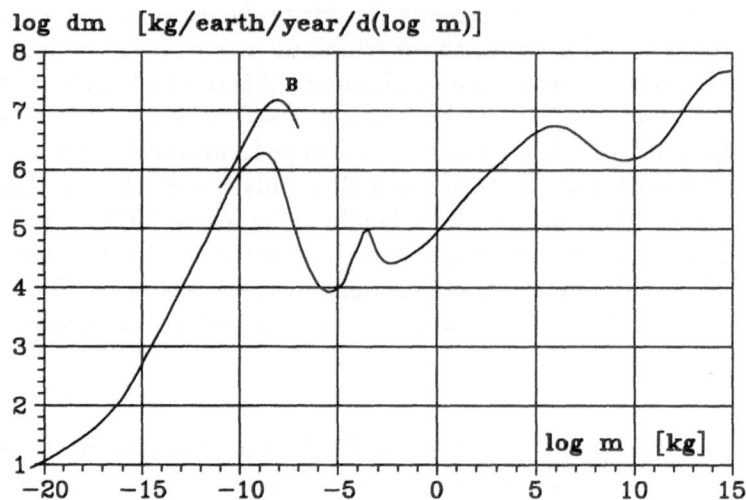

Figure 2. Increment of mass, log dm, per one order of mass per the entire Earth's surface per year is plotted against the logarithm of mass, log m. The dust range contains also new measurements by Love and Brownlee (1993) denoted **B**.

4.1. INCREMENTAL MASS INFLUX

Incremental fluxes dm derived from Table 4 are given in Figure 2 for the whole mass range from 10^{-21} kg to 10^{15} kg. There are three mass ranges with unstable populations: log $m = (-9.5, -6), (-4, -2.5)$ and $(5, 9)$. The first region corresponds to the situation where the losses of particles caused by the Poynting-Robertson effect outweigh the collisional gains of particles from bigger bodies (Grün *et al.*, 1985; Leinert and Grün, 1989). The second region at log $m = -3$ corresponds to the mass range where meteor showers are the most distinct phenomena. Perhaps a good part of sporadic meteors in this mass range originates from shower meteors perturbed and spread to such an extent that they cannot be distinguished from the sporadic background. The third region corresponds perhaps to a majority of small cometary fragments (small inactive comets) (Ceplecha, 1992; Ceplecha *et al.*, 1997) as already proposed by Kresák (1978) to explain the comet discovery statistics. Perhaps comets need to have masses of at least 10^9 kg to be active.

The total mass influx of all interplanetary bodies to the Earth over the mass range 10^{-21} kg to 10^{15} kg is 1.3×10^8 kg per year for the entire Earth surface. This value is one order of magnitude higher than those derived from extrapolations of fluxes of small bodies (*e.g.* Hughes, 1978). But one should be aware that this flux is mostly due to very large bodies: one body

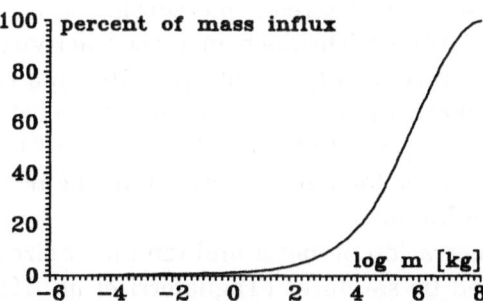

Figure 3. Percent of total mass influx as function of mass. Assumption of averaging over 100 years is incorporated. Because one body with mass of 10^8 or larger comes in 100 years on average, this is equivalent to omission of all bodies larger than 10^8 kg. Thus 100% is achieved just at 10^8 kg. Bodies between 10^3 kg and 10^7 kg form the most important source of mass influx of interplanetary bodies to Earth during time intervals of the order of 100 years.

of 10^{14} kg to 10^{15} kg coming at an average interval of 10^7 to 10^8 years. Masses less than 10^{-7} kg contribute to the influx by 4×10^6 kg per year per Earth's surface; masses between 10^{-7} kg and 10^2 kg contribute by 0.8×10^6 kg; masses between 10^2 kg and 10^9 kg contribute by 20×10^6 kg; masses between 10^9 kg and 10^{15} kg contribute by 100×10^6 kg per year per Earth's surface.

We see that the mass influx in perspective of a human life span is significantly lower then the total influx. If one takes 100 years as a period for averaging, one has to omit all bodies with masses over 10^8 kg and the total influx becomes 2.4×10^7 kg per year per Earth. The interplanetary dust component of the influx is 4×10^6 kg per year per Earth, i. e. about 17% of the influx originates in the interplanetary dust and 83% in bodies mostly inside a mass range of 10^5 kg to 10^8 kg (10 to 100 m sizes). We see that these large bodies form the most important source of mass influx onto the Earth from the perspective of human life span (Figure 3). Because most of this incoming mass is converted to small dust grains (due to fragmentation and ablation), these bodies are also the most important source of meteoric dust in the middle atmosphere (Ceplecha, 1976). This situation would change, if one assumes that the significantly larger dust fluxes resulting from the Long Duration Exposure Facility (Love and Brownlee, 1993) are real. Then the original dust component of the influx would be about 3× larger than the flux from the 10 to 100 m size bodies.

The size range of visual meteors is located in a minimum of incremental flux. The flux of visual sporadic meteors is the least important component from the point of view of the total influx. However, we should take into consideration also the flux of stream meteoroids seen as meteor showers. There are periods of time with much increased activity called meteoroid storms increasing the flux by a factor of up to 10^5 on a time scale of hours. A flux increase of 1000× or higher is already the most efficient source of Earth influx for a short period of tens of minutes up to few hours. These bodies are the real danger for human activity in the near Earth space and on the surface of the Moon.

Atmospheric penetration of meter and ten meter size bodies is regularly and globally observed by satellites (Tagliaferri et al., 1994; McCord et al., 1995). Bodies of 10 and 100 m sizes imperil the Earth's biosphere on a time scale of hundreds and thousands of years. A good example from this century is the Tunguska body explosion. An atmospheric explosion of a body of about 5× larger size over ocean waters would give rise to an enormous tsunami wave already endangering coastal ranges. Bodies in the size range of 10 and 100 meters are the least known bodies of the solar system. Their exploration is of prime importance. Satellite systems should be improved to be able to yield precise data on trajectories, velocities and orbits, and also on spectral radiation at penetration of these bodies into the Earth's atmosphere. This can help to reveal the composition and structure of these bodies, as well as new insights into the atmospheric interaction processes for such large bodies. Also observing these bodies in interplanetary space by systems of sensitive telescopes using reflected sunlight, either from ground or from space stations, can help to place these bodies in their planetary context.

Acknowledgements

This work has been supported by Contracts 205/94/1862 and 205/97/0700 of the Grant Agency of the Czech Republic and by the Sandia National Laboratories Contract AJ-4706.

References

Ayers, W. G., McCrosky, R. E. and Shao, C.-Y. (1970) Photographic observations of 10 artificial meteors, *Smithsonian Astrophys. Obs. Spec. Rep.* **317**, 1–40.

Borovička, J. and Spurný, P. (1996) Radiation study of two very bright terrestrial bolides, *Icarus* **121**, 484–510.

Ceplecha, Z. (1958) On the composition of meteors, *Bull. Astron. Inst. Czechosl.* **9**, 154–159.

Ceplecha, Z. (1961) Multiple fall of Příbram meteorites Photographed: Double-station photographs of the fireball and their relations to the found meteorites, *Bull. Astron. Inst. Czechosl.* **12**, 21–47.

Ceplecha, Z. (1967) Classification of meteor orbits, *Smithsonian Contr. Astrophys.* **11**, 35–60.

Ceplecha, Z. (1968) Discrete levels of meteor beginning heights, *Smithsonian Astrophys. Obs. Spec. Rep.* **279**, 1–54.

Ceplecha, Z. (1976) Fireballs as an atmospheric source of meteoric dust, in *Interplanetary Dust and Zodiacal Light*, IAU Coll. 31, eds. H. Elsässer, and H. Fechtig, Springer-Verlag, Lecture Notes in Physics 48, 385–388.

Ceplecha, Z. (1977 Meteoroid Populations and Orbits, in *Comets, Asteroids, Meteorites: interrelations, evolution and origins*, ed. A. H. Delsemme, Univ. Press, Toledo, pp. 143–152.

Ceplecha, Z. (1983) New aspects in classification of meteoroids, in *Asteroids, Comets, Meteors*, eds. C.I. Lagerkvist and H. Rickman, Astron. Obs. Univ, Uppsala, pp. 435–438.

Ceplecha, Z. (1985) Fireball information on meteoroids and meteorites, *Bull. Astron. Inst. Czechosl.* **36**, 237–241.

Ceplecha, Z. (1988) Earth's influx of different populations of sporadic meteoroids from photographic and television data, *Bull. Astron. Inst. Czechosl.* **39**, 221–236.

Ceplecha, Z. (1992) Influx of interplanetary bodies onto Earth, *Astron. Astrophys.* **263**, 361–366.

Ceplecha, Z. (1994) Meteoroid properties from photographic records of meteors and fireballs, in *Asteroid, Comets, Meteors 1993, IAU Sympos. 160*, eds. A. Milani, M. Di-Martino, and A. Celino, Kluwer, Dordrecht, pp. 343–356.

Ceplecha, Z. (1996) Luminous efficiency based on photographic observations of the Lost City fireball and implications for the influx of interplanetary bodies onto Earth, *Astron. Astrophys.* **311**, 329–332.

Ceplecha, Z. (1997) Influx of large meteoroids onto Earth, *Proceedings of SPIE* **3116**, 134–143.

Ceplecha, Z. and McCrosky, R. E. (1976) Fireball end heights: a diagnostic for the structure of meteoric material, *J. Geophys. Res.* **81**, 6257–6275.

Ceplecha, Z., Spurný, P., Borovička, J. and Keclíková, J. (1993) Atmospheric fragmentation of meteoroids, *Astron. Astrophys.* **279**, 615–626.

Ceplecha, Z., Spalding, R.E., Jacobs, C. and Tagliaferri, E. (1996) Luminous efficiencies of bolides, *Proceedings of SPIE* **2813**, 46–56.

Ceplecha, Z., Jacobs, C. and Zaffery C. (1997) Correlation of ground- and space-based bolides, in *Near-Earth Objects*, ed. J.L. Remo, *Annals of the New York Acad. of Sci.* **822**, 145–154.

Ceplecha, Z., Borovička, J., Elford W.G., ReVelle D.O., Hawkes R.L., Porubčan V. and Šimek M. (1998) Meteor phenomena and bodies, *Space Science Reviews* **84**, 327–471.

Elford, W.G. and Hawkins, G.S. (1964) Meteor echo rates and the flux of sporadic meteors, *Radio Meteor Project Res. Rep.* **9**, Harvard and Smithsonian Obs., Cambridge.

Farinella P. and Vokrouhlický D. (1999) Semimajor axis mobility of asteroidal fragments, *Science* **283**, 1507–1510.

Grün, E., Zook, H.A., Fechtig, H. and Giese, R.H. (1985) Collisional balance of the meteoric complex, *Icarus* **62**, 244–272.

Halliday, I., Griffin, A.A. and Blackwell A.T. (1981) The Innisfree Meteorite Fall, *Meteoritics* **16**, 153–170.

Halliday I., Blackwell A.T., and Griffin A.A. (1984) The frequency of meteorite falls on the Earth, *Science* **223**, 1405–1407.

Hawkes, R.L., Jones, J. and Ceplecha, Z. (1984) The populations and orbits of double-station TV meteors, *Bull. Astron. Inst. Czechosl.* **35**, 46–64.

Hoffmann, H.J., Fechtig, H., Grün, E. and Kissel, J. (1975a) First results of the micrometeoroid experiment S215 on HEOS 2 satellite, *Planet. Space Sci.* **23**, 215–224.

Hoffmann, H.J., Fechtig, H., Grün, E. and Kissel, J. (1975b) Temporal fluctuations and anisotropy of the micrometeoroid flux in the Earth-Moon system, *Planet. Space Sci.* **23**, 985–991.

Hughes, D.W. (1978) in J.A.M. McDonnell (ed.), Meteors, *Cosmic Dust*, John Willey, Chichester, pp. 123–186.

Jacchia L.G. (1958) On two parameter used in the physical theory of meteors, *Smithsonian Contr. Astrophys.* **2**, 181–187.

Jones, J., Sarma, T. and Ceplecha, Z. (1985) Double-station observations of 454 TV meteors III: Populations, *Bull. Astron. Inst. Czechosl.* **36**, 116–122.

Kresák, Ľ. (1978) The comet and asteroid population of the Earth's environment, *Bull. Astron. Inst. Czechosl.* **29**, 114–135.

Leinert, C. and Grün E. (1989) Interplanetary dust, in *Physics of the Inner Heliosphere*, eds. R. Schwenn and E. March, Springer, Heidelberg, pp. 207–260.

Love, S.G. and Brownlee, D.E. (1993) A direct measurement of the terrestrial mass accretion rate of cosmic dust, *Science* **262**, 550–553.

McCord, T.B., Morris J., Persing, D., Tagliaferri, E., Jacobs, C., Spalding, R.E., Grady, L. and Schmidt R. (1995) Detection of a meteoroid entry into the Earth's atmosphere on February 1, 1994, *J. Geophys. Res.* **100**, 3245–3249.

McCrosky R.E., Posen A., Schwartz, G. and Shao, C.-Y. (1971) Lost City meteorite – its recovery and a comparison with other fireballs, *J. Geophys. Res.* **76**, 4090–4108.

Nauman, R.J. (1966) The near Earth meteoroid environment, *NASA TND* **3717**.

Rabinowitz D.L. (1993) The size distribution of the Earth-approaching asteroids, *Astrophys. J.* **407**, 412–427.

Rabinowitz D.L. (1994) The size and shape of the near-Earth asteroid belt, *Icarus* **111**, 364–377.

Rabinowitz D.L. (1996) Observations constraining the origins of Earth-approaching asteroids, in *Completing the Inventory of the Solar System*, eds. T.W. Rettig and J.M. Hahn, *ASP Conference Series* **107**, pp. 13–28.

Rabinowitz, D.L., Gehrels, T., Scotti, J.V., McMillan, R.S., Perry, M.L., Winslewski W., Larson, S.M., Howel, E.S. and Mueller, B.E.A. (1993) Evidence for a near-Earth asteroid belt, *Nature* **363**, 704–706.

ReVelle, D.O. (1997) Historical Detection of Atmospheric Impacts by Large Bolides Using Acoustic-Gravity Waves, in *Near-Earth Objects*, ed. J.L. Remo, *Annals New York Acad. Sci.* **822**, 284–302, (also available as *LA-UR-95-1263*, National Laboratory, Los Alamos, pp. 26.

Sekanina Z. (1983) The Tunguska event: no cometary signature in evidence, *Astron. J.* **88**, 1382–1484.

Shoemaker, E.M., Wolfe, R.F. and Shoemaker, C.S. (1990) Asteroid and comet flux in the neighborhood of Earth, in *Global Catastrophes in Earth History*, eds. V.L. Sharpton and P.D. Ward, Geological Soc. of America, pp. 155–170.

Tagliaferri E., Spalding R., Jacobs C., Worden S.P. and Erlich A. (1994) Detection of meteoroid impacts by optical sensors in Earth orbit, in *Hazards due to comets and asteroids*, eds. T. Gehrels, Univ. of Arizona Press, Tucson, pp. 199–220.

Verniani, F. and Hawkins, G.S. (1965) Masses, magnitudes, and densities of 320 radio meteors, *Radio Meteor Project* **12**, 1–35, Harvard and Smithsonian Obs., Cambridge.

Vokrouhlický D. and Farinella P. (1998) Orbital evolution of asteroidal fragments into the ν_6 resonance via Yarkovsky effects, *Astron. Astrophys.* **335**, 351–362.

Wetherill, G.W. (1985) Asteroidal source of ordinary chondrites, *Meteoritics* **30**, 1–22.

Wetherill, G.W. and ReVelle, D.O. (1981a) Which fireballs are meteorites? A study of the Prairie Network photographic meteor data, *Icarus* **48**, 308–328.

Wetherill, G.W. and ReVelle, D.O. (1981b) Relationship between comets, large meteors and meteorites, in *Comets*, ed. in L. Wilkening, Univ. of Arizona Press, Tucson, pp. 297–319.

THE IMPACT RECORD ON ASTEROID SURFACES

W.F. BOTTKE, JR.
Cornell University
Ithaca, NY 14853-6801

AND

R. GREENBERG
University of Arizona
Tucson, AZ 85721

Abstract. The surfaces of main-belt and near-Earth asteroids have been carved over the age of the solar system by violent collisions with other asteroids. The craters and related features left behind can be used to interpret each asteroid's unique impact history, which depends on several factors: (i) the dynamical and physical properties of the asteroid (*e.g.*, size, orbit, composition, internal structure), (ii) the nature of the projectile population striking the asteroid over time, (iii) the asteroid's response to a high-velocity impact, and (iv) the capacity of impact erosion processes and tidal disruption to modify preexisting craters. In this chapter, we describe how numerical hydrocodes, dynamical calculations, and simulations of surface histories allow us to unscramble the complex crater records seen on main belt asteroids like 951 Gaspra and 243 Ida and constrain the processes that have created the impact records. We also present a template for interpreting crater history results from near-Earth asteroids like 433 Eros, which will be visited soon by the NEAR spacecraft.

1. Introduction

As several chapters in this book attest, there has been a long and rich history of interpretation of the impact record on planets and of the impact processes that created that record as small bodies bombarded the planets. We also know that the asteroids, which represent a major component of the impactors, indeed the dominant component for the terrestrial planets, have

M. Ya. Marov and H. Rickman (eds.), Collisional Processes in the Solar System, 51–71.
© 2001 *Kluwer Academic Publishers. Printed in the Netherlands.*

also bombarded one another. Such mutual collisions have ground down the earlier population of asteroids by disrupting the bodies, thus processing the populations into smaller and smaller objects.

In addition to being comminuted, asteroids are scarred by impacts as well. The nature of the size distribution of the bombarding asteroidal population is such that numbers increase strongly as size decreases. A given target asteroid must experience numerous surface-scarring impacts before eventually being disrupted by a more energetic impact. Here we use the term scarring as a generic modification of the surface morphology. Generally such scars are expected to be craters. However, for impacts near the limit of disruption, the surface may be scarred in ways that are somewhat different from the morphology of craters found in larger planets or satellites, perhaps for example forming more amorphous concavities or facets.

In recent years we have begun to obtain images of the surfaces of asteroids, which record the history of impacts that span this wide range of possible effects on the target. Interpretation of those records has the potential of providing important information about the target asteroid as well as the population that bombarded it.

However, we still lack the requisite level of understanding of the impact processes themselves, and of the effect of a given impact into a particular size and type of target. Thus, in order for interpretation of the impact record on asteroids to reach its full potential, we will need to investigate further the effects of impacts. Several lines of research are underway, including numerical simulations and laboratory experiments. In addition, constraints on the nature of the impact processes can come from the asteroids themselves, as we fit the observational data with theoretical scenarios based on models of the impacting populations and impact processes.

This topic is one of active on-going research. Here we summarize accomplishments to date, and identify the many areas of uncertainty and issues requiring additional investigation.

2. Modeling the Crater History of Asteroids

The surface of an asteroid tells the story of what has happened to that body over time. To read that story, we must document an asteroid's morphological, orbital, and spectral properties, interpret how these factors relate to one another, and place them in context. Remote sensing observations via spacecraft allow us to measure an asteroid's shape, surface spectral signature, size, rotation rate, cratering record, and bulk density. Other asteroid properties (*e.g.,* internal composition, internal structure, chemical history) cannot be determined unless we land on the asteroid.

Faced with a paucity of information, scientists interpret asteroid histories using deductive methods. In the following sections, we will describe how to interpret the history of (a) a main belt asteroid, and (b) a near-Earth asteroid. For reference, a near-Earth asteroid (NEA) is defined by perihelion distance $q \leq 1.3$ AU and aphelion distance $Q \geq 0.983$ AU (Rabinowitz et al., 1994).

2.1. ASTEROID LIFETIMES

From the moment of their formation, asteroids are subjected to a continuous rain of small bodies (predominantly asteroids), causing both cratering and catastrophic disruption events. Primordial objects, defined as those few that have avoided disruption since the formation of the solar system, should have a record of impact events stretching ~ 4.6 Gyr, while the majority, created as collisional products, have abbreviated crater records. Determining whether an asteroid is primordial or a collisional product via remote sensing is difficult, even with high resolution surface images. For this reason, it is useful to estimate an asteroid's mean lifetime against collisional disruption. This value gives us a statistical measure of how long a given asteroid has been collecting craters. Assuming a target asteroid with diameter D is capable of being disrupted by the impact of a projectile of diameter d, a target asteroid's collisional lifetime is:

$$\tau = P_i \left(D^2/4 \right) N \left(> d \right).$$ (1)

$N \left(> d \right)$ is the cumulative number of bodies with diameter larger than d in the impactor population and P_i is the "intrinsic collision probability" that projectiles will strike the target asteroid. The "intrinsic collision probability" is defined (Wetherill, 1967) as the probability that a single member of the impacting population will hit a unit area of the target surface in a unit of time. Gravitational focussing is generally neglected here because asteroid escape velocities are $\sim$ m s^{-1} whereas asteroid impact velocities are $\sim$ km s^{-1} (Bottke et al., 1994). Methods for calculating P_i will be discussed in Sect. 2.2. For main belt objects striking one another, $P_i \approx 3 \times 10^{-18}$ km^{-2} yr^{-1} (Farinella and Davis, 1992; Bottke et al., 1994).

Eq. (1) can be used to determine the transition diameters between primordial asteroids and collisional products in the main belt. If $\tau < 4.6$ Gyr, the asteroid is most likely a collisional product. If $\tau > 4.6$ Gyr, the ratio of 4.6 Gyr / τ yields the probability that the asteroid is a collisional product. These calculations require the value of the minimum projectile diameter d that can disrupt a given target diameter D. This value d is given by the catastrophic disruption criterion defined by Melosh et al. (1994) and Melosh and Ryan (1997):

$$D = 0.7 \left(\frac{d^2 v}{\sqrt{G}\rho} \right)^{1/3}, \qquad (2)$$

where G is Newton's gravitational constant, ρ is asteroid density, and v is the impact velocity. Assuming some typical main belt values (*e.g.*, $\rho = 2700$ kg m^{-3} and $v = 5.3$ km s^{-1}; Bottke *et al.*, 1994), we find that $D = 200$ km, 100 km, 50 km, 20 km, and 10 km target asteroids are disrupted when they are struck by $d = 43.2$ km, 15.3 km, 5.4 km, 1.4 km, and 0.5 km projectiles, respectively. Large asteroids, with significant gravitational fields, are more difficult to disrupt.

To estimate $N (> d)$ for main belt asteroids with diameters larger than 1 km, we use the size distribution presented in Durda *et al.* (1998), who combined cataloged main belt asteroids with $d > 30$ km with population estimates of 5 km$< d < 30$ km bodies derived from debiased observational data (Van Houten *et al.*, 1970; Jedicke and Metcalfe, 1998). Estimating the size distribution of $d < 5$ km bodies is problematic because of a paucity of observational data. Rather than use the model dependent size distribution of $d < 5$ km bodies of Durda *et al.* (1998), and to keep things simple, we extend the size distribution to small bodies using the incremental slope index $p = 2.343$ (*i.e.*, $dN/dd \propto d^{-p}$), derived using a weighted least-squares fit to Palomar-Leiden Survey data (Van Houten *et al.*, 1970). We caution that this approximation is only valid for asteroids larger than 1 km or so; we use it because most of the target asteroids discussed in this chapter are disrupted by $d > 1$ km projectiles.

Substituting values into Eq. 1, we find that a 30 km asteroid has a collisional lifetime near 4.6 Gyr, such that objects with $D < 30$ km are almost certainly collisional products. For reference, a $D = 30$ km asteroid is disrupted by a $d = 2.5$ km projectile, and there are roughly $N (> 2.5\text{km}) \sim 320,000$ main belt objects. On the other hand, the collisional lifetime of a 100 km object is 2.4×10^{10} years, about 5 times the age of the solar system. Hence, the probability that a given $D = 100$ km object is a collisional product is $\sim 20\%$. For $D = 200$ km, we find a collisional lifetime of 4.4×10^{10} years and a probability of being a collisional product of 11%.

Similar results come from the consideration of asteroid families, groups of asteroids with similar orbital and spectral properties. Asteroid families are believed to be remnants of catastrophic disruption events among large asteroids, with 4-5 prominent families formed over several Gyr from $D > 200$ km parent bodies (Tanga *et al.*, 1999). Since there are roughly 35 asteroids with $D > 200$ km, it appears that $\sim 10\%$ have disrupted over the age of the Solar System, matching the estimate above for 200 km bodies.

Since the main belt is the primary reservoir of NEAs (Bottke *et al.*, 2000), we can use the above results to place constraints on NEA surface ages. NEAs are believed to be fragments of main belt asteroids that have reached Earth-approaching orbits via mean-motion resonances with the planets and/or secular resonances, where orbital frequencies are commensurate with the solar system's natural frequencies (Williams, 1973; Wisdom, 1983). This hypothesis is supported by the affinity between (i) NEA orbits and the results of N-body codes simulating the evolution of test bodies in main belt resonances, (ii) main belt and NEA spectral characteristics, and (iii) main belt collisional products with $D < 30$ km and NEA sizes, all but one which are $D < 30$ km. Accordingly, we predict that NEAs, like sub-30 km main belt asteroids, should have surface ages shorter than the age of the Solar System.

2.2. ASTEROID COLLISION PROBABILITIES AND IMPACT VELOCITIES

In the previous section, we estimated the likely age of a "generic" main belt asteroid. To determine the likely surface age of a particular main-belt asteroid (*e.g.*, 951 Gaspra) or NEA, however, requires more precise values of P_i and v. These values are obtained by (i) establishing the orbital distribution of the impactor population, (ii) computing the P_i and v between the target asteroid and each body of the impactor population, and (iii) determining total and average values from the P_i and v distributions.

2.2.1. *Main-belt target*

We first consider the orbital distribution of the projectile population capable of striking the target over time. According to Eq. (1), most of the craters observed on $D < 100$ km main-belt asteroids are made by projectiles smaller than 10 km. The orbital distribution of $D < 10$ km objects in the main belt, however, are biased by observational selection effects (Jedicke and Metcalfe, 1998), such that many more small asteroids have been discovered in the inner main-belt than the outer main-belt (*e.g.*, Bowell *et al.*, 1994). To avoid this problem, we use asteroid orbits which do not suffer from bias (*i.e.*, larger bodies), and assume their distribution is representative sample of all projectile orbits. One such representative set is 682 asteroids with $D \geq 50$ km (Farinella and Davis, 1992; Bottke *et al.*, 1994). We caution that this method may introduce some errors. According to Eq. (1), roughly half of these larger asteroids are primordial, whereas all asteroids with $D < 30$ km are collisional products. It is unclear whether the former set of objects, not yet in collisional equilibrium, can fully represent the latter set of objects, which has been heavily processed by collisions. For example, this approach does not account for unusual local projectile populations corresponding to asteroid families (Tanga *et al.*, 1999). As more small asteroid data become available, we will be able to better compensate for these factors.

Once we have our representative main-belt orbital sample, we can calculate P_i and v for each projectile-target pair (Bottke *et al.*, 1994). The method calculates collision probabilities and impact velocities between bodies on fixed (a, e, i) orbits, integrating over uniform distributions of longitudes of apsides and nodes to account for secular precession, which is quick ($\sim$ 0.01 Myr) compared with the collision timescale ($>$ 1 Myr for objects capable of making an observable crater). For this reason, all possible orbital intersection positions between each pair must be evaluated and weighed. The characteristic values of P_i and v for the target body are determined from the distribution of P_i and v values.

2.2.2. *Near-Earth asteroid*

Calculating meaningful P_i and v values for asteroids outside the main-belt is problematic. The population of inner Solar System asteroids (*e.g.*, Mars-crossing asteroids, NEAs) is heavily biased, such that it is difficult to choose a set of known objects capable of representing the projectile population. Moreover, NEAs followed complicated dynamical pathways for millions of years before reaching their current orbits. Thus, mean-motion resonances, secular resonances, and planetary close encounters are modifying the orbits of NEAs at the same time that main-belt and inner Solar System asteroids are colliding with them. This chaotic evolution means that NEAs cannot be tracked backwards in time more than a few hundred years, too short to understand where a particular NEA originated, how it arrived at its current location, or how long the entire trip took. It also means that P_i and v change as NEAs evolve in (a, e, i) space.

To overcome these problems, work is proceeding along several fronts. The debiased projectile population may be determined from recent estimates of the debiased orbital and size distribution of the NEA and MC populations (*e.g.*, Bottke *et al.*, 2000). The orbital pathways of particular NEAs may be estimated by numerically integrating suites of test bodies emerging from various main belt resonances. If enough test bodies are used, this technique must eventually produce trajectories passing through every possible NEA orbit. Assuming the NEA we are interested in resides at a given (a, e, i) position, we can collect all the test bodies passing through that position and weigh their orbital paths by the flux of material coming from their source resonance. This will give us some idea of which source resonance is most likely to produce a particular NEA. Additional constraints may come from comparing the NEA's spectra to the spectra of asteroids near possible source resonances. When a most probable orbital path is found, we can calculate P_i and v by comparing its evolving (a, e, i) values against our sample objects representing the orbits of the projectile population.

2.3. ASTEROID CRATERING RATE

The production rate of L diameter craters (or larger) on our target asteroid per square km per year is given, in general, by:

$$R(> L) = \frac{P_i \; (D^2 + l^2) \; N \, (> l)}{4 \pi D^2}. \qquad (3)$$

where l is the diameter of the projectile which makes a crater L when it strikes the target asteroid at v. $N \, (> l)$ is the cumulative number of projectiles with diameter l or larger in the impactor population. Estimates of $N \, (> l)$ for the main belt are discussed above, though the shape of the size distribution for $l < 3$ km projectiles is unknown. For this reason, models of crater populations often use power-law size distributions with adjustable log-log slopes (sometimes with breaks in the slope) in the small asteroid range. Estimates of $N \, (> l)$ for the NEA region are found in Rabinowitz et al. (2000) and Bottke et al. (2000), yielding an absolute magnitude (H) distribution for $15 < H < 22$ NEAs of $N(< H) \sim 10^{0.35H}$. In Bottke et al. (2000), this distribution is fixed to the estimated set of NEAs with $13 < H < 15$ (roughly 66 bodies). Bottke et al. (2000) use the conversion formula $D = 4365 \times 10^{-H/5}$ to transform H into D.

At this time, the population of Mars-crossing asteroids adjacent to the main belt which have not yet entered NEA orbits (defined as "IMC"s by Bottke et al., 2000) can be estimated as follows. The number of observed IMCs with $H < 15$ is ~ 200, while the observed number of NEAs with $H < 15$ is ~ 50 (Bowell et al., 1994). If we assume that both values suffer from a comparable degree of observational bias, we can postulate that the IMC population is roughly 4 times (i.e., 200/50) the size of the NEA population.

Since many IMCs and NEAs remain coupled to the main belt throughout their evolution, and the main belt size distribution is ~ 250-1000 times more numerous at a given l value than the IMC or NEA populations, respectively, we predict that most craters on IMCs or NEAs are produced by main belt projectiles. If we eventually find a IMC or NEA with a lightly cratered surface, it would suggest that (i) the surface was reset by some process after the asteroid became collisionally decoupled from the main belt, or (ii) the object is a new collisional product which spent little time exposed to main belt projectiles.

The relationship between impactor size l and crater size L have often been determined using scaling laws, which extrapolate high-velocity impact experiments in the laboratory up to the energies of main belt asteroid collisions (a review of the subject can be found in Melosh, 1989). Complementary, and probably more realistic results, are obtained via numerical hydrocode models, which are calibrated using both laboratory experiments and nuclear weapons tests (*e.g.*, Nolan *et al.*, 1992; 1996; Asphaug *et al.*, 1996).

The inadequacies of analytical scaling laws became apparent from analysis of the first images of an asteroid's surface, which were obtained when the Galileo spacecraft flew past Gaspra in 1991. At that time, conventional wisdom from scaling laws predicted that Gaspra could not have sustained large (multi-km) crater-forming impacts without being catastrophically fragmented. Initial crater counts from the Galileo data (Belton *et al.*, 1992) identified craters only up to 1.5 km in diameter. However, the presence of larger concavities on Gaspra suggested that scars of larger impacts had been recorded. Given that these features were up to 5 km across, comparable to the dimensions of the asteroid itself, it was not surprising that their morphologies were not those of conventional craters. Nolan *et al.* (1992) then performed hydrocode simulations of impacts, demonstrating that Gaspra could have sustained such energetic impacts, leaving scars without disruption (see also Nolan, 1994; Nolan *et al.*, 1996). Those results have now been confirmed and extended by numerous researchers.

Hydrocode results show that asteroids react far differently to collisions than predicted by scaling laws. When a hypothetical km-sized homogenous asteroid is hit by an object tens of meters in diameter, a shock wave is launched from the impact site which shatters the body into multiple fragments. These pieces usually have very low relative velocities and thus remain gravitationally bound. The asteroid becomes a "rubble pile", a collection of large and small components held together by gravity, rather than physical strength. For this reason, most km-sized bodies are probably rubble piles covered by regolith (Love *et al.*, 1996; Benz and Asphaug, 1999).

The shock front also pulverizes material at the impact site, such that the crater excavation occurs in virtually strengthless material. Thus, the scaling relationship between l and L changes as we move from small to large craters. Small craters form in the "strength-scaling" regime, where the final size of the crater is controlled by the strength of the target surface. Large craters, however, have their formation controlled by the target's gravity, such that their sizes are given by "gravity-scaling". Since weak material does not transport energy efficiently, the size of the largest crater in the gravity regime is significantly larger than suggested by extrapolation of strength-scaling laws. This process explains why observed asteroid surfaces (*i.e.*, Gaspra, Ida, Mathilde, Vesta) and other small bodies (*e.g.*, Phobos, Deimos, Amalthea, Thebe, Janus, Epimetheus, Hyperion, and Proteus) have such huge craters on their surface (Thomas, 1999). The transition between small and large craters (and strength- and gravity-scaling) is best determined via direct hydrocode experimentation, which can also account for important factors like the target asteroid's shape and internal structure.

3. Crater Erasure Processes

Assuming the above components are well understood, Eq. (3) can be used to interpret fresh and well-preserved cratered surfaces. Older and more degraded surfaces, however, cannot be examined until we account for physical processes capable of erasing asteroid craters.

Beginning with the hydrocode simulations of Nolan *et al.* (1992), which were interpreted by Greenberg *et al.* (1994; 1996), it has been possible to describe how impacts destroy old craters as well as create new ones. We describe several of these processes below.

3.1. GLOBAL JOLTING OF THE SURFACE

Numerical hydrocode experiments approaching the limit of catastrophic disruption have shown that the impact of a sufficiently large projectile on a target asteroid will cause material to jump up over the asteroid's entire surface (Greenberg *et al.*, 1994; 1996). As long as the projectile is not large enough to disrupt the target, when this material settles back to the asteroid, it buries large scale topography on a scale commensurate with the impact energy. This process is analogous to striking the bottom of a pan of sand with a hammer. If we increase the amount of sand in the pan (*i.e.*, big asteroids), or we decrease the velocity of the hammer striking the pan (*i.e.*, small projectiles), global jolt is reduced. For Gaspra-sized asteroids ($D \sim 14$ km), the impact of a projectile which creates a $L \sim 8$ km crater will erase all craters smaller than 1 km (Greenberg *et al.*, 1994). For larger asteroids like Ida, ($D \sim 28$ km), a $L \sim 20$ km crater is needed to do the same job (Greenberg *et al.*, 1996). Factors like asteroid shape and internal structure also modify the propagation of seismic waves and the degree of shaking for a given impact.

3.2. SANDBLASTING

When small projectiles strike the target asteroid, they break up surface material, create regolith, and form tiny craters. The cumulative effect of these projectiles on the underlying topography is analogous to the industrial process of sandblasting. Thus, old craters may be smoothed and degraded until they are no longer distinguishable from the background surface. Though this effect has not been quantitatively modeled, conservative estimates suggest that a crater L which has been saturated three times over by craters between 0.1–1.0 L is erased. Sandblasting is a more important effect on

large asteroids than small asteroids, since their surface is less susceptible
to crater erasure via global jolting.

3.3. THE "COOKIE-CUTTER" EFFECT

When craters are formed on a planetary surface or on an asteroid, they
erase all preexisting smaller craters at the impact site. This process cuts
circular holes in the surface distribution of craters, much like a cookie cutter does when it removes dough. The size of the cut consists of the new
crater size, augmented by seismic shaking (local jolt) and burial by crater
ejecta in the vicinity of the impact. Conservative models of this effect assume that all craters within one radius of the center of the new crater are
obliterated by the "cookie cutter" (Greenberg et al., 1996). This process
becomes more effective as impact energy increases, both because of crater
size and augmented local or regional jolt.

3.4. TIDAL DISTORTION OR DISRUPTION

The spectacular breakup of comet D/Shoemaker-Levy 9 by Jupiter's tidal
forces in 1992 has fueled speculation that rubble-pile asteroids passing near
a planet may be disrupted. This hypothesis has been tested via N-body simulations of encounters between ellipsoid-shaped rubble-pile asteroids and
the Earth (Bottke et al., 1997; Richardson et al., 1998; Bottke et al., 1999).
Their results indicate that tidal forces set individual asteroid components
into motion, producing a landslide moving towards the ends of the body
which modifies or buries craters. One possible outcome is a "SL9-type" disruption, with the asteroid pulled into a train of fragments. Less dramatic
are mass shedding events, where clumps and/or solid fragments escape off
the ends of the object, or gross asteroid shape distortions, where the tidal-
induced landslide stops short of mass shedding. 1620 Geographos, a NEA
with a fast rotation rate (i.e., $P \sim 5.2$ h) and a distinctive shape (i.e.,
highly elongated with a single convex side, tapered ends, and small pro-
tuberances swept back against the rotation direction) may be an example
of an object which has shed mass via tidal forces (Bottke et al., 1999).
Thus, we suspect that 1620 Geographos and some other planet-crossing
objects have strongly-modified crater records. This process, like global jolt,
may create a blank slate, resetting and simplifying the subsequent impact
record.

3.5. OTHER FACTORS

There are several additional factors which modify the cratering history of
asteroids. Asteroids which are members of dynamical families have similar

(proper) (a, e, i) values, such that the local projectile population has a higher P_i and lower characteristic v than background projectiles. These bodies also have more stringent constraints on their lifetime, since they were formed during the family-forming event itself. Cratering records may also be modified to some degree by secondaries. For example, several small craters on 243 Ida show evidence that they were produced by the impact of boulders reaccreted after an impact event (Geissler et al., 1996).

4. Applications to Specific Cases

4.1. 951 GASPRA

Initial crater counts on images of the surface of asteroid Gaspra obtained by the Galileo spacecraft (Fig. 1) were based on identification of traditional morphologies and the expectation that such a small asteroid could not have survived a hit energetic enough to create a crater much larger than about a kilometer across (Belton et al., 1992). In addition, the relatively low numbers, relative to what was expected for a crater-saturated surface, were interpreted to imply that the crater population was representative of the production function of the impacting population. That was to say that the number of craters of any size is directly proportional to the number of impactors that make craters of that size. On the basis of that population, combined with a strength-scaling-based production law (crater diameter proportional to impactor diameter), Belton et al. (1992) concluded that the impacting population was an extrapolation of the Palomar-Leiden Survey population (PLS; Van Houten et al., 1970) with incremental power-law index −2.95 down to a diameter of 175 m, at which size the distribution bends to an index of −3.5. That population differs from the "simple" size distribution we introduced in Sect. 2.1 above because (a) it did not include statistical corrections to the PLS (Jedicke and Metcalfe, 1998), and (b) it introduced the very steep distribution for small bodies as seemingly directly required by the steep crater population on Gaspra.

Greenberg et al. (1994) took a different approach, recognizing the possibility that a small asteroid might survive impacts that would leave very large concavities (if not the crater or basin morphologies familiar from the surfaces of larger bodies). Their crater counts thus included features up to 5 km across, and an estimate that the entire surface may have about 8 craters larger than 4 km across. These are very large features on a body less than 20 km long and 10 km across.

In order to interpret this impact record, Greenberg et al. (1994) developed an impact history model based on the numerical simulations of impact events by Nolan et al. (1992) and Nolan (1994). Those simulations showed how the processes of crater erasure and modification discussed in

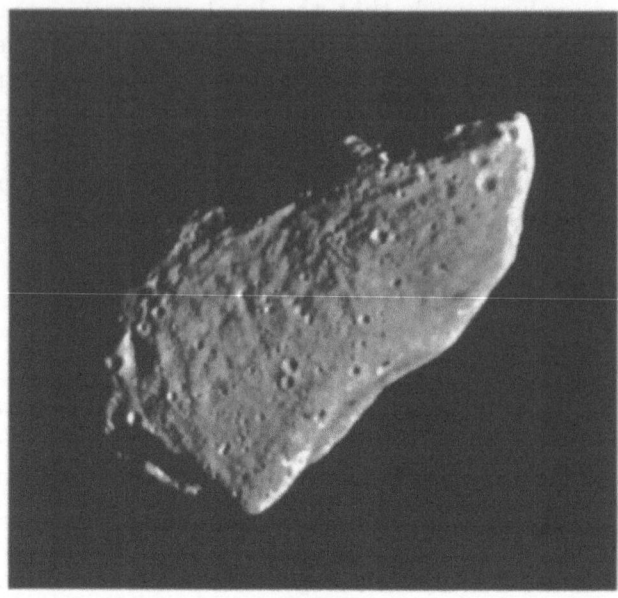

Figure 1. An image of asteroid 951 Gaspra taken by the Galileo spacecraft. The resolution is ~ 54 meters/pixel. Gaspra is an irregular body with dimensions about $19 \times 12 \times 11$ km. The large concavity on the lower right limb is about 6 km across, while the the prominent crater on the terminator, center left, is ~ 1.5 km. Note the abundance of small craters; more than 600 craters, 100-500 m in diameter are visible here. The number of such small craters compared to larger ones is much greater for Gaspra than for previously studied small bodies. Gaspra's irregular shape suggests is a collision product.

Sect. 3 could explain what had been interpreted as an unsaturated appearance by Belton *et al.* (1992). With that new insight into the effects of major impact events, Greenberg *et al.* (1994) found that the current appearance of Gaspra does not directly reflect the production of craters, but rather the interplay between crater production and erasure. Even taking into account considerable uncertainty as to the exact quantitative nature of the jolt, cookie-cutter, ejecta blanketing, sandblasting and crater production processes, Greenberg *et al.* (1994) were only able to fit the observed distribution with an impacting population with considerably more small bodies than had been inferred by Belton *et al.* (1992): The size distribution was found to bend upward to a power-law index of −4 below 100 m.

Another important result was that a major impact had probably jolted away nearly all craters only about 50 Myr ago, even though such an event is to be expected only about every 500 Myr. The significance of that result is that it serves as a reminder that the specific impact record on any particular asteroid is likely affected strongly by the stochastic events of the largest impacts. Gaspra-sized bodies probably have diverse appearances even given

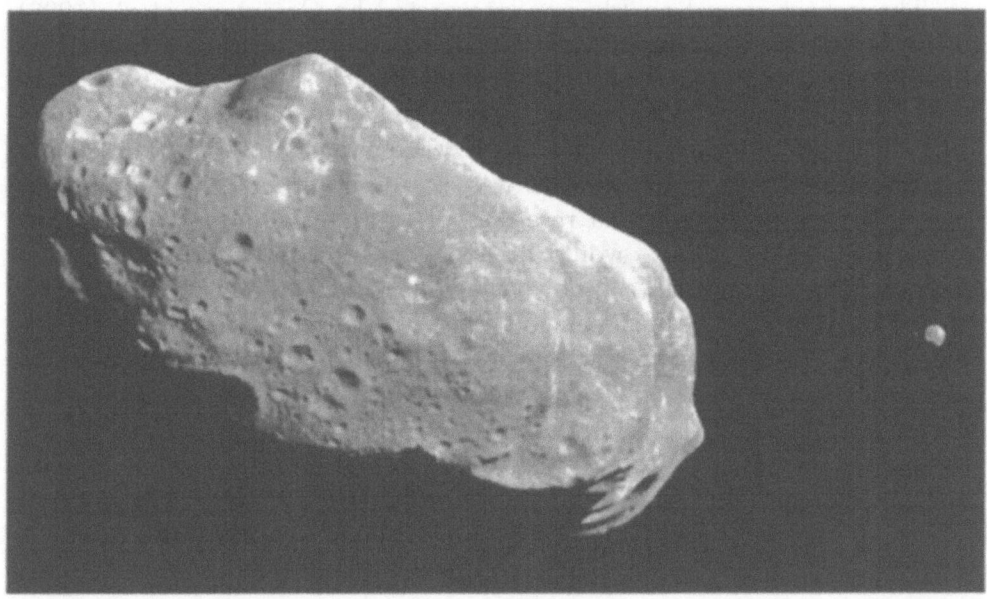

Figure 2. An image of asteroid 243 Ida and its moon Dactyl taken by the Galileo spacecraft. Ida is 58 km long and 23 km wide, more than twice as large as Gaspra. It is a member of the Koronis family, produced by the catastrophic disruption of its parent asteroid. This view shows numerous degraded craters on Ida, suggesting its surface is geologically old. Dactyl has a mean diameter of 1.4 km. At the time of this image, Dactyl was 85 km from the center of Ida.

the same impacting population. Our understanding of the record of impacts will improve greatly as we obtain a much larger sample of information on asteroid surfaces.

4.2. 243 IDA

When images of 243 Ida were returned from the Galileo spacecraft in 1994 (Fig. 2), they contained two types of information relevant to its impact history. First, we obtained good images of the cratered surface, and second, we discovered the presence of a small satellite, Dactyl, with diameter 1.4 km. A body of Dactyl's size has a likely lifetime against disruption of only 100 Myr. Also, the most likely mode of formation was as an accompaniment to the formation of Ida in the break-up of the precursor parent body of the Koronis family of asteroids (Durda, 1996). Thus, either (a) Ida, Dactyl, and the Koronis family formed less than about 100 Myr ago, or (b) Dactyl is a fragment of a larger, earlier disrupted satellite of Ida. Either of those possibilities is consistent with the constraint that the Koronis family formation event occurred ∼ 1.5 Gyr ago (Davis *et al.*, 1996).

The impact record on Ida was interpreted by Greenberg *et al.* (1996), using techniques similar to those that had been applied to Gaspra, and based on a more extensive set of numerical impact simulations. They found that the same main-belt population that had been inferred to have bombarded Gaspra appeared to have created the impact record on Ida, and in fact the record was closer to what was expected without the anomalously recent large impact that had reset Gaspra's crater counts. The observed crater population is what would be expected after only about 50 Myr of impacts (consistent with (a) above, the young Ida/Dactyl/Koronis model), or after a steady state between jolt erasure and production is reached in more than 1 Gyr (consistent with case (b) above). In either case, the result is consistent with the 1.5 Gyr upper limit to the age of the Koronis family derived by Davis *et al.* (1996). In any case, this work demonstrated again the importance of both crater erasure and formation during bombardment of asteroidal surfaces.

Studies of the distribution of ejecta have proven to be significant on Ida. First, Geissler *et al.* (1996) discovered and explained the numerous boulders lying on the surface of Ida as ejecta debris. The locations of the boulders correlate with the computed expected re-impact sites, as the leading surfaces of the elongated asteroid sweeps ejecta from the space around it. Also, computations of the redistribution of ejecta from specific craters was shown to correlate with somewhat spectrally bluer areas on the surface. Thus we can begin to address issues of composition and compositional mixing on the basis of studies of impacts into asteroids. This issue also becomes important in the context of 4 Vesta, which we will discuss in Sect. 4.4.

4.3. 253 MATHILDE

The NEAR spacecraft recently imaged 253 Mathilde (Fig. 3), a heavily-cratered C-type main belt asteroid whose mean diameter is $D = 53$ km (Veverka *et al.*, 1997). Its chief characteristics include a very low bulk density (~ 1.3 g cm^{-3}), such that its porosity is probably near 50% (Veverka *et al.*, 1999), and a surface dominated by huge craters, 5 of which have diameters L between 20 and 33 km (Chapman *et al.*, 1999). These craters, many which are adjacent to one another, do not appear to have noticeably modified one another, suggesting that Mathilde's porous and/or broken-up interior efficiently damp out the propagation of impact energy. This evidence confirms the inference first made for Gaspra (Greenberg *et al.*, 1994) that such damping allows a target to survive impacts that create craters comparable to the size of the target. This may be especially true for such a low density object as Mathilde, where damping is especially effective. The nearly pristine topography near the largest craters, however, suggests

Figure 3. An image of C-type asteroid 253 Mathilde taken from the Near Earth Asteroid Rendezvous (NEAR) spacecraft. Mathilde has a mean diameter $D = 53$ km and a low mean density near 1.3 g cm^{-3}. Four craters have been observed with diameters equal to or greater than Mathilde's mean radius. Their relatively undamaged appearance suggests they did not cause significant damage to the asteroid when they formed. Accordingly, Mathilde's interior must be a poor transmitter of impact energy, such that it can withstand large impacts without disruption.

that global jolt is probably ineffective at erasing small craters on this asteroid. For this reason, crater numbers should be dominated by effects like sandblasting and cookie-cutter (*e.g.*, Chapman *et al.*, 1999).

Conventional crater scaling laws cannot be applied to Mathilde because its internal structure is very different from the typical target bodies used in laboratory impact experiments. For this reason, large-scale cratering events on porous asteroids do not match those observed in laboratory experiments with less-porous geological materials. Instead, some insights into Mathilde's craters come from impact experiments of high velocity projectiles into porous under-dense material and from nuclear cratering tests in Pacific atolls. These results suggest that craters in porous media form primarily by compaction (with little deposition of ejecta exterior to the crater) rather than traditional excavation and ejection (Housen *et al.*, 1999). The small amount of ejecta launched away from the compaction site have low enough velocities that nearly all of the material is redeposited within the crater bowl. This scenario would explain how craters near an impact site avoided damage. We caution, however, that the material properties used in the Housen *et al.* experiments may not be good analogues for Mathilde. On the other hand, recent numerical hydrocode modeling suggests that highly porous, rubble-pile asteroids can readily damp out the energy produced by large collisions; in these simulations, compaction is not a factor (Asphaug *et al.*, 1996; 1999; Asphaug, 2000). It is clear that further modeling of impacts into porous or shattered objects will be needed to interpret Mathilde's cratering history.

Despite this, we can glean some useful information from standard modeling efforts. The intrinsic collision probability P_i and mean impact velocity v for projectiles with Mathilde are 3.82×10^{-18} km^{-2} yr^{-1} and 5.3 km s^{-1}, respectively. Using Eqs. (1) and (2), we find that Mathilde's collisional lifetime against $d \sim 5$ km projectiles is 5 Gyr, slightly longer than the age of the Solar System. Based on this, the probability that Mathilde is a collisional product is $\sim 90\%$ (*i.e.*, 4.6 Gyr / 5.0 Gyr) using the present population of main belt projectiles.

Mathilde's crater rate was recently estimated by Davis (1999) using gravity-scaling laws and the assumption that Mathilde's internal structure was similar to sand. He suggested that $L = 20$ km and 33 km craters on Mathilde could have been made by $l = 0.75$ and 1.3 km projectiles, respectively. Using the size distribution from Sect. 2.1, $N(> 1.3\,\text{km}) \approx 1 \times 10^6$ and $N(> 0.75\,\text{km}) \approx 5 \times 10^6$. Thus, the inverse frequency of forming $L = 20$ and 33 km craters being formed on Mathilde is 80 Myr and 400 Myr, respectively, far shorter than the age of the Solar System. These ages would imply that Mathilde is ~ 1 Gyr old. On the other hand, if a $l = 2\text{-}3$ km projectile was needed to make a $L = 33$ km crater, the time to form this crater on Mathilde would increase to ~ 2 Gyr. Further work is needed to understand how to accurately transform l to L on Mathilde.

4.4. 4 VESTA

Vesta, a $D = 530$ km main belt asteroid, is the smallest surviving planetary body to have undergone the terrestrial-planet-like processes of heating and differentiation (Fig. 4). The fact that Vesta's basaltic crust is more-or-less intact after ~ 4.5 Gyr of collisional exposure and that it retains significant geologic diversity and hemispheric dichotomy (Binzel *et al.*, 1997) supports the idea that Vesta is primordial. Images from the Hubble Space Telescope show that a ~ 450 km impact basin exists on Vesta's southern hemisphere (Thomas *et al.*, 1997). Hydrocode modeling of the basin-forming event suggests it may have been created when a $l \sim 42$ km diameter asteroid struck Vesta at 5.4 km s^{-1} (Asphaug *et al.*, 1997).

We can use the methods described in this chapter to examine this event in more detail. The intrinsic collision probability P_i for projectiles with

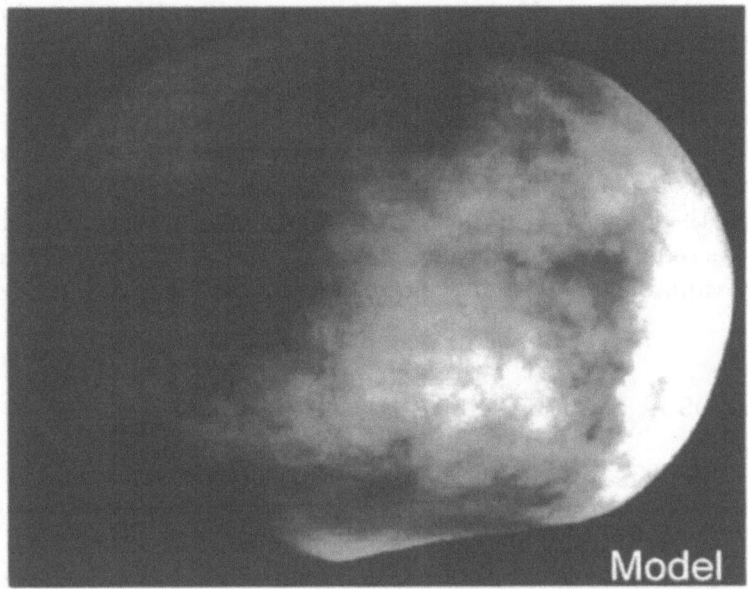

Figure 4. A model of 4 Vesta derived from Hubble Space Telescope images. The resolution of the images used to create this shape was 36 km/pixel. Vesta is 530 km across. A 450 km diameter crater containing a large central peak is seen at the bottom of the image. The crater is about 8 km deep with a rim that rises in places an additional 8 to 14 km. Its central peak is about 13 km high. Other craters on Vesta are up to 150 km in diameter and a few km deep. At least part of the impact crater has a deep mafic absorption band, consistent with plutonic basalts and/or the excavation of olivine mantle material (Thomas *et al.*, 1997).

Vesta is 2.9×10^{-18} km^{-2} yr^{-1} while $N \, (> 42 \, \text{km}) \sim 800$. Putting these components together, we estimate that a 450 km basin forms on Vesta every 6 Gyr. Thus, the probability that a basin-forming event occurred over the age of the Solar System is 75% (*i.e.,* 4.6 Gyr / 6.0 Gyr), large enough that we can have confidence in our procedure, but not so large that Vesta should have more than one big impact basin.

Observations indicate that olivine mantle material was excavated when the crater was formed (Thomas *et al.*, 1997). In addition, several geologic units on Vesta have a substantial olivine component (Binzel *et al.*, 1997). From a theoretical point of view, these results are not surprising for the following reason: The thickness of a basalt layer on a differentiated body the size of Vesta could not be more than about 15 km, so the larger basin-forming impact events would certainly excavate well into the olivine mantle so that ejecta would include larger olivine fractions. If the ejecta are well-mixed and widely distributed over the surface, the surface would be uniform, with spectra reflecting the olivine-basalt mix. However, the diversity of spectral units on Vesta's surface (reported by Binzel *et al.*), argues against such uniform mixing and distribution. The observations seem to suggest that olivine is exposed within the largest basins, or that olivine-rich ejecta have been non-uniformly distributed, and that ejecta from smaller craters have not re-blanketed the surface uniformly. Clearly, the impact record on Vesta places constraints on the impacting population, and especially on the physics of impact excavation and ejecta distribution. The implications require careful consideration. These constraints will become increasingly valuable as more data become available regarding the asteroid's surface.

4.5. 433 EROS AND OTHER NEAR-EARTH OBJECTS

In February 2000, the NEAR spacecraft will begin orbiting 433 Eros, a $40.5 \times 14.5 \times 14.1$ km S-type NEA. This mission promises to give us a close look at an asteroid roughly the same shape and size as 243 Ida. Extensive imaging and scientific measurements will help us understand Eros's surface composition, geology, physical properties, and internal structure. This data will be fed into sophisticated hydrocode simulations, allowing us to better model the transformation between projectile sizes (l) and crater sizes (L) and important effects like global jolt. Given the similarity between Eros and Ida, the study of Eros may help us place Ida's cratering history in better context (*e.g.*, Greenberg *et al.*, 1996). The biggest problem interpreting Eros's cratering history (and comparing it with 243 Ida) is that Eros left the main belt some time ago and has followed a unique dynamical path to reach its current orbit in the Amor region ($a = 1.46$ AU, $e = 0.22$, $i = 10.8°$). As discussed above, the unknown provenance and chaotic evolution of NEAs make it difficult to constrain the appropriate projectile population and parameters like P_i and v. On the other hand, as we begin to understand better how to use the impact record on an asteroid's surface to characterize the bombarding population, differences between main belt targets (*e.g.*, Ida) and near-Earth asteroids (*e.g.*, Eros) may shed light on the experience of all bodies that come to Earth from the main belt. Data from the NEAR spacecraft should tell us whether this issue can be realistically attacked using available tools.

Additional images revealing the impact histories of near-Earth asteroids are being obtained from Earth-based delay-Doppler radar observations. The highest resolution images from these investigations often reveal craters, concavities, and faceted surfaces, undoubtedly fashioned by collisions (*e.g.*, images of 4179 Toutatis; Ostro *et al.*, 1999). As the resolution improves and the numbers of bodies imaged increases, it will be possible to gain improved understanding of the collision histories and impacting population to which these bodies have been exposed over the duration of their surface ages. Their collision histories and their orbital histories are intimately interwoven: Collisions have contributed to resetting of orbital parameters, while the orbital evolution has determined the degree of exposure to main-belt or other impacting populations. Thus the impact record on the surface of small bodies may provide an important constraint as we unravel the history of these populations.

5. Conclusions

Because asteroids have travelled through various portions of the Solar System with a variety of ages, sizes, and orbital histories, the impact record on their surfaces provide potentially useful probes of the environments through which they have passed.

Interpreting this record in a meaningful way will require scientific advances along two fronts. First, we require more asteroid surface images. Progress is underway toward that goal as we improve our techniques and instruments for radar imaging and as spacecraft make small bodies their primary objectives and, increasingly, high priority ancillary targets of opportunity. Second, we need to know precisely what happens when one asteroid strikes another. The role of parameters such as impact velocity, projectile size, and the target's physical properties, size, and shape will have to be understood before we can adequately analyze an asteroid's cratered surface. Advances are underway in this area as well, thanks to improving numerical codes and computational hardware, grounded in on-going impact experiments and theoretical studies.

So far, limited data on a few asteroids, and imperfect (but improving) understanding of impact processes has required fairly speculative components in each analysis. Nevertheless, the exercise of interpreting the data in hand has identified the key issues that need further study and has blocked out the general approach that will be useful as both our observational data and theoretical understanding improve in the near future.

References

Asphaug, E. (2000) The large, undisturbed craters of Mathilde: Evidence for structural porosity, *Lunar Planet. Sci.* **31**, 1864.

Asphaug, E. (1997) Impact origin of the Vesta family, *Meteor. Planet. Sci.* **32**, 965–980.

Asphaug, E. and Thomas, P.C. (1999) Modeling mysterious Mathilde, *Lunar Planet. Sci.* **30**, 2028.

Asphaug, E., Ostro, S.J., Hudson, R.S., Scheeres, D J. and Benz, W. (1996) Disruptive impacts into small asteroids, *Nature* **393**, 437–439.

Belton, M.J.S., Veverka, J., Thomas, P., Helfenstein, P., Simonelli, D., Chapman, C., Davies, M.E., Greeley, R., Greenberg, R. and Head, J. (1992) Galileo encounter with 951 Gaspra First pictures of an asteroid, *Science* **257**, 1647–1652.

Benz, W. and Asphaug, E. (1999) Catastrophic disruptions revisited, *Icarus* **142**, 5–20.

Binzel, R.P., Gaffey, M.J., Thomas, P.C., Zellner, B.H., Storrs, A.D. and Wells, E.N. (1997) Geologic mapping of Vesta from 1994 Hubble Space Telescope images, *Icarus* **128**, 95–103.

Bottke, W.F., Jedicke, R., Morbidelli, A., Petit, J.–M., Gladman B. (2000) Understanding the distribution of near-Earth asteroids, *Science* **288**, 2190–2194.

Bottke, W.F., Richardson, D.C. and Love, S.G. (1997) Can tidal disruption of asteroids make crater chains on the Earth and Moon? *Icarus* **126**, 470–474.

Bottke, W.F., Jr., Richardson, D.C., Michel, P. and Love, S.G. (1999) 1620 Geographos and 433 Eros: Shaped by planetary tides? *Astron. J.* **117**, 1921–1928.

Bottke, W.F., Nolan, M.C., Greenberg, R. and Kolvoord, R.A. (1994) Velocity distributions among colliding asteroids, *Icarus* **107**, 255–268.

Bowell, E., Muinonen, K., and Wasserman, L.H. (1994) A public-domain asteroid orbit data base, in *Asteroids, Comets, Meteors 1993*, eds. A. Milani, M.Di Martino, and A. Cellino, Kluwer, pp. 477–481.

Chapman, C.R., Merline, W.J. and Thomas, P. (1999) Cratering on Mathilde, *Icarus* **140**, 28–33.

Davis, D.R. (1999) The collisional history of asteroid 253 Mathilde, *Icarus* **140**, 49–52.

Davis, D.R., Chapman, C.R., Durda, D.D., Farinella, P. and Marzari, F. (1996) The formation and collisional/dynamical evolution of the Ida/Dactyl system as part of the Koronis family, *Icarus* **120**, 220–230.

Durda, D.D., Greenberg, R., and Jedicke, R. (1998) Collisional models and scaling laws: A new interpretation of the shape of the main belt asteroid distribution, *Icarus* **135**, 431–440.

Durda, D.D. (1996) The formation of asteroidal satellites in catastrophic collisions, *Icarus* **120**, 212–219.

Farinella, P. and Davis, D.R. (1992) Collision rates and impact velocities in the main asteroid belt, *Icarus* **97**, 111–123.

Geissler, P., Petit, J.–M., Durda, D.D., Greenberg, R., Bottke, W., Nolan, M and Moore, J. 1996 Erosion and ejecta reaccretion on 243 Ida and its moon, *Icarus* **120**, 140–157.

Greenberg, R., Nolan, M.C., Bottke, W.F., Kolvoord, R.A. and Veverka, J. (1994) Collisional history of Gaspra, *Icarus* **107**, 84–97.

Greenberg, R., Bottke, W.F. Nolan, M., Geissler, P., Petit, J.–M., Durda, D.D., Asphaug, E. and Head, J. (1996) Collisional and dynamical history of Ida, *Icarus* **120**, 106–118.

Housen, K.R., Holsapple, K.A. and Voss, M.E. (1999) Compaction as the origin of the unusual craters on the asteroid Mathilde, *Nature* **402**, 155–157.

Jedicke, R., and Metcalfe, T.S. (1998) The orbital and absolute magnitude distributions of main belt asteroids, *Icarus* **131**, 245–260.

Love, S.G. and Ahrens, T.J. (1996) Catastrophic impacts on gravity dominated asteroids, *Icarus* **124**, 141–155.

Melosh, H.J. and Ryan, E.V. (1997) Asteroids: Shattered but not dispersed, *Icarus* **129**, 562–564.

Melosh, H.J., Nemtchinov, I.V., Zetzer, Yu.I. (1994) Non-nuclear strategies for deflecting comets and asteroids, in *Hazards Due to Comets and Asteroids*, ed. T. Gehrels, Univ. of Arizona Press, Tucson, pp. 1111-1132.

Melosh, H.J. (1989) *Impact Cratering: A Geologic Process*, Oxford University Press, New York.

Nolan, M.C., Asphaug, E., Melosh, H.J. and Greenberg, R. (1996) Impact craters on asteroids: Does gravity or strength control their size?, *Icarus* **124**, 359–371.

Nolan, M.C. (1994) *Delivery of meteorites from the asteroid belt*, Ph.D Thesis., Univ. Arizona, Tucson.

Nolan, M.C., Asphaug, E. and Greenberg, R. (1992) Numerical simulation of regolith production on 951 Gaspra, *Meteoritics* **27**, 270.

Rabinowitz, D., Helin, E., Lawrence, K., Pravdo, S. (1999) A reduced estimate of the number of kilometer-sized near-Earth asteroids, *Nature* **403**, 165–166.

Rabinowitz, D., Bowell, E., Shoemaker, E.M. and Muinonen, K. (1994) The population of Earth-crossing asteroids, in *Hazards Due to Comets and Asteroids*, ed. T. Gehrels, Univ. of Arizona Press, Tucson, pp. 285–312.

Richardson, D.C., Bottke, W.F. and Love, S.G. (1998) Tidal distortion and disruption of Earth-crossing asteroids, *Icarus* **134**, 47–76.

Tanga, P., Cellino, A., Michel, P., Zappala, V., Paolicchi, P. and Dell'Oro, A. (1999) On the size distribution of asteroid families: The role of geometry, *Icarus* **141**, 65–78.

Thomas, P.C. (1999) Large craters on small objects: occurrence, morphology, and effects, *Icarus* **142**, 89–96.

Thomas, P.C., Binzel, R.P., Gaffey, M.J., Storrs, A.D. Wells, E.N. and Zellner, B.H. (1997) Impact excavation on asteroid 4 Vesta: Hubble Space Telescope results, *Science* **277**, 1492–1495.

Van Houten, C.J., Van Houten-Groeneveld, I., Herget, P. and Gehrels, T. (1970) The Palomar-Leiden survey of faint minor planets, *Astron. Astrophys. Suppl.* **2**, 339.

Veverka, J., and 13 co-authors (1999) NEAR encounter with asteroid 253 Mathilde: Overview, *Icarus* **140**, 3–16.

Veverka, J., and 16 co-authors (1997) NEAR's flyby of 253 Mathilde: Images of a C asteroid, *Science* **278**, 2109–2114.

Wetherill, G.W. (1967) Collisions in the asteroid belt, *J. Geophys. Res.* **72**, 2429–2444.

Williams, J.G. (1973) Meteorites from the asteroid belt? *Eos Trans. AGU* **54**, 233.

Wisdom, J. (1983) Chaotic behavior and the origin of the 3/1 Kirkwood gap, *Icarus* **56**, 51–74.

COMETARY DYNAMICS

HAROLD F. LEVISON
Department of Space Studies
Southwest Research Institute, Boulder, Colorado, USA 80302

AND

MARTIN J. DUNCAN
Department of Physics, Queen's University
Kingston, Ontario, Canada K7L 3N6

1. Introduction

Comets have played an important role in the history of the solar system. It is generally accepted that comets were among the first macroscopic objects to form in the outer solar nebula (Weidenschilling & Cuzzi 1993) and they contain critical information about giant planet formation (Lissauer & Stewart 1993). Comets therefore represent an important resource that allow us to study conditions that existed in parts of the solar nebula before the planets formed. In addition, comets have played a significant role in the surface geology of the solid planets and satellites because they are responsible for a significant fraction of the impact flux (Weissman 1985,1990, Zahnle et al. 1998), and may have even contributed to the late veneering of volatiles on the terrestrial planets (Chyba 1990; Chyba et al. 1995).

The comets that we see today are, in general, in a very short-lived phase of their lives. That is, the dynamical and physical lifetimes of observed comets are short compared to the age of the solar system. Thus, they must be coming from some reservoir or reservoirs that slowly allow comets to leak out to regions where they can be detected. These reservoir(s) must have the following characteristics. First, they must be long-lived. That is, they must be stable enough so that a significant number of objects survive for the age of the solar system. If not, these regions would currently be empty. Second, there must be some processes or mechanisms that allow objects to escape these regions. Since the lifetimes of the observed comets are very short, these mechanisms must be active in the solar system today.

M. Ya. Marov and H. Rickman (eds.), Collisional Processes in the Solar System, 73–90.
© 2001 *Kluwer Academic Publishers. Printed in the Netherlands.*

It is currently believed that there are three main sources of the known comets: the Oort cloud, the Kuiper belt, and the scattered comet disk. Each of these will be discussed in more detail below. Much of the research into the dynamics of comets that has been performed in recent years has had the main goal of understanding the structure and evolution of these mostly invisible reservoirs. These reservoirs were formed at the same time as the planets. Indeed, they can be viewed as the dregs of planet formation. As such, the dynamical structure of the cometary reservoirs can supply very important clues to the process of planet formation.

This chapter is organized as follows: Section 2 discusses the comet classification scheme we will adopt, Section 3 discusses the dynamics of the nearly-isotropic comets and Section 4 discusses the dynamics of the ecliptic population.

2. Cometary Taxonomy

There is a wide variety of cometary orbits observed in the solar system. This is just the tip of the iceberg since there are many more comets that do not get close enough to the Sun to be detected. Thus, in order to make a discussion of comet dynamics tractable, we must first classify the orbits of comets in some meaningful way. Ideally, this taxonomy should take into account our current understanding of where comets come from. For this chapter, we will adopt a scheme based on that of Levison (1996).

The dots in Figure 1 represent the orbits of all known comets in Marsden's comet catalog of 1992 as a function of their semi-major axis, a, and Tisserand parameter, T. The Tisserand parameter is usually defined as

$$T = a_J/a + 2\sqrt{(1 - e^2)a/a_J} \, \cos(i), \tag{1}$$

where a_J is Jupiter's semi-major axis, and a, e, and i refer to an object's semi-major axis, eccentricity, and inclination, respectively. It is an approximation to the Jacobi constant, which is an integral of the motion in the circular restricted three-body problem. It is also a measure of the relative velocity between a comet and Jupiter during close encounters, $v_{rel} = v_c\sqrt{3 - T}$, where v_c is Jupiter's velocity about the Sun. Objects with T close to, but smaller than, 3 have very slow, and thus very strong, encounters with Jupiter. Objects with $T > 3$ cannot cross Jupiter's orbit in the circular restricted case, being confined to orbits either totally interior or totally exterior to Jupiter. The shaded regions in the figure represent classes in our taxonomy scheme. Figure 2 represents this scheme in a more graphical form.

The most significant division is based on the Tisserand parameter. In this scheme comets with $T > 2$ are designated *ecliptic* comets because most

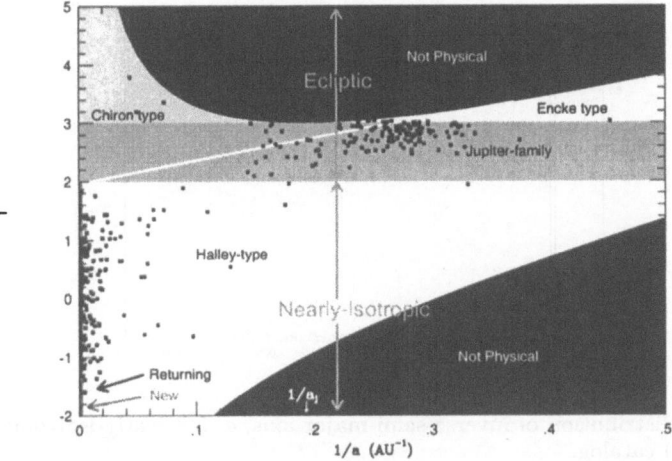

Figure 1. The location of the taxonomy classes defined in Levison (1996) as function of the Tisserand parameter (T) and semi-major axis (a). The major classes of ecliptic and nearly-isotropic comets are defined by T and are independent of a. The ranges of these two classes are thus shown with arrows only. The extent of the subclasses is shown by different shadings. Also shown is the location of all the comets with $1/a > 0$ in Marsden's catalog of cometary orbits and the location of the first 3 discovered Centaur 'asteroids'. The white curve shows the relationship of T versus a for a comet with $q = 2.5$ AU and $i = 0$. Comets above and to the left of this line have $q > 2.5$ AU and thus are difficult to detect. Reproduced from Levison (1996).

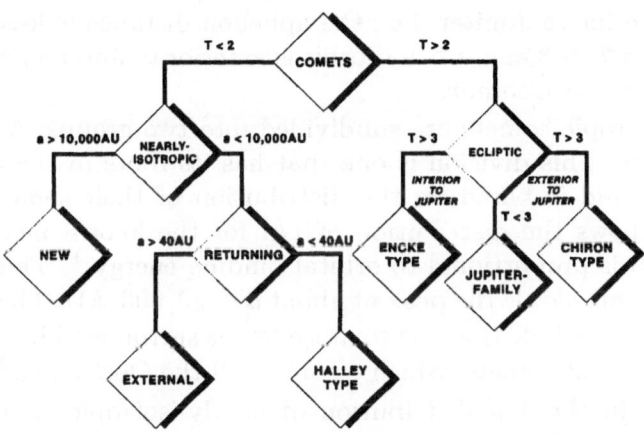

Figure 2. The family tree for comets. Reproduced from Levison (1996).

members have small inclinations. These objects most likely originate in the Kuiper belt (Levison & Duncan 1997) or the *scattered disk* (Duncan & Levison 1997), see §4. Comets with $T < 2$, which are mainly comets from the Oort cloud, are designated *nearly isotropic* comets, reflecting their

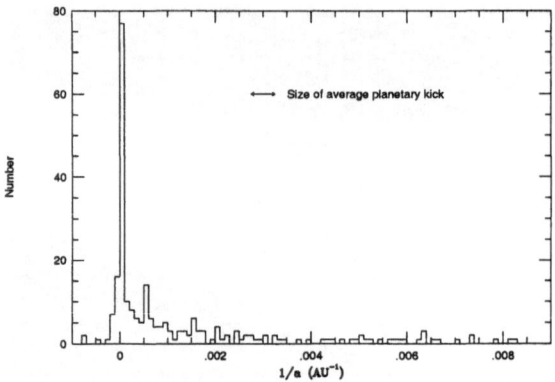

Figure 3. The distribution of inverse semi-major axis, a, for nearly-isotropic comets in Marsden's (1992) catalog.

inclination distribution, see §3.

Ecliptic comets can be further subdivided into three groups. Comets with $2 < T < 3$ are mainly on Jupiter-crossing orbits and are dynamically dominated by that planet. We call these *Jupiter-family* comets. Comets with $T > 3$ (not Jupiter-crossing) are not considered members of the Jupiter family. A comet that has $T > 3$ and $a < a_J$ is designated an *Encke-type*. Note that this combination of T and a implies that the orbit of this object is entirely interior to Jupiter, i.e., the aphelion distance is less than a_J. A comet that has $T > 3$ and $a > a_J$ (orbit is exterior to Jupiter) is designated as *Chiron-type* or a *Centaur*.

Nearly-isotropic comets are subdivided into two groups: *New* and *Returning* comets. This division is one that has its roots in the dynamics of these objects and is based on the distribution of their semi-major axes, a. Figure 3 shows the distribution of $1/a$ for the known nearly-isotropic comets, which is proportional to orbital binding energy [1]. The most striking feature of this plot is the peak at about $a \sim 20,000$ AU. This feature led Oort (1950) to conclude that the solar system is surrounded by a spherically symmetric cloud of comets, which we now call the Oort cloud[2].

The peak in the $1/a$ distribution of nearly-isotropic comets is fairly narrow. And yet the typical kick that a comet receives when it passes through the planetary system is approximately $\pm 0.0005\,\mathrm{AU}^{-1}$ (Oort 1950, cf. Figure 3). Thus it is unlikely that a comet that is in the peak when it first passes through the solar system will be in the peak during successive passes. It is concluded from this argument (Oort & Schmidt 1951) that comets in

---

[1]Recall that $1/a < 0$ implies that the comet is unbound.

[2]It is interesting to note that there were only 19 comets in Oort's original histogram.

the peak ($a \gtrsim 10,000\,\text{AU}$) are dynamically 'New' in the sense that this is the first time that they have passed through the planetary system.

Comets not in the peak are most likely objects that have been through the planetary system before and thus are called *Returning*. The class of Returning comets covers a large range of objects — from comets with $a \approx 10,000\,\text{AU}$ down to three comets with periods less than 20 years ($a = 7.4\,\text{AU}$). Having a class as large as the Returning comets may be somewhat awkward. Therefore, this class is further divided into two groups based on their dynamics. Carusi et al. (1987a, 1987b) have found that a significant fraction of Returning comets with small semi-major axes are trapped (at least temporally) in mean motion resonances with Jupiter (also see Chambers 1992). They suggest that many of the comets that are currently not librating in such a resonance may have done so in the past or may do so in the future if their semi-major axes are small enough. In our classification scheme, comets that have a small enough semi-major axis ($a < 40\,\text{AU}$) to be able to be trapped in a mean motion resonance with a giant planet are designated as *Halley-type* comets and those that have semi-major axes larger than this as *external* comets.

3. Nearly-Isotropic Comets

As described above, nearly-isotropic comets most likely originate in the Oort cloud. Due the fact that New comets have inclinations which are isotropic in the sky, the Oort cloud is believed to be a nearly spherical distribution, at least in the outer regions (we address this issue below in more detail), of comets, centered on the Sun. The position of its inner edge is very uncertain. The position of its outer edge is defined by the solar system's tidal truncation radius at about $1 - 2 \times 10^5\,\text{AU}$ from the Sun (Smoluchowski & Torbett 1984).

Oort cloud comets have been stored at great distances from the Sun for nearly the age of the solar system. However, it is most likely that they were formed in the outer regions of the planetary system (Kuiper 1951). That is, the comets now in the Oort cloud are thought to have formed in nearly circular orbits between the giant planets and to be the remnants of planetary formation (Duncan, Quinn, & Tremaine 1987; Gaidos 1995; Fernández 1997). At first, their semi-major axes evolved in a random walk because of repeated gravitational scatterings by the giant planets. When the semi-major axes of these bodies reached about $10^4\,\text{AU}$, galactic perturbations (i.e. the galactic tidal field and passing stars) lifted their perihelia out of the planetary system where they were stored for billions of years. Although other theories for the origin of the Oort cloud have been proposed (a complete discussion is beyond the scope of this chapter, see Fernández 1985

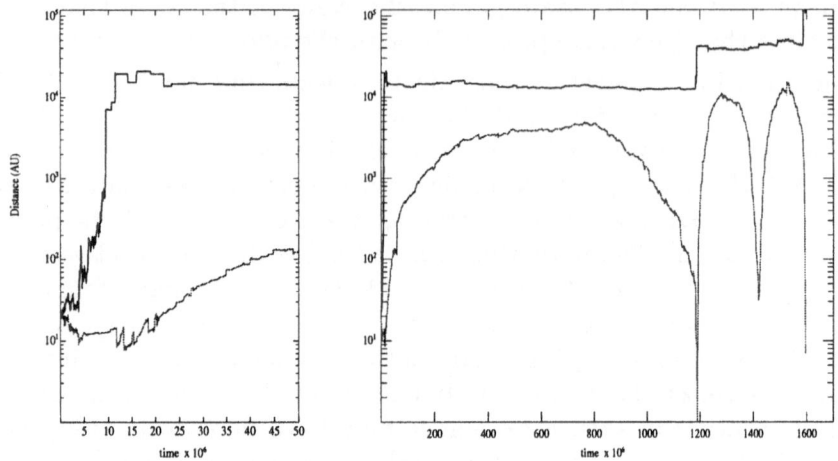

Figure 4. The temporal evolution of the barycentric semi-major axis (upper) and peri-helion distance (lower) of a typical particle from the integrations of Levison et al. (2000a). The two panels simply show the integrations on different temporal scales.

or Weissman 1996 for a review), this theory has become the dominant one.

The orbits of comets stored in the Oort cloud evolve due to the gravitational effects of the galactic disk (Heisler & Tremaine 1986, Bailey 1986), passing stars (Oort 1950, Hills 1981), and passing giant molecular clouds (Biermann & Lüst 1978, Hut & Tremaine 1985, Weissman 1990). Since these perturbations to first order act as a torque, the semi-major axis of a comet remains fixed while its perihelion distance varies. Occasionally, a comet will evolve so that its perihelion distance falls to within the planetary region, after which it can become a visible comet.

Figure 4 shows the temporal evolution of a typical particle taken from an integration by Levison at al. (2000a, in preparation). This simulation started with objects on nearly circular orbits between the giant planets and included the gravitational effects of the Sun, giant planets, galactic tides, and passing stars. Initially the particle in Figure 4 was on a nearly-circular orbit with a semi-major axis (shown in black) of 22AU. The particle's eccentricity grows until it is gravitationally interacting with Uranus. At this point Uranus hands the particle off to Saturn, as can be seen in the figure as a decrease of the perihelion distance (shown in red) to less than 10AU. Between roughly 5 and 20 million years, Saturn then forces the semi-major axis to increase to about 10,000 AU, after which the galactic tides become important. The tides lift the particle beyond the reach of the planetary system. As this point, the object has a semi-major axis $\sim 14,000$AU.

Between 20 million years and 1.2 billion years the particle is mainly under the gravitational control of the galactic tide. Passing stars play only a small role causing the small discontinuous jumps seen in the figure. After 1.2 billion years, the particle completes one oscillation in perihelion distance and reenters the planetary system. An encounter with Saturn on the way in increases its semi-major axis to 45,000AU. After the encounter, the particle evolves onto an orbit with $q = 0.85AU$. Interestingly, it never becomes a visible comet, for although it spends 1.2 million years with $q < 2AU$, its orbital period is almost 10 million years. So, during the phase when $q < 2AU$ the particle is far away from the Sun. As a result of the increase in a, the perihelion oscillation period decreases significantly ($\propto a^{-2}$; Duncan, Quinn, & Tremaine 1987). The particle completes two more oscillations between 1.2 and 1.6 billion years, after which it again reenters the planetary region. This time, it encounters Jupiter and is ejected from the solar system.

There are several paths that an Oort cloud comet can take in order to become a visible comet. As we showed above, galactic tides cause the perihelion distances of Oort cloud comets to change, feeding them directly into the planetary region. Following Duncan, et al. 1987, the typical change in a comet's perihelion distance, q, in time t is approximately:

$$\Delta q \approx \frac{1}{\sqrt{GM_\odot}} 5\pi G\rho_0 q^{1/2} a^2 t \sin^2 i, \qquad (2)$$

where $M_\odot$ is the mass of the Sun, ρ_0 is the mean density of the galactic disk, and a and i are the comet's semi-major axis and inclination with respect to the galactic plane, respectively. If the comet's semi-major axis is large enough, then the galactic tide can deliver the comet to a small enough perihelion distance to become visible without encountering the giant planets. This occurs if the time it takes to evolve from q beyond 10AU (near Saturn's orbit) to $q \sim 1AU$ is less than an orbital period. This, in turn, occurs for comets with $a \gtrsim 20,000AU$ according to Eq(2), also see Hills (1981). New comets should come only from this region. These comets have $1/a \approx 5 \times 10^{-5}$ and are responsible for the spike seen in the semi-major axis distribution of long-period comets, see Figure 3.

As described above, as New comets come through the planetary system, they receive a kick in energy that is, on average, larger than its binding energy (or $1/a$, see Figure 3). Thus depending on the sign of the kick, they are either ejected from the solar system or evolve onto orbits that have much smaller semi-major axes. If the latter is the case, the new semi-major axis is often small enough so that galactic tides are again unimportant. For these objects, their future evolution will again appear as a random walk in semi-major axis while the perihelion distribution remains relatively constant (see Wiegert & Tremaine 1999 for a more detailed discussion). These objects produce the 'Returning' nearly-isotropic comets.

Over the years, many researchers have modeled the dynamical behavior of New comets as they evolve into Returning comets (Oort 1950; Whipple 1962; Weissman 1980; Emel'yanenko & Bailey 1998; Wiegert & Tremaine 1999) [3]. They have always found that the ratio of the number of Returning comets to the number of New comets in their models is significantly larger than observed. Despite valiant efforts to explain this discrepancy with dynamics alone, it was found that the only way to change the models so that they match the observations is to allow the comets in the models to physically age. That is, comets must fade and become extinct (or disintegrate) as a function of time. The most recent attempt at this (Wiegert & Tremaine 1999) found a good match if: (1) the fraction of comets remaining visible after m apparitions is proportional to $m^{-0.6\pm0.1}$, or (2) if $\sim 95\%$ of comets life for only ~ 6 returns and the remainder last indefinitely (also see Weissman 1980).

As described above, New comets sample only the outer regions of the Oort cloud since comets from the inner region cannot penetrate the 'Jupiter Barrier'. According to Wiegert & Tremaine (1999), 98% of the External Returning comets were once visible New comets. Thus, these comets are also only sampling the outer Oort cloud. It appears that this may not be the case for the Halley-type comets (hereafter HTCs), however. Oort cloud comets with semi-major axes interior to 20,000 AU should be encountering the giant planets. These comets contribute significantly to the distribution of HTCs.

The most complete work on this subject was been performed by Levison at al. (2000b, in preparation). These integrations have not yet been fully analyzed, so the results that follow should be viewed as preliminary. Since these simulations are not yet published, we briefly describe the procedures used to produce them. Levison at al. (2000b) followed the dynamical evolution of 27,700 massless test particles under the gravitational effects of the Sun, giant planets, galactic tides, and passing stars using the RMVS3 integrator (Levison & Duncan 1994). These test particles were initially uniformally distributed in: (1) semi-major axes from 5000AU to 50,000AU, (2) perihelion distance from the maximum of 0.1 AU and $(10AU-\Delta q)$ to 30AU, where Δq is a function of the comets initial semi-major axis and is taken from Equation (2)[4], and (3) cosine of the inclination. The remaining orbital elements were chosen at random. The orbit for each test particle was integrated for up to 1 billion years or until it was ejected from the system, hit the Sun or a planet, or evolved back into the Oort cloud.

One important diagnostic for the HTCs is their inclination distribution.

---

[3]To be more precise, the efforts have concentrated on what Levison (1996) calls 'external' comets. Halley-type comets will be discussed in more detail below.

[4]Levison at al. (2000b) assumed $\rho_\circ = 0.1 M_\odot/pc^3$ and i is set to $60°$, which is the inclination of the galactic plane with respect to the ecliptic. They also set t to the orbital period of the comet.

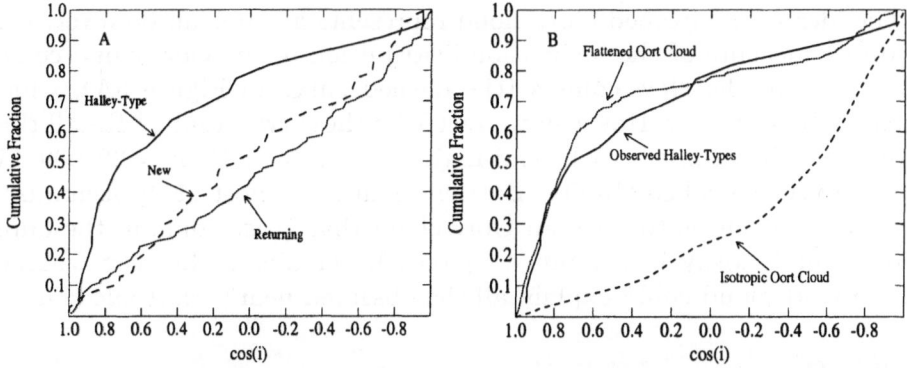

Figure 5. The cumulative distribution of the cosine of inclination, cos (i), for various observed or hypothetical comet populations. In this figure, a population that is isotropic in inclination would be represented by a line starting at (0,0) and going to (1,1). A) The three subclasses of nearly-isotropic comets: New (dashed curve), External Returning (dotted curve), and HTCs (solid). B) A comparison between the results of Levison et al. (2000b)'s simulations and the real HTCs (solid curve). The dashed curve shows the results of a simulation assuming an isotropic Oort cloud. The dotted curve shows the results from an inner Oort cloud that is flattened with a median inclination of 25°.

Figure 5A shows the cumulative inclination distribution of New comets (dashed), External Returning comets (dotted), and HTCs (solid) taken from the Marsden catalog. While the New and External comets are isotropic, the HTCs are mainly prograde with a median inclination of only 45°. It has generally been assumed that this asymmetry was due to the dynamics of the capture process. However, the results of Levison et al. (2000b) indicates that this may not be the case.

The dashed curve in Figure 5B shows the cumulative inclination distribution of the HTCs captured from an isotropic Oort cloud as predicted by the Levison el al. (2000b) simulations. Note that it does not produce a good match to the observations (solid curve in Figure 5B). Indeed, the capture dynamics from an isotropic Oort cloud tends to produce more retrograde comets than prograde[5]. Only 25% of the HTCs produced in these simulations were prograde. This can be compared to the value of 80% for the real HTCs.

The only way that Levison et al. (2000b) found to reproduce the observed inclination distribution of the HTCs was to start with a flattened Oort cloud. The dotted curve in Figure 5B shows the results of a simulation in which the Oort cloud was flattened with a median inclination of 25°. This appears to match the observations fairly well. This model also matches the other orbital elements well.

---

[5]Prograde objects are more likely to be captured, but have shorter dynamical lifetimes. The latter is the dominant of these two competing effects. As a result, the steady-state distribution of the Levison el al. (2000b) simulations has a majority of retrograde objects.

The idea of a flattened Oort cloud represents a problem, however. Our one direct measure of the inclination distribution of the Oort cloud comes in the form of the New comets (the dashed curve in Figure 5A), which appear to be isotropic. How can we reconcile these two results? Recall that we only see New comets with semi-major axes greater than $\sim 20,000$ AU because of the so-called 'Jupiter Barrier'. Thus, the most likely scenario to solve this problem is to have an Oort cloud that is isotropic in the outer regions, but is disky in the inner regions. It remains to be seen whether such an Oort cloud could explain all the observed nearly-isotropic comets.

4. The Dynamics of Ecliptic Comets

We turn now to the study of the origin and dynamical evolution of the class of objects which we referred to in §2 as 'ecliptic comets'. The observed ecliptic comets have a very flattened inclination distribution with a median inclination of only 11°. In the last 10 years or so, research attempting to explain this inclination distribution has been extremely active. Indeed, attempting to understand these comets has led to one of the most important discoveries in planetary science in the last half of the twentieth century — the discovery of the Kuiper belt.

Ecliptic comets[6] were originally thought to originate from nearly isotropic comets that had been captured into short-period orbits by gravitational enounters with the planets (Newton 1891; 1893; Everhart 1977). Fernández (1980) argued that this process is too inefficient and suggested that a belt of distant icy planetesimals beyond Neptune could serve as the source of most of these comets. Duncan et al. (1988) strengthened this argument by performing dynamical simulations which showed that a cometary source beyond Neptune with a low initial inclination distribution (which they named the Kuiper belt) was far more consistent with the observed orbits of most of these comets than the randomly distributed inclinations of comets in the Oort cloud (see also Quinn et al. 1990). The Kuiper belt was discovered in 1992 (Luu & Jewitt 1993).

Levison & Duncan (1997: hereafter LD97) have presented the most comprehensive simulations to date of the dynamical evolution of objects leaking from the Kuiper belt. They performed numerical orbital integrations of 2200 massless particles as they evolved from Neptune-encountering orbits in the Kuiper belt for times up to a billion years or until they either impacted a massive body or were ejected from the solar system. The initial orbits for these particles were chosen from a previous set of integrations of objects

---

[6]They were not known as ecliptic comets at this time. Historically, researchers were interested in 'short-period' comets, which were defined to have periods less than 200 years.

that were initially in low-eccentricity, low-inclination orbits in the Kuiper belt but evolved onto Neptune-crossing orbits on timescales between 1 and 4 billion years (Duncan et al. 1995).

LD97 found that as objects evolve inward from the Kuiper belt, they tend to be under the dynamical control of just one planet. That planet will scatter the comets inward and outward in a random walk, typically handing them off to the planet directly interior or exterior to it. Therefore, the comets tend to have eccentricities of about 25% between handoffs. However, once they have been scattered into the inner solar system by Jupiter, they can have much larger eccentricities as they evolve outward.

Figure 6 shows the evolution of a typical particle in perihelion distance (q) - aphelion distance (Q) plane, as it evolves from the Kuiper belt $(q > 30AU)$ to a visible Jupiter-family comet (the most populous of the visible ecliptic comets; hereafter JFCs). The positions are joined by blue lines until the particle first became 'visible' $(q < 2.5\,AU)$ and are linked in red thereafter. Initially, the particle spent considerable time with perihelion near the orbit of Neptune (30 AU) and aphelion well beyond the planetary system. However, once its perihelion dropped to Uranus' location, this particle, which was chosen at random from LD97's integrations, clearly shows the handoff behavior described above. It evolved at relatively small eccentricity to visibility (cf. the lines of constant eccentricity e =0.2 and 0.3 on the figure) and it spent considerable time with perihelion or aphelion near the semi-major axis of one of the three outer planets. Its post-visibility phase is reasonably typical of Jupiter-family comets, with much larger eccentricities than the pre-visibility comets and perihelion distances near Jupiter or Saturn.

The median dynamical lifetime of the ecliptic comets was found to be 4.5×10^7 years. (This is the time from the first encounter with Neptune to ejection from the solar system, placement in the Oort cloud, or impact with the Sun or a planet). LD97 found that about 30% of the objects in the integrations became visible comets $(q < 2.5\,AU)$. Of those that became visible, 99.7% were Jupiter-family comets (see §2) at the time of first visibility. The very narrow range in the Tisserand parameter, T, with respect to Jupiter observed in the JFCs is mainly due to the fact that when they were on their initial Neptune-encountering orbits, the comets had a narrow range in the Tisserand parameter with respect to Neptune.

By comparing the orbital element distribution of the simulated JFCs to that of the observed JFCs, LD97 estimated that the physical lifetime of JFCs is between 3000 and 25, 000 years. The most likely value is 12, 000 years. From this they estimate that the total number of ecliptic comets is 1.2×10^7. The latter estimate refers to objects that produce active comets with total absolute magnitudes brighter then 9 (roughly 1km in radius).

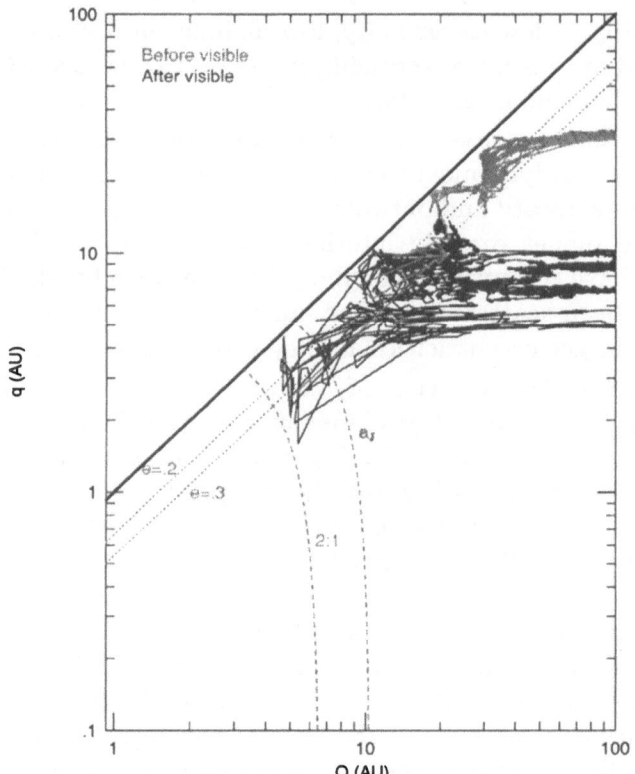

Figure 6. The evolution of a representative particle originating in the Kuiper belt from the integrations of LD97. Locations of the particle's orbit in the q–Q (perihelion-aphelion) plane are joined by lightly shaded lines until the particle became 'visible' ($q < 2.5$ AU) and are linked in darker shading thereafter. By definition, comets cannot have orbits with $q > Q$, so they cannot lie in the upper left of the diagram. The sampling interval was every 10^4 years in the previsibility phase and every 10^3 years thereafter. Also shown in the figure are three lines of constant eccentricity at $e = 0$, 0.2, and 0.3. In addition, we plot two dashed curves of constant semi-major axis, one at Jupiter's orbit and one at its 2:1 mean motion resonance.

LD97's estimate of the total number of comets should be viewed as approximate because it is dependent on physical properties of comets that are not well understood (see Levison et al. 2000c).

Although when the ecliptic comets first become visible they are almost entirely members of the Jupiter-family, a small fraction of these comets switch to visible Halley-type comets (HTCs). Since those that switch have longer dynamical lifetimes, on average about 13% of all visible ($q < 2.5\,AU$) comets derived from the Kuiper belt with $a < 40\,AU$ will be HTCs. However, the orbital element distribution of the simulated HTCs is not consistent with the observed distribution. In particular, their semi-major axes are too small. Since LD97 found that it takes at least 10^5 years and usually over

10^6 years to become a visible HTC after the comet first becomes visible, most of these comets have likely become extinct since this is longer then the typical physical lifetime estimated above. Thus, although the Kuiper belt can be the source of at least some of the HTCs, initially low-inclination bodies encountering Neptune are unlikely to provide a significant number of them.

Perhaps the most interesting result of the LD97 simulations was that about 5% of the particles survived the length of the integration (10^9 years). All of the survivors had semi-major axes outside the orbit of Neptune. This result implied that there may be a significant population of objects with highly eccentric orbits in an extended disk beyond the orbit of Neptune – a 'scattered' comet disk. Although, the idea of the existence of a scattered comet disk dates to Fernández & Ip (1983), it was not until the modern work of LD97 and the followup paper, Duncan & Levison (1997; hereafter DL97), that demonstrated that the scattered disk should exist. The first scattered disk object was discoverd shortly after this prediction (Luu et al. 1997).

DL97 is the only published investigation of the scattered disk which uses modern direct numerical integrations. DL97 extended the LD97 integrations to 4 billion years. It was found that 1% of the particles remained in orbits beyond Neptune after 4 billion years. So, if at early times, there was a significant amount of material from the region between Uranus and Neptune or the inner Kuiper belt that evolved onto Neptune-crossing orbits, then there could be a significant amount of this material remaining today. What is meant by 'significant' is the main question when it comes to the current importance of the scattered comet disk.

DL97 found that some of the long-lived objects were scattered to very long-period orbits where encounters with Neptune became infrequent. However, at any given time, the majority of them were interior to 100 AU. Their longevity is due in large part to their being temporarily trapped in or near mean-motion resonances with Neptune. The 'stickiness' of the mean motion resonances, which was first mentioned by Holman & Wisdom (1993), leads to an overall distribution of semi-major axes for the particles that is peaked near the locations of many of the mean motion resonances with Neptune. Occasionally, the longevity is enhanced by the presence of the Kozai resonance (Kozai 1962).

In all long-lived cases, particles had their perihelion distances increased so that close encounters with Neptune no longer occurred. Frequently, these increases in perihelion distance were associated with trapping in a mean motion resonance, although in many cases it has not yet been possible to identify the exact process that was involved. On occasion, the perihelion distance can become large, but 81% of scattered disk objects have perihelia

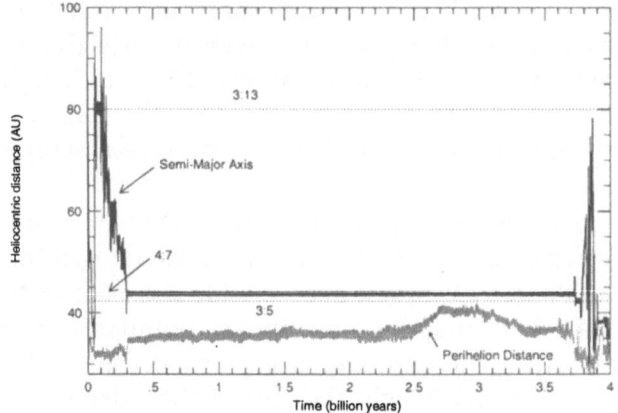

Figure 7. The temporal behavior of a long-lived member of the scattered disk. The black curve shows the behavior of the comet's semi-major axis. The gray curve shows the perihelion distance. The three dotted curves show the location of the 3:13, 4:7, and 3:5 mean motion resonances with Neptune. Reproduced from DL97.

between 32 and 36 AU.

Figure 7 shows the dynamical behavior of a typical long-lived particle. This object initially underwent a random walk in semi-major axis due to encounters with Neptune. At about 7×10^7 years it was temporarily trapped in Neptune's 3:13 mean motion resonance for about 5×10^7 years. It then performed a random walk in semi-major axis until about 3×10^8 years, when it was trapped in the 4:7 mean motion resonance, where it remained for 3.4×10^9 years. Notice the increase in the perihelion distance near the time of capture. While trapped in this resonance, the particle's eccentricity became as small as 0.04. After leaving the 4:7, it was trapped temporarily in Neptune's 3:5 mean motion resonance for $\sim 5 \times 10^8$ yr and then went through a random walk in semi-major axis for the remainder of the simulation.

DL97 estimated an upper limit on the number of possible scattered disk objects by assuming that they are the sole source of the Jupiter-family comets. DL97 computed the simulated distribution of comets throughout the solar system at the current epoch (averaged over the last billion years for better statistical accuracy). They found that the ratio of scattered disk objects to visible Jupiter-family comets (those with a perihelion distance less than 2.5 AU) is 1.3×10^6. Since there are currently estimated to be 500 visible JFCs (LD97), there are $\sim 6 \times 10^8$ comets in the scattered disk if it is the sole source of the JFCs. Figure 8 shows the spatial distribution for this model.

It is quite possible that the scattered disk could contain this much material. It was noted above that $\sim 1\%$ of the objects in the scattered disk

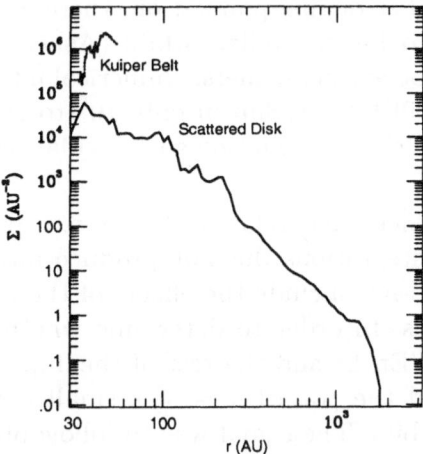

Figure 8. The surface density of comets beyond Neptune for two different models of the source of Jupiter-family comets. The dotted curve is a model assuming that the Kuiper belt is the current source (Levison & Duncan 1997). There are 7×10^9 comets in this distribution between 30 and 50 AU. This curve ends at 50 AU because the models are unconstrained beyond this point and not because it is believed that there are no comets there. The solid curve is DL97's model assuming the scattered disk is the sole source of the Jupiter-family comets. There are 6×10^8 comets currently in this distribution.

remain after 4 billion years, and that 6×10^8 comets are currently required to supply all of the Jupiter-family comets. Thus, a scattered comet disk requires an initial population of only 6×10^{10} comets (or $\sim 0.4 M_{\oplus}$, Weissman 1990) on Neptune-encountering orbits. Since planet formation is unlikely to have been 100% efficient, the original disk could have resulted from the scattering of even a small fraction of the tens of Earth masses of cometary material that must have populated the outer solar system in order to have formed Uranus and Neptune.

There is now convincing observational evidence for the existence of the scattered disk. The first scattered disk object discovered was 1996 TL$_{66}$, found in October 1996 by Jane Luu and colleagues (Luu et al. 1997). Current observations indicate that it has a semi-major axis of 85 AU, a perihelion of 35 AU, and an inclination of 24°. Several other candidates have subsequently been reported in other surveys and such objects generally have orbital parameters in agreement with those found in the simulations of DL97. The total mass in 100-km sized scattered disk objects is estimated to be $\sim 0.05 M_{\oplus}$ (Trujillo et al. 2000), comparable to the mass in similar sized Kuiper belt objects interior to 50 AU. The mass in comet-sized bodies in both of these trans-Neptunian populations remains to be determined.

Finally, we consider the Encke-type comets, which are low inclination comets totally interior to Jupiter's orbit (see §1). Although comet 2P/Encke is the only active member of this population, there are several kilometer-size asteroids known to be in similar orbits (Asher et al. 1993). These small 'asteroids' could be extinct comets. Numerical integrations of its orbit show that 2P/Encke will hit the Sun in only 10^5 to 10^6 years (Levison & Duncan 1994) due to its close association with secular resonances (Valsecchi et al. 1995).

The origin of 2P/Encke and related objects is a puzzle. It was a surprise that the LD97 integrations did not produce any comets similar to 2P/Encke, but they did not include the effects of the terrestrial planets or non-gravitational effects. In order to determine whether dynamical pathways exist between 2P/Encke and the rest of the Jupiter family, Valsecchi et al. (1995) integrated the trajectories of a small number of real asteroids with Encke-like orbits. Their goal was to follow objects 'backward' in time to map out possible dynamical pathways. They included the terrestrial planets in their simulations, but not non-gravitational forces. They found that some objects evolved onto orbits similar to those of the JFCs, but the timescales were longer than the physical lifetime of a comet. Thus, it is unlikely that a comet evolving in the 'forward' direction would be active after the process was complete. Steel & Asher (1996) performed a similar set of integrations in which they crudely included non-gravitational effects. They also found that some of their objects evolved onto JFC-like orbits. In addition, they found more reasonable timescales. So, dynamical pathways between the Jupiter family and 2P/Encke exist, but it is not clear whether either process is efficient enough to explain 2P/Encke's existence, since both simulations integrated 'backward' from Encke-type orbits to the Jupiter family. These questions can be answered only with 'forward' integrations including nongravitational effects.

5. Closing Comments

In the last 10 or 12 years there has been a revolution in our understanding of the formation and dynamical evolution of comets. This revolution has occurred due to the ability of scientists to perform orbital integrations of a large number of comets for timescales on the order of the age of the solar system. The study of the dynamics of comets has led to important discoveries about the structure of the Solar System: the Kuiper belt and the scattered disk. These in turn supply us with important clues about the formation of the planets.

However, important mysteries still exist. In particular, we believe that attempts to explain the observed orbital element distribution of the Halley-type comets may lead to important clues to the structure of the Oort cloud. This, in turn, may supply important information about not only the planet formation process, but the galactic environment in which the Sun and Solar System formed (for example see Gaidos 1995). Only the future will tell.

References

Asher, D.J., Clube, S.V.M., & Steel, D.I.: 1993, Asteroids in the Taurid Complex, *MNRAS*, **264**, 93.

Bailey, M.E.: 1986, The mean energy transfer rate to comets in the Oort cloud and implications for cometary origins, *MNRAS*, **218**, 1.

Biermann, L., & Lüst, R.: 1978, *Sitz. ber. Bayer. Akad. Wiss. Mat. Naturw. Kl.*

Carusi, A., Kresák, & Valsecchi, G.: 1987a, High-Order Librations of Halley-Type Comets, *Astron. & Astrophy*, **187**, 899.

Carusi, A., Kresák, Perozzi, E., & Valsecchi, G.: 1987b, Long-term resonances and orbital evolutions of Halley-type comets, European Regional Astronomy Meeting of the IAU, 10th, Prague, Czechoslovakia, Aug. 24-29, 1987, 2, 29.

Chambers, J.: 1992, Why Halley-types resonate but long-period don't: A dynamical distinction between short- and long-period comets, *Icarus*, **125**, 32.

Chyba, C.F.: 1990, Impact delivery and erosion of planetary oceans in the early inner solar system, *Nature*, **343**, 129.

Chyba, C.F, Owen, T.C., & Ip W.-H.: 1995, Impact delivery of volatiles and organic molecules to earth, in *Hazards Due to Comets and Asteroids*, ed. T. Gehrels (Tucson: Univ. of Arizona Press), 9.

Duncan, M., Quinn, T., & Tremaine, S.: 1987, The formation and extent of the solar system comet cloud, *Astron. J.*, **94**, 1330.

Duncan, M., Quinn, T., & Tremaine, S.: 1988, The origin of short-period comets, *Astrophy. J. Lett.*, **328**, L69.

Duncan, M., Levison, H., & Budd M.: 1995, The Dynamical Structure of the Kuiper Belt, *Astron. J.*, **110**, 3073.

Duncan, M. & Levison, H.: 1997, A scattered comet disk and the origin of Jupiter family comet, *Science*, **276**, 1670.

Emel'yanenko, V.V, & Bailey, M.E.: 1998, Capture of Halley-type comets from the near-parabolic flux, *MNRAS*, **298**, 212.

Everhart, E.: 1977, The evolution of comet orbits as perturbed by Uranus and Neptune, in *Comets–Asteroids–Meteorites*, ed. A.H. Delsemme, (Ohio: University of Toledo Press), 99.

Fernández, J. A.: 1980, On the existence of a comet belt beyond Neptune, *MNRAS*, **192**, 481.

Fernández, J.A.: 1985, The Formation and Dynamical Survival of the Comet Cloud, in *Dynamics of Comets: Their Origin and Evolution*, eds. A. Carusi & G. Valsecchi (Dordrecht: Reidel), 45.

Fernández, J.A.: 1997, The Formation of the Oort Cloud and the Primitive Galactic Environment, *Icarus*, **129**, 106.

Fernández, J.A. & Ip, W.-H.: 1983, On the time evolution of the cometary influx in the region of the terrestrial planets, *Icarus*, **54**, 377.

Gaidos, E.J.: 1995, Paleodynamics: Solar system formation and the early environment of the sun, *Icarus*, **114**, 258.

Heisler, J., & Tremaine, S.: 1986, The influence of the galactic tidal field on the Oort comet cloud, *Icarus*, **65**, 13.

Hills, J.G.: 1981, Comet showers and the steady-state infall of comets from the Oort cloud, *Astron. J.*, **86**, 1730

Holman, M.J., & Wisdom, J.: 1993, Dynamical stability in the outer solar system and the delivery of short period comets, *Astron. J.*, **105**, 1987.

Hut, P., & Tremaine, S.: 1985, Have interstellar clouds disrupted the Oort comet cloud?, *Astron. J.*, **90**, 1548.

Kozai, Y.: 1962, Secular perturbations of asteroids with high inclination and eccentricity, *Astron. J.*, **67**, 591.

Kuiper, G.P.: 1951, On the origin of the Solar System, in *Astrophysics: A Topical Symposium*, ed. J.A. Hynek (New York: McGraw Hill), 357.

Levison, H., & Duncan, M.: 1994, The long-term dynamical behavior of short-period comets, *Icarus*, **108**, 18.

Levison, H., Duncan, M: 1997, From the Kuiper Belt to Jupiter-Family Comets: The Spatial Distribution of Ecliptic Comets, *Icarus*, **127**, 13.

Levison, H.: 1996, in *Completing the Inventory of the Solar System*, eds. T.W. Rettig & J.M. Hahn. (San Francisco: ASP), 173.

Levison, H.F., Dones, L., Duncan, M.J., & Weissman, P.R.: 2000a, in preparation.

Levison, H.F., Dones, L., & Duncan, M.J.: 2000b, in preparation.

Levison, H.F., Duncan, M.J., Zahnle, K., Holman, M., & Dones, L.: 2000c, Planetary Impact Rates from Ecliptic Comets, *Icarus*, in press.

Lissauer, J., & Stewart, G.: 1993, Growth of planets from planetesimals, in *Protostars and Planets III*, eds. E. H. Levy, J. I. Lunine, M. S. Mathews (Tucson: Univ. of Arizona Press.), 1061.

Luu, J. & Jewitt, D.: 1993, Discovery of the candidate Kuiper belt object 1992 QB1, *Nature*, **362** 730.

Luu, J., Jewitt, D., Trujillo, C.A., Hergenrother, C.W., Chen, J., & Offutt, W.B.: 1997, A New Dynamical Class in the Trans-Neptunian Solar System, *Nature*, **387**, 573.

Marsden, B.G.: 1992, *Catalog of Cometary Orbits* (Cambridge: Smithson. Astrophy. Obs).

Newton, H.A.: 1891, Capture of comets by planets, *Astron. J.*, **11**, 73.

Newton, H.A.: 1893, *Mem. Natl. Acad. Sci.*, **6**, 7.

Oort, J.H.: 1950, The structure of the cloud of comets surrounding the Solar System and a hypothesis concerning its origin, *Bull. Astron. Inst. Neth.*, **11**, 91.

Oort, J.H. & Schmidt, M.: 1951, Differences between new and old comets, *Bull. Astron. Inst. Neth.*, **11**, 259.

Quinn, T.R., Tremaine, S., Duncan, M.J: 1990, Planetary perturbations and the origins of short-period comets, *Astrophys. J.*, **355**, 667.

Smoluchowski, R., & Torbett, M.: 1984, The boundary of the solar system, *Nature*, **311**, 38.

Steel, D.I. & Asher, D.J.: 1996, On the origin of Comet Encke, *MNRAS*, **281**, 937.

Trujillo, C.A., Jewitt, D.C., & Luu, J.X.: 2000, Population of scattered Kuiper belt, to appear in *Astrophy. J. Lett.*, **529**.

Valsecchi, G.B., Morbidelli, A., Gonczi, R., Farinella, P., Froeschlé, Ch., & Froeschlé, C.: 1995, The dynamics of objects in orbits resembling that of P/Encke, *Icarus*, **118**, 169.

Weidenschilling, S., & Cuzzi, J.: 1993, Formation of planetesimals in the solar nebula, in *Protostars and Planets III*, eds. E. H. Levy, J. I. Lunine, M. S. Mathews (Tucson: Univ. of Arizona Press.), 1031.

Weissman, P.R.: 1980, Physical loss of long-period comets, *Astron. & Astrophy.*, **85**, 191.

Weissman, P.R.: 1985, Dynamical evolution of the Oort Cloud, in *Dynamics of Comets: Their Origin and Evolution*, eds. A. Carusi and G.B. Valsecchi (Dordrecht: Reidel), 87.

Weissman, P.R.: 1990, The cometary impactor flux at the Earth, in *Global Catastrophes in Earth History*, eds. V.L. Sharpton and P.D. Ward, GSA Special Paper 247, 171.

Weissman, P.R.: 1996, in *Completing the Inventory of the Solar System*, eds. T.W. Rettig & J.M. Hahn. (San Francisco: ASP), 265.

Whipple, F.L.: 1962, On the distribution of semimajor axes among comet orbits, *Astron. J.*, **67**, 1.

Wiegert, P. & Tremaine, S. 1999, The Evolution of Long-Period Comets, *Icarus*, **137** 84.

Zahnle, K., Dones, L., & Levison, H.F.: 1998, Cratering Rates on the Galilean Satellites, *Icarus*, **136**, 202.

VARIABLE OORT CLOUD FLUX DUE TO THE GALACTIC TIDE

JOHN J. MATESE
Department of Physics, University of Louisiana
Lafayette, Louisiana, 70504-4210 USA

KIMMO A. INNANEN
Department of Physics and Astronomy, York University
Toronto, Ontario, Canada, M3J IP3

AND

MAURI J. VALTONEN
Tuorla Observatory, University of Turku
FIN-21500 Piikkiö, Finland

Abstract. We review the subject of the time dependence of the component of Oort cloud comet flux due to the adiabatic Galactic tide, including the possibility of detecting such a signal in the terrestrial cratering record.

1. Introduction

Over long time scales the flux of new comets coming from the outer Oort cloud is likely to be dominated by the near-adiabatic tide due to the Galactic matter distribution (Heisler, 1990). As the Solar System moves in its Galactic orbit, this tide is substantially modulated for all models of the Galactic mass distribution that are consistent with stellar dispersion studies. Therefore a quasi-periodic variability of the tidally induced component of the Oort cloud flux having significant amplitude is to be expected (Matese *et al.*, 1995). Valtonen *et al.* (1995) have shown that any periodicity in the flux from the Oort cloud is reflected in its contribution to the terrestrial cratering rate, though with a small time delay and added noise. If Shoemaker *et al.* (1990, 1998) were correct in their estimate that 80% of terrestrial craters having diameter > 100 km are produced by long-period comets (and 50% of craters > 50 km), then the phase and period of the So-

M. Ya. Marov and H. Rickman (eds.), Collisional Processes in the Solar System, 91–102.
© 2001 *Kluwer Academic Publishers. Printed in the Netherlands.*

lar System oscillation about the Galactic disk should be consistent with the
ages of the accurately dated largest craters. The phase is well constrained,
but the dynamically predicted plane crossing period has been sufficiently
uncertain (30–45 My) to preclude a meaningful comparison with the best
fit measured cratering period of 34–37 My. If the mean plane crossing period is ultimately determined to exclude this interval, we will be able to
confidently reject Shoemaker's hypothesis of the dominance of cometary
impacts in the production of the largest terrestrial craters.

2. Tidal Dynamics

We are primarily interested in understanding how the Galactic tide makes
an Oort cloud comet observable. A widely accepted model (Duncan *et al.*,
1987) of the formation of the Solar System comet cloud considers unaccreted
comets in the giant planetary region being gravitationally pumped by these
planets into more extended orbits. When semimajor axes grow to values
$a \approx 5000$ AU, Galactic tidal torques can sufficiently increase the angular
momentum of a comet, $\mathbf{H} = \mathbf{r} \times \mathbf{v}$, and therefore its perihelion distance
$q = a(1 - e)$,

$$H = \left[\mathrm{GM}_\odot a(1 - e^2)\right]^{1/2} = \left[\mathrm{GM}_\odot (2q - q^2/a)\right]^{1/2}, \qquad (1)$$

so that it becomes detached from the planetary zone. The near-adiabatic
nature of the Galactic tide will keep a essentially constant during this stage.
Episodically, passing molecular clouds and stars impulsively pump comet
energies as well as effectively randomize the phase space of semimajor axis
orientations and angular momenta, thus forming the Oort cloud (Bailey,
1986).

   To make a comet observable, it must be injected into the inner planetary
region so that it is sufficiently insolated to form a coma. Here we discuss
the dynamical mechanism that is predominantly responsible for doing so,
the quasi-adiabatic tidal interaction with the Galaxy. The same mechanism
that increased angular momentum and detached the comet orbit from the
planets can also decrease angular momentum and bring it back into the
planetary zone. But now the planets Saturn and Jupiter provide a dynamical "barrier" to the migrating perihelia of Oort cloud comets.

   Semimajor axes $a > 10000$ AU denote the energy range commonly referred to as the outer Oort cloud. As we shall see, comets with smaller
semimajor axes are inefficiently torqued by the tide so that they are dynamically captured by Saturn or Jupiter before becoming observable. Only
if the tidal interaction is sufficiently strong to make the perihelion distance
migrate from beyond this "loss cylinder" barrier at ≈ 15 AU to the observ-

able zone interior to ≈ 5 AU in a single orbit, will we recognize the comet as having originated in the outer Oort cloud – a "new" comet is observed.

If we ignore modest mass inhomogeneities due to local molecular clouds and voids (Frisch and York, 1986), we can take the gravitational potential generated by the smoothed Galactic mass density to be axisymmetric (the analysis presented here is adapted from Heisler and Tremaine (1986)),

$$\nabla^2 U(R, Z) = \frac{1}{R}\frac{\partial}{\partial R}R\frac{\partial U}{\partial R} + \frac{\partial^2 U}{\partial Z^2} = 4\pi G\rho(R, Z), \qquad (2)$$

where $R = 0$ defines the Galactic center and $Z = 0$ the Galactic midplane. From this potential we can obtain the R-dependent azimuthal velocity, Θ, and angular velocity, Ω. Values at the Solar location (Merrifield, 1992) are denoted by the subscript $\circ$, and the present epoch is $t \equiv 0$,

$$R_\circ(0) \approx 8\text{kpc}, \quad \Theta_\circ(0) = \left[\frac{\partial U}{\partial \ln R}\bigg|_{R_\circ}\right]^{1/2} \approx 200\text{kms}^{-1} \qquad (3)$$

with uncertainties of approximately 10% (Kuijken and Tremaine, 1994).

The Galactic tidal field in an orbiting, but non-rotating Solar reference frame O is

$$\mathbf{F}_O(t) = -\nabla U(\mathbf{R}_\circ + \mathbf{r}) + \nabla U(\mathbf{R}_\circ) = -(\mathbf{r} \cdot \nabla)\nabla U(R_\circ(t), Z_\circ(t)). \qquad (4)$$

Here $\mathbf{r}$ is the comet position vector relative to the Sun with components (x, y, z) in frame O, and components

$$(x', y', z') \equiv (x\cos\Omega_\circ t + y\sin\Omega_\circ t, y\cos\Omega_\circ t - x\sin\Omega_\circ t, z) \qquad (5)$$

in a rotating frame with x' axis pointing toward the Galactic center, y' axis opposite to Θ and z' along the Galactic normal.

In this notation, the tidal force is expressible as

$$\mathbf{F}_O(t) = \Omega_\circ^2(1 - 2\delta)\mathbf{x}' - \Omega_\circ^2\mathbf{y}' - (\Omega_z^2 + 2\delta\Omega_\circ^2)\mathbf{z}'. \qquad (6)$$

Here $\Omega_\circ = \frac{\Theta}{R}\big|_\circ$ and $\delta \equiv \frac{d\ln\Theta}{d\ln R}\big|_\circ$ are simply related to the Oort constants A and B (Heisler and Tremaine, 1986), while $\Omega_z^2 \equiv 4\pi G\rho_\circ$. It can be shown (Binney and Tremaine, 1987) that the perigalactic radial frequency of the Solar motion is larger than its orbital frequency by a factor $(2 + 2\delta)^{1/2}$. The time dependence of $\mathbf{F}_O(t)$ is contained not only implicitly in the components (x, y, z), but explicitly in the transformation to (x', y', z') of Eq. (5) and in Ω_z, $\Omega_\circ$, and δ through their dependencies on $\rho_\circ = \rho(R_\circ(t), Z_\circ(t))$.

Our Galaxy has a nearly flat rotational velocity curve at the Solar location (Merrifield, 1992), $\delta \approx -0.1$. The various time scales are then the comet orbital period ($P \approx 5$ My), the oscillatory Solar Z period about the Galactic midplane ($P_z \approx 2\pi/\Omega_z \approx 60$–90 My), the Solar azimuthal period about the Galactic center ($P_\circ = 2\pi/\Omega_\circ \approx 240$ My), and the radial period of the Solar orbit $\approx P_\circ/\sqrt{2} \approx 180$ My. The larger value of P_z corresponds to the no-dark-disk-matter case while the smaller period is obtained in model calculations (Matese et al., 1995) with modest amounts of CDDM (dark disk matter distributed over compact scale heights, comparable to the interstellar medium).

The angular momentum evolves in accord with the Newtonian equation

$$\frac{d\mathbf{H}}{dt} = \mathbf{r} \times \mathbf{F}_O \equiv \boldsymbol{\tau}_O. \tag{7}$$

Also of interest is the Laplace-Runge-Lenz eccentricity vector which instantaneously points to the osculating perihelion point

$$\mathbf{e} = \frac{\mathbf{v} \times \mathbf{H}}{GM_\odot} - \frac{\mathbf{r}}{r}, \quad \frac{d\mathbf{e}}{dt} = \frac{\mathbf{v} \times \boldsymbol{\tau}_O - \mathbf{H} \times \mathbf{F}_O}{GM_\odot}. \tag{8}$$

Since we are primarily interested in the tidal mechanism as it relates to making a comet observable during a single orbit, we consider the various terms in $\mathbf{F}_O$ in the context of a time average over a single comet period. Comparing terms, we see that $(\Omega_\circ/\Omega_z)^2 \approx 0.1$ so that setting $\delta \to 0$ introduces only a modest formal error of $\approx 2\%$.

To proceed further in the spirit of doing an orbital average of the equations of motion, two distinct approximations could be made. If we set $\Omega_\circ \to 0$, which formally introduces errors of $\approx 10\%$, the problem simplifies substantially and a complete analytic solution to the orbital averaged equations of motion can be obtained for all of the orbital elements in the adiabatic limit (Matese and Whitman, 1989; Breiter et al., 1996).

Alternatively, we can concentrate on near-parabolic comets since they are most easily made observable by the Galactic tide in a single orbit. In this approximation, the coordinates in $\mathbf{F}_O$ are replaced by those of a comet freely falling along its semimajor axis with $e \to 1$ and position vector $\mathbf{r} \to -r\hat{\mathbf{q}} = -r\hat{\mathbf{e}}$, i.e.,

$$(x', y', z') \to -r \left(\cos b \, \cos(l - \Omega_\circ t), \, \cos b \, \sin(l - \Omega_\circ t), \, \sin b \right) \tag{9}$$

where b, l are the Galactic latitude and longitude of perihelion. In this case we introduce relative errors of order $1 - e = q/a \approx 10^{-3}$ in the equations of motion. Further, from Eq. (8), eccentricity changes during an orbit are of order $\Delta \mathbf{e} \approx PF_O H/GM_\odot \approx 4\pi\sqrt{2q/a}(P/P_z)^2 \approx 10^{-3}$, so we can treat the perihelion angles b, l as constants during the averaging of the torque to the same level of approximation. Note that the conventional orbital angles, the longitude of the ascending node (Ω), the argument of perihelion (ω), and the inclination (i), are all rapidly changing for a near-parabolic comet and thus cannot be held constant in a perturbative orbital average analysis.

Defining an azimuthal unit vector, $\hat{\phi} \equiv \hat{z} \times \hat{q}/\cos b$, we construct a near-constant set of orthogonal unit vectors ($\hat{q}, \hat{\phi}, \hat{\theta} \equiv \hat{\phi} \times \hat{q}$). Since $\mathbf{H} \perp \hat{q}$, it has two components in this basis (H_ϕ, H_θ) which can be changed by the torque. Performing the time average over an orbit, $\langle r^2 \rangle = a^2(4 + e^2)/2 \to 5a^2/2$, we obtain

$$
\begin{aligned}
\langle \tau_\mathbf{O} \rangle &= \tfrac{5}{2}a^2 \cos b \left[\hat{\phi} \sin b \left(\Omega_z{}^2 + \Omega_o{}^2 \cos 2(l - \Omega_o t) \right) \right. \\
&+ \left. \hat{\theta} \left(\Omega_o{}^2 \sin 2(l - \Omega_o t) \right) \right].
\end{aligned}
\tag{10}
$$

We are not presently interested in the long term evolution of an individual comet orbit, but in the single-orbit evolution of all near-parabolic comets, so we set $t = 0$. The relative error introduced in replacing the torque by its time average over a single orbit is of magnitude $\approx \left[\frac{\Omega_o{}^2 P}{\Omega_z} \right]^2 < 10^{-2}$ when we set $t = 0$ in the sinusoidal functions. The largest error is made when we treat $\Omega_z{}^2 = 4\pi G \rho(R_o(t), Z_o(t))$ as adiabatically constant. A standard analysis in perturbation theory shows that in using the adiabatic approximation we make an error of order

$$
Max \left[\left(\frac{\dot{\Omega}_z P}{\Omega_z} \right)^2, \frac{\ddot{\Omega}_z P^2}{\Omega_z} \right] \approx \frac{1}{6} \left(\frac{Z_{max} \Omega_z P}{Z_\rho} \right)^2.
\tag{11}
$$

Here $Z_{max}/Z_\rho = Order(1)$ is the ratio of the Solar amplitude to the scale height of the disk density. Today $Z_o \approx 10$ pc $< Z_{max} \approx 80$ pc (Reed, 1997) and the acceleration term dominates. This a-dependent error is $\approx 5\%$ for $a = 30000$ AU.

When both torque components are included, we find that the tidally induced change in angular momentum during a single orbit is

$$
\Delta \mathbf{H} \approx \frac{5}{2} \Omega_z{}^2 P a^2 \cos b \left(\hat{\phi} \sin b [1 + \varepsilon \cos 2l] + \hat{\theta} \varepsilon \sin 2l \right)
\tag{12}
$$

where $\varepsilon \equiv (\Omega_o/\Omega_z)^2$. However, the assumption of azimuthal symmetry in the local tidal field is probably in error by an amount comparable to ε, so that a more nearly self consistent final result, onto which we should append an $\approx 20\%$ uncertainty in the l, b dependence, would be

$$
\Delta \mathbf{H} \approx \hat{\phi} \frac{5}{2} \Omega_z{}^2 P a^2 \sin b \cos b.
\tag{13}
$$

The minimum value of the semimajor axis that can enable a comet to leap the loss cylinder barrier in a single orbit is ≈ 25000 AU for the observed no-dark-matter disk density of $\rho_0 \approx 0.1 M_\odot pc^{-3}$ (Flynn and Fuchs, 1994) which is in good agreement with the observed inner edge of the Oort cloud energy distribution, when we account for uncertainties in the determination of the original value of the cometary semimajor axis, prior to its entry back into the planetary region.

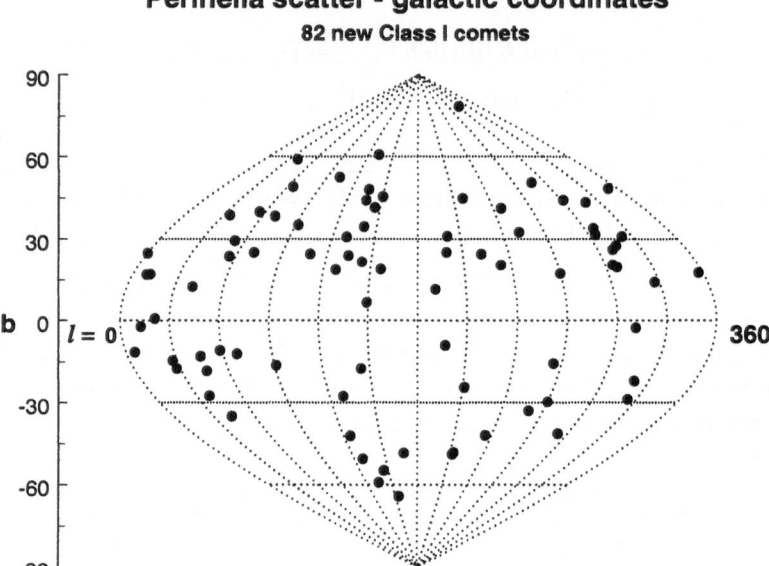

Figure 1. An equal area scatter on the celestial sphere of the perihelion directions of 82 new Class I comets. Results are distributed in Galactic latitude, b, and longitude, l.

The change in **H**, and so the change in q, is predicted to be minimal at $\sin b \cos b = 0$, *i.e.*, at the Galactic poles and equator, once we recognize that b is essentially constant during an orbit for near-parabolic comets. Therefore, if the tide dominates in making Oort cloud comets observable, we should see minima in the distributions at these values. Randomly oriented perihelia would be uniformly distributed on the celestial sphere. In Fig. 1 we show an equal area scatter in perihelia directions of new Class I Oort cloud comets (Marsden and Williams, 1996) which illustrates this characteristic signature of the Galactic tide.

Byl (1983) was the first to recognize the importance of the adiabatic Galactic tide in making Oort cloud comets observable. He also noted the prediction of smallest changes in q for perihelia near the poles and equator and pointed out that the observed distributions had such depletions. But Byl modeled the Galactic interaction as a point mass and several others, most notably Heisler and Tremaine (1986), observed that the Galactic disk dominates the core interaction and performed the appropriate analysis.

The first comprehensive modeling of observable Oort cloud comet orbital element distributions was given later (Matese and Whitman 1989, 1992). A Monte Carlo procedure was employed in which the *in situ* Oort cloud population was modeled to have an energy distribution determined by Bailey (1986). This model has an angular momentum distribution which is empty inside the loss cylinder (but otherwise random), and a $\hat{q}$ distribution that is also randomly oriented. Elements describing comets as they left the planetary region on their prior orbit were selected from this population, *i.e.*, $(a, l, b, H_\phi{}^{prior}, H_\theta{}^{prior})$ are chosen. $\Delta \mathbf{H}$ is then computed from Eqs. (12-13), and comet orbital properties recorded if the change made the comet "observable". Distributions of orbital elements are then obtained for the theoretically "observable" population, and correlations between orbital elements were studied. Predicted distributions were found to compare reasonably with those actually observed. These results have been recently confirmed (Wiegert and Tremaine, 1999).

The most significant evidence that the Galactic tide dominates stellar impulses in making comets observable during the present epoch is the three-fold correlation between orbital elements that is predicted by tidal theory and is observed (Matese *et al.*, 1999). The correlations are embodied in Eq. (13). We leave the reader to investigate the discussion there.

3. Time Dependent Oort Cloud Comet Flux

In a similar manner Matese *et al.* (1995) have modeled the Solar motion $\mathbf{R}_o(t)$ to estimate the time dependence of the tidally-dominated Oort cloud flux over time scales of hundreds of My. In Eq. (13) the time dependence of the tidal strength was obtained by replacing $\Omega_z \rightarrow \Omega_z(\mathbf{R}_o(t - P/2))$ so that the predicted observations at time t were appropriately retarded.

In Fig. 2 we show the modeled flux for a single case with a mean plane crossing period of 36 My, the period that best fits the observed periodicity of large craters (Matese *et al.*, 1998). The peak flux times lag the Galactic plane crossing times by ≈ 2 My and are not precisely periodic because of decreasing Galactic density as the Sun recedes from the Galactic core. The phase of the oscillations is restricted by observations which place the last previous plane crossing at ≈ 1.5 Ma in the past (and the next flux peak ≈ 1 Ma in the future). A perigalactic period of ≈ 180 My is observable in the data. The background shown is a combination of the adiabatic tidal effects

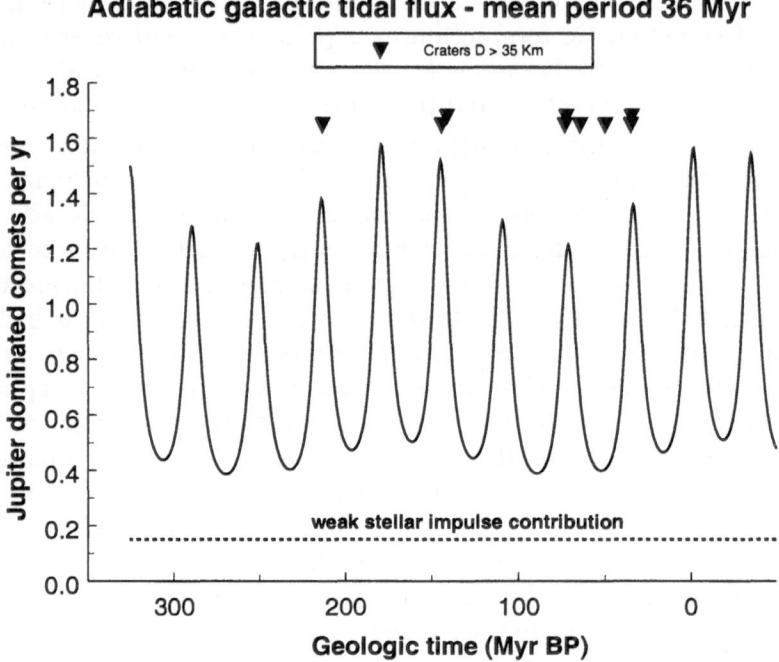

Figure 2. A model of the variable Oort cloud comet flux as modulated by the adiabatic Galactic tide with a mean plane crossing period of 36 My. Past time is positive here. Markers for the 9 accurately dated large craters listed in Table 1 are shown.

of the large-scale-height old stellar population as well as an assumed steady state contribution from stellar impulses of the outer Oort cloud affecting the phase space not accessible to the Galactic tide. We do not show the larger but rarer contributions due to random stellar-induced showers from the inner Oort cloud which are estimated to add $\approx$ 20% to the total, but which dominate only 2% of the time (Heisler, 1990).

The peaks above the background are attributable to the adiabatic tide of the compact component of the disk composed of molecular clouds, dust, young stars and a modest amount of CDDM. Random molecular cloud impulses of the Oort cloud are not included, but will have their probabilities modulated is a manner similar to that shown. The standard deviations of the peaks are $\approx$ 4–5 My, but if the background is included, the formal standard deviation of a complete cycle is closer to 7–8 My. That is, roughly 2/3 of the model's flux occurs in a time interval of $\approx$ 15 My. Random strong stellar impulses will broaden this interval, while stochastic, strong

TABLE 1. Large Accurately Dated Craters

Crater	D(km)	T(My BP)	ΔT(My)
Chesapeake	85	35.5	0.5
Popigai	100	35.7	0.8
Montagnais	45	50.5	0.8
Chicxulub	170	64.98	0.05
Kara	65	73	3
Manson	35	73.8	0.3
Mjølnir	40	142	3
Morokweng	100	145	3
Manicouagan	100	214	1

molecular cloud impulses will narrow it, but not below a value of $\approx$ 9 My. Phase jitter in the Solar motion due to impulses should be substantially less than these widths in a 240 My interval.

Also shown in Fig. 2 are markers for the 9 largest accurately dated craters. The data, listed in Table 1, are taken from Grieve and Pesonen (1996), as modified by Shoemaker (1998) and Rampino and Stothers (1998). The crater name, diameter (D), age (T) and dating uncertainty ($\Delta T \leq 3$ My) are given. Figure 3 shows the results for oscillation models having a range of plane crossing periods from 25–45 My. Statistical analyses have been performed (Matese et al., 1998; Rampino and Stothers, 1998). It is concluded that only if the mean Solar plane crossing period is ultimately found to be in the interval 34–37 My will we be able to say that there is a statistically significant correlation between the Galactic oscillation cycle of the Solar System and periodicity in the formation of large terrestrial impact craters. The statistical significance ($> 2\sigma$) would hold independent of whether one considers only the largest 5 craters or all 9 craters. Including smaller, accurately dated craters will degrade the significance level.

A recent analysis using Hipparcos observations of A and F star distributions (Holmberg and Flynn, 2000) is the first that effectively sheds light on the question of the Solar period. They conclude that there is no evidence for enough CDDM to significantly reduce the Solar oscillation plane crossing period below 45 My. Prior analyses could not definitively exclude modest amounts of CDDM and therefore could not effectively constrain this period.

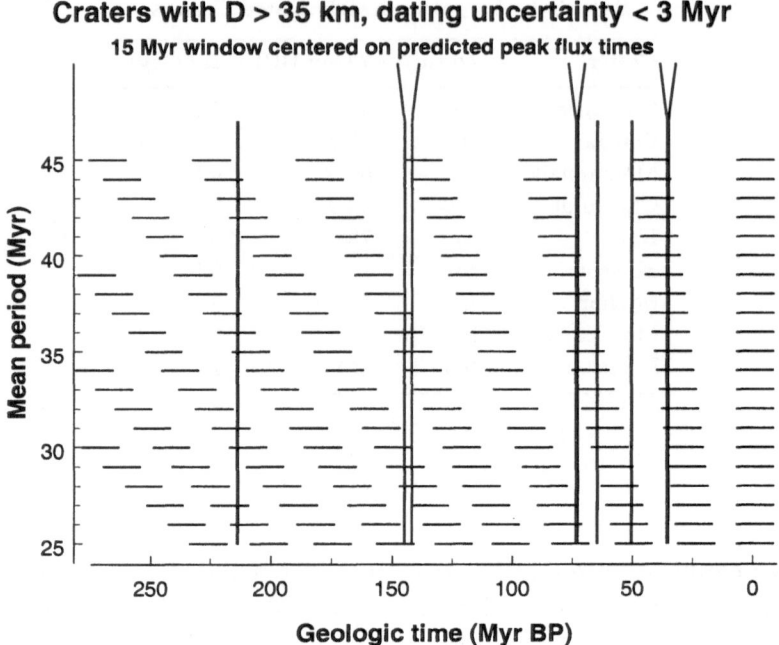

Figure 3. A comparison of 9 accurately dated, large crater ages with a sequence of peak flux times for various mean plane crossing periods ($\frac{1}{2}\bar{P}_z$), as predicted by the Galactic oscillations model. A 15 My window is centered on each peak flux time as a reference.

4. Summary

The Galactic tide dominates in making Oort cloud comets enter the planetary region during the present epoch, and likely over long time scales. Substantial modulation of the tidally induced comet flux must occur independent of the existence or non-existence of CDDM. This is due to the adiabatic variation of the local disk density during the Solar cycle. A Galactic oscillations model in which the Solar cycle is manifest in the terrestrial cratering record will only be sustainable as a working hypothesis if the mean cycle period is found to be in the interval 34–37 My. Should the Holmberg-Flynn result be confirmed, we can reject Shoemaker's suggestion that impacts from long-period comets dominate large terrestrial crater formation if the axisymmetic Galactic disk model is appropriate. But it is evident that non-axisymmetric, periodic perturbations such as spiral density waves, or a central, bar-like mass concentration, may affect the classical periods discussed in this paper, especially if resonances should play a role.

We leave these interesting possibilities for future consideration.

5. Acknowledgements

KAI is grateful for partial financial support from NSERC (Canada), and for the support and hospitality of MJV at the Tuorla Observatory in the spring of 1999.

References

Bailey, M.E. (1986) The Mean Energy Transfer Rate to Comets in the Oort Cloud and Implications for Cometary Origins, *MNRAS* **218**, 1-30.

Binney, J. and Tremaine, S. (1986) Chap. 3.2.3 in *Galactic Dynamics*, Princeton University Press (Princeton NJ), pp. 120-126.

Breiter, S., Dybczynski, P.A. and Elipe, A. (1996) The Action of the Galactic Disk on the Oort Cloud Comets; Qualitative Study, *Astron. Astrophys.* **315**, 618-624.

Byl, J. (1983) Galactic Perturbations of Nearly-Parabolic Cometary Orbits, *Moon Planets* **29**, 121-137.

Duncan, M., Quinn, T. and Tremaine, S. (1987) The Formation and Extent of the Solar System Comet Cloud, *Astron. J.* **94**, 1330-1338.

Flynn, C. and Fuchs, B. (1994) Density of Matter in the Galactic Disc, *MNRAS* **270**, 471-479.

Frisch, P.C. and York, D.G. (1986) Interstellar Clouds Near the Sun, in *The Galaxy and the Solar System* (R.N. Smoluchowski, J.N. Bahcall, and M.S. Matthews, Eds.), University of Arizona Press, pp. 83-102.

Grieve, R.A.F. and Pesonen, L.J. (1996) Terrestrial Impact Craters: Their Spatial and Temporal Distribution and Impacting Bodies, *Earth, Moon, Planets* **72**, 357-376.

Heisler, J. (1990) Monte Carlo Simulations of the Oort Comet Cloud, *Icarus* **88**, 104-121.

Heisler, J.S. and Tremaine, S. (1986) Influence of the Galactic Tidal Field on the Oort Cloud, *Icarus* **65**, 13-26.

Holmberg J., and Flynn, C. (2000) The Local Density of Matter Mapped by *Hipparcos*, *MNRAS* **313**, 209-216.

Kuijken, K., and Tremaine, S. (1994) On the Ellipticity of the Galactic Disk, *Astroph. J.* **421**, 178-194.

Marsden B.G., and Williams, G.V. (1996) *Catalogue of Cometary Orbits, 11th Ed.*, Cambridge: Smithsonian Astrophysical Observatory.

Matese, J.J. and Whitman, P.G. (1989) The Galactic Disk Tidal Field and the Nonrandom Distribution of Observed Oort Cloud Comets, *Icarus* **82**, 389-402.

Matese, J.J. and Whitman, P.G. (1992) A Model of the Galactic Tidal Interaction with the Oort Comet Cloud, *Celest. Mech. Dynam. Astron.* **54**, 13-36.

Matese, J.J., Whitman, P.G., Innanen, K.A. and Valtonen, M.J. (1995) Periodic Modulation of the Oort Cloud Comet Flux by the Adiabatically Changing Galactic Tide, *Icarus* **116**, 255-268.

Matese, J.J., Whitman, P.G., Innanen, K.A. and Valtonen, M.J. (1998) Variability of the Oort Cloud Comet Flux: Can it be Manifest in the Cratering Record?, in *Highlights in Astronomy* (J. Andersen, Ed.) **11A**, pp. 252-256.

Matese, J.J., Whitman, P.G. and Whitmire, D.P. (1999) Cometary Evidence of a Massive Body in the Outer Oort Cloud, *Icarus* **141**, 354-366.

Merrifield, M.R. (1992) The Rotation Curve of the Milky Way to 2.5 R_o from the Thickness of the HI Layer, *Astron J.* **103**(5), 1552-1563.

Rampino, M.R. and Stothers, R.B. (1998) Mass Extinctions, Comet Impacts, and the Galaxy, in *Highlights in Astronomy* (J. Andersen, Ed.) **11A**, pp. 246-251.

Reed, B.C. (1997) The Sun's Displacement from the Galactic Plane: Limits from the Distribution of OB-Star Latitudes, *Pub. Astron. Soc. Pac.* **109**, 1145-1148.

Shoemaker, E.M., Wolfe, R.F. and Shoemaker, C.S. (1990) Asteroid and Comet Flux in the Neighborhood of the Earth, *Geol. Soc. of America Special Paper 247*, 155-170.

Shoemaker, E.M. (1998) Impact Cratering Through Geologic Time, *J. Roy. Astron. Soc. Canada* **92**, 297-309.

Valtonen, M.J., Zheng, J.Q., Matese, J.J. and Whitman, P.G. (1995) Near-Earth Populations of Bodies coming from the Oort Cloud and Their Impacts with Planets, *Earth, Moon, Planets* **71**, 219-223.

Wiegert, P. and Tremaine, S. (1999) The Evolution of Long-Period Comets, *Icarus* **137**, 84-122.

GALACTIC TRIGGERING OF PERIODIC COMET SHOWERS AND MASS EXTINCTIONS ON EARTH

MICHAEL R. RAMPINO

Earth & Environmental Sciences Program, New York University
100 Washington Square East, New York, NY 10003, USA;

NASA, Goddard Institute for Space Studies
2880 Broadway, New York, NY 10025, USA

Abstract. Geological and astronomical studies can be synthesized to provide a hypothesis connecting the history of life on Earth with the dynamics of the Galaxy. Impacts of comets and asteroids on the Earth are believed to be capable of causing severe environmental effects and mass extinction of life. The size-frequency distribution of earth-crossing objects is in general agreement with the hypothesis that collisions with large bodies (mostly comets) $\gtrsim 5$ km in diameter could explain the record of ~ 25 extinction episodes in the last 545 Ma. The five major mass extinctions could be related to impacts of the largest comets ($\gtrsim 10$ km in diameter). Several of these extinction events have thus far been correlated or tentatively correlated with concurrent impact ejecta layers and/or large impact craters, suggesting times of increased flux of impactors. Smaller comet and asteroid impacts might be responsible for lesser global and sub-global faunal and floral turnover events.

Periods of $\sim 30 - 36$ My have been found in sets of well-dated large impact craters, and extinction shows a periodic component of ~ 30 My. These results could be explained by periodic or quasi-periodic showers of Oort Cloud comets. The pacemaker for such comet showers may involve the Sun's vertical oscillation through the plane of the Milky Way Galaxy, suggesting that evolutionary changes on the Earth are partly related to simple galactic dynamics.

M. Ya. Marov and H. Rickman (eds.), Collisional Processes in the Solar System, 103–120.

1. Introduction

In the geologic picture that has been standard for more than 160 years, the important events in the planet's history, including mass extinction of living species, are engendered largely by causes internal to the Earth. In the early 1980's, that view was challenged. Alvarez *et al.* (1980) provided hard physical evidence for a link between the end-Cretaceous mass extinction (65 million years ago[1]) and the impact of a large asteroid or comet. Subsequently, Raup and Sepkoski (1984) presented analyses suggesting that extinctions showed an approximately 26 to 30 million year period, and thus might have a common cause, most likely extraterrestrial. In response to those findings, Rampino and Stothers (1984a) proposed a model in which the periodic or quasi-periodic extinction events were related to comet showers caused by the periodic passage of the Solar System through the central plane of the Milky Way Galaxy. Rampino and Haggerty (1996a) later applied the name Shiva Hypothesis to the galactic comet shower hypothesis in reference to the Hindu deity of cyclical destruction and renewal (see also Rampino, 1998).

Rampino and Stothers (1984a) suggested that the probability of encounters with molecular clouds that could perturb the Oort comet cloud and cause comet showers is modulated by the Sun's oscillation about the galactic disk. The distribution of clouds in the galactic disk suggests that an encounter would be more likely as the Sun passes through the plane region, and hence the encounters would be quasi-periodic, with a period equal to the time between plane crossings.

The time between plane crossings (the half-period of vertical oscillations) is given by:

$$P_{1/2} = (\pi/4G\rho)^{1/2} \tag{1}$$

where ρ is the mean volume density of matter near the galactic plane, and G is the constant of gravity. The density of the known visible matter in the plane is 0.10 ± 0.01 $M_\odot pc^{-3}$ (Matese *et al.*, 1995; Stothers, 1998), but some additional dark matter is commonly believed to be present. Recently, Stothers (1998) utilized all recent determinations of local mass density to estimate that the average value of ρ is 0.15 ± 0.01 $M_\odot pc^{-3}$, implying about 30% dark matter (probably in the form of cold interstellar clouds) (see also Gould *et al.*, 1996; Crézé *et al.*, 1998). This would give a half-period in the order of 34 ± 3 My; more extreme estimates of dark matter (e.g., Bahcall *et al.*, 1992) lead to half-periods as short as 27 to 28 My (see Matese *et al.*, 1995; Clube and Napier, 1996).

---

[1] In the following, 'Ma' will denote specific time intervals into the past, while 'My' will denote general time intervals

The Rampino/Stothers model was criticized on the grounds that the amplitude of the Sun's present excursions from the galactic plane was too small to produce a quasi-periodic modulation of comets that would show up over the background of non-periodic impacts (Thaddeus and Chanan, 1985). Numerical simulations of the galactic modulation using the best available astronomical data, however, suggested that this effect should be detectable in the terrestrial record of cratering (Stothers, 1985).

Rampino and Stothers (1984a) also argued that there could be enough scatter in the cloud encounters to explain the apparent offset between the most recent extinction recognized by Raup and Sepkoski (in the mid-Miocene at $\sim$ 11 Ma) and the most recent plane crossing in the last few million years. Subsequent work, however, has identified an extinction event in the Pliocene ($\sim$ 2.3 Ma), although both the Mid-Miocene and Pliocene extinctions were minor events as compared with the Late Cretaceous (65 Ma) or even the Late Eocene (35 Ma) extinctions (Raup and Sepkoski, 1986; Sepkoski, 1995).

Tidal forces produced by the overall gravitational field of the Galaxy can also cause perturbations of cometary orbits. Since these forces vary with the changing position of the Solar System in the Galaxy, they provide a mechanism for the periodic variation in the flux of Oort cloud comets into the inner Solar System (Napier, 1987; Bailey et al., 1990; Matese et al., 1995, 1996, 1998). The cycle time and degree of modulation also depend critically on the mass distribution in the galactic disk.

Again, if the extreme range of the various recent estimates in the disk matter density is considered (Bahcall et al., 1992), then the period could be as short as $\sim$ 28 My, with a comet flux ratio of $\sim$ 4 to 1. The standard deviations of the flux peaks were 4 to 5 My, and the times of peak comet flux lag the times of galactic plane crossing by about 1 to 2 My (Valtonen et al., 1995). The amplitude of the comet pulses and the length of the cycle interval are modulated somewhat by the epicyclic motion of the Sun about the Galactic center, with a period of about 180 My (Matese et al., 1995, 1998)

Differences in the mass-frequency distributions of large asteroids and comets and the higher velocity of comets in general suggest that the largest and most energetic impactors are most likely comets (Shoemaker et al., 1990). Thus, the largest craters should preferentially show the galactic modulation of comet flux. Smaller impacts may also show a periodic signal, if comets commonly fragment. A comet shower in the Late Eocene ($\sim$ 35 Ma) (Rampino, 1995) is suggested by enriched extraterrestrial ^{3}He in sediments covering about 2 to 3 My, apparently in dust from a pulse of comets or comet breakup in the inner Solar System (Farley et al., 1998). This interval brackets the dates of two known large impacts (Popigai and Chesapeake) and several less well-dated smaller craters, and could represent a 4-fold increase in the impact rate (Montanari *et al.*, 1998). Similar pulses of increased comet flux could also explain the stepped nature of some extinction events, and clusters of similar-age craters and impact layers seen in the geologic record (Hut *et al.*, 1987; Shoemaker and Wolfe, 1986; Montanari *et al.*, 1998). Furthermore, McEwen *et al.* (1997) have proposed that the cratering rate in the inner Solar System increased in the past 300 Ma due to an increased flux of comets.

2. Impact and Extinction

The major mass extinction at the Cretaceous/Tertiary (K/T) boundary (65 Ma) coincided with the impact of a comet or asteroid $\sim$ 10 km in diameter which created the $\sim$ 180 km diameter Chicxulub impact structure in the Yucatán region of Mexico (e.g., Hildebrand *et al.*, 1995; Pope *et al.*, 1998). Such a large impact is calculated to have caused a global catastrophe related to dense clouds of fine ejecta, production of nitric oxides and acid rain, and smoke clouds from fires triggered by atmospheric re-entry of ejecta (Toon *et al.*, 1997). Other effects include tsunami, enhanced greenhouse effect from atmospheric water vapor derived from an ocean impact, CO_2 released by impact into carbonate rocks, and cooling and acid rain from large amounts of sulfuric acid aerosols derived from calcium sulfate target rocks (Pope *et al.*, 1997, 1998; Toon *et al.*, 1997).

The size vs. frequency curve for impacts on the Earth, based on observations of earth-crossing asteroids and comets, and the cratering record of the inner planets (Shoemaker *et al.*, 1990), predict that in an $\sim$ 100 My period, the Earth should be hit by several comets larger than a few km in diameter ($\sim 10^{23}$ J or 10^7 Mt TNT equivalent) and perhaps one $\sim$ 10 km diameter ($\sim 10^{24}$ J, 10^8 Mt) object (most likely cometary) (Shoemaker *et al.*, 1990; Gehrels, 1999).

A $\gtrsim 10^8$ Mt event would produce a more severe and widespread environmental disaster than a $\sim 10^7$ Mt event (Melosh *et al.*, 1990; Toon *et al.*, 1997), and these results lead to the prediction of $\sim$ 5 major mass extinctions, and about 25 ± 5 less severe pulses of extinction during the Phanerozoic Eon (the last 545 Ma) resulting from impacts. The independent paleontological record of extinctions for that interval clearly shows 5 major mass extinctions and $\sim$ 20 less severe extinction pulses, in agreement with the estimates for impact-induced extinctions (Fig. 1; note that overall extinction rates were probably higher during the period prior to $\sim$ 500 Ma, and the lack of resolution of distinct peaks in that interval is a result of poor knowledge of the fossil record, e.g., Sepkoski, 1995). The agreement is an unexpected congruence of astronomy and paleontology.

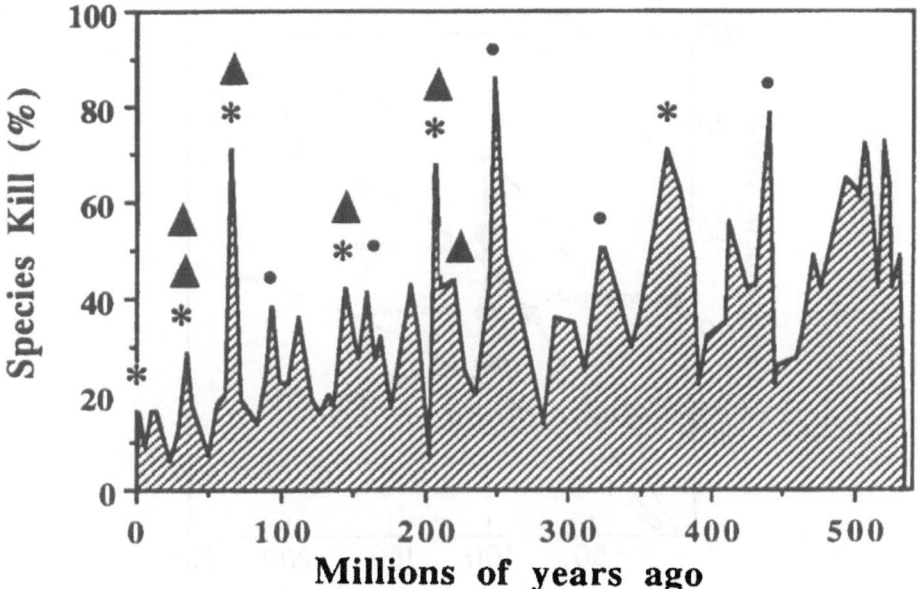

Figure 1. Percent extinction of marine species per geologic stage (or substage) during the Phanerozoic (species data derived from genus-level data of Sepkoski, 1995, treated with the reverse rarefaction estimates of Raup, 1979), and compiled record of proven and possible impact evidence (triangles = large, dated impact structures; asterisks = impact ejecta; dots = iridium spikes above background; see text).

Proposed "kill curve" relationships between mass extinctions and impacts of various magnitudes (Fig. 2; Raup, 1992) can be compared with data representing the largest (> 80 km diameter) known impact craters with well-defined ages that overlap the ages of mass-extinction boundaries (Table 1). The observed points in Fig. 2 (using Raup's original curve) agree with the predicted curves within the errors permitted by the geologic data, supporting at least a first-order relationship between large impacts and mass extinctions.

The age of the Triassic/Jurassic boundary (Rhaetian/Hettangian) as evidenced by a palynological break in the Newark Basin (Fowell *et al.*, 1994), suggests a revision in the age of that boundary from ~ 205.7±4.0 Ma (Gradstein and Ogg, 1996) to a younger age of ~ 201 Ma. Thus, the ~ 100 km diameter Manicouagan impact structure, dated by U-Pb methods at 214±1 Ma, may be too old to mark the end of the Rhaetian (76% marine species extinction). The crater might, in fact, be related to the somewhat older Norian/Rhaetian boundary (~ 40% species extinction) dated at ~ 209.6 ± 4.1 Ma, or the Carnian/Norian boundary at 220.7 ± 4.4 Ma (~ 42% species extinction) (Fig. 1). This interpretation supports suggested

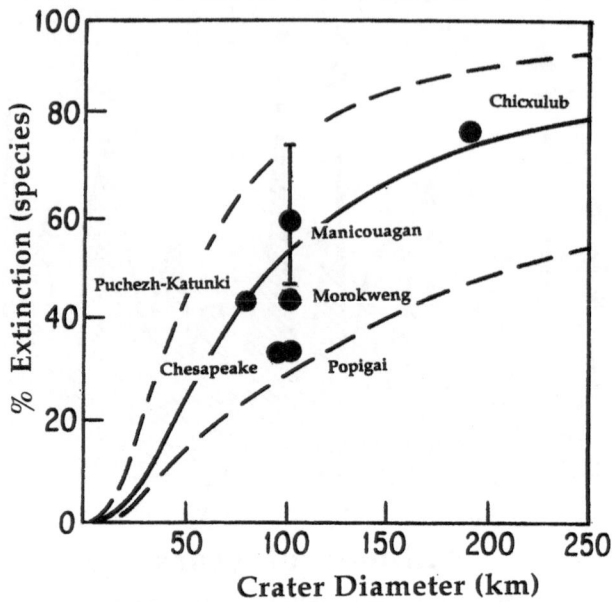

Figure 2.   Proposed kill curve for Phanerozoic marine species plotted against estimated
size of impact craters associated with extinctions of various magnitudes (assuming that
the two are related) (solid line with dashed estimated error lines) (after Raup, 1992).
Largest well-dated impact craters with ages overlapping mass extinction times are plotted
for comparison (see Table 1). New dating of the extinction record suggests that the
Manicouagan crater may be associated with the end-Carnian extinction event ($\sim$ 40%
species kill). The kill curve of Poag (1997) suggests a step function in the curve, with a
rapid rise in global species kill from very low values, at a crater size somewhat greater
than 100 km in diameter.

modifications of the original kill curves that incorporate a threshold level for
significant global mass extinctions at impacts that create craters somewhat
larger than 100 km in diameter ($\sim 10^7$ Mt events) (Jansa *et al.*, 1990; Poag,
1997, Montanari *et al.*, 1998).

3.   Direct Evidence of Impacts at Times of Extinction

Shocked minerals (including shocked quartz, stishovite, and shocked zir-
cons), impact glass (microtektites/tektites), microspherules with structures
indicating high-temperature origin, and Ni-rich spinels are diagnostic of
hypervelocity impact. These material have thus far been reported in stratigraphic horizons close to the times of six recorded extinction pulses (Table 2) (Rampino *et al.*, 1997; Grieve, 1997). Recent preliminary reports of
shocked quartz from the Permian/Triassic boundary ($\sim$ 248 Ma) in Antarc-

TABLE 1. Large dated impact craters (Grieve and Pesonen, 1996, with updates) with associated marine extinctions (Sepkoski, 1995).

Name	Diameter (km)	Age (Ma)	Extinction	% Species
Popigai	100	35.7 ± 0.8	Late Eocene	30
Chesapeake	90	35.2 ± 0.3	Late Eocene	30
Chicxulub	180	65.2 ± 0.4	K/T	76
Morokweng	100?	145 ± 2	J/K	42
Manicouagan	100	214 ± 1	Late Triassic:	
			end-Rhaetian	76
			or end-Norian	40
			or end-Carnian	42
Puchezh-Katunki	80	220 ± 10	end-Carnian?	42

TABLE 2. Stratigraphic evidence of impact debris at or near extinction events (for references see text).

Age	Evidence
Pliocene (2.3 Ma)	impact debris
Late Eocene (35 Ma)	microtektites (multiple), tektites, microspherules, shocked quartz, spinels
Cretaceous/Tertiary (65 Ma)	microtektites, tektites, shocked minerals, stishovite, spinels, iridium
Jurassic/Cretaceous (~ 142 Ma)	shocked quartz
Late Triassic (~ 2014 Ma)	shocked quartz (multiple)
Late Devonian (~ 368 − 365 Ma)	microtektites (multiple)

tica and Australia (Retallack *et al.*, 1998), and the Triassic/Jurassic boundary (~ 201 Ma?) in eastern North America (Mossman *et al.*, 1998) require further data for confirmation.

In total, at least six of the ~ 25 extinction peaks in Fig. 1 (Pliocene ~ 2.3 Ma, Late Eocene ~ 35 Ma, Cretaceous/Tertiary 65 Ma, Jurassic/Cretaceous ~ 144 Ma, Late Triassic ~ 201 Ma, Late Devonian ~ 365 Ma) are correlative with large impact craters and/or stratigraphic evidence of impact ejecta (Tables 1 and 2). Several other extinction events are associated with "possible" evidence of impact consisting of iridium concentrations somewhat elevated with respect to background values or suspected shock minerals (see tables in Rampino et al., 1997 and Grieve 1997).

Direct comparisons of impact crater ages and mass extinction show significant statistical correlations (Matsumoto and Kubotani, 1996; Yabushita, 1998), and smaller craters also tend to match stratigraphic stage boundaries defined by lesser faunal changes (Stothers, 1993; Rampino and Schwindt, 1999). Clusters of impacts could explain the problems in global correlation of some geologic time boundaries. For example, the Jurassic/Cretaceous boundary ($\sim$ 142 to 145 Ma) does not yet have an internationally accepted definition (Gradstein and Ogg, 1996), stemming in part from the difficulty of correlating faunal changes in the northern (Boreal) and equatorial (Tethyan) regions of the time. Recent work has shown that the age of the 40 km diameter Mjølnir impact structure in the Barents Sea matches the time of the Boreal J/K boundary at about 142 Ma (Dypvik et al., 1996; see also Zakharov et al., 1993). The smaller Gosses Bluff impact in Australia (142.5 $\pm$ 0.5 Ma) also dates from the same interval. The age of the larger ($>$ 100 km diameter) Morokweng impact structure in South Africa (145 $\pm$ 2 Ma), however, better matches the timing of the slightly older Tethyan J/K boundary. These data suggest that several impacts and extinction events over a period of several million years were involved in the Jurassic/Cretaceous transition.

4. Smaller Impacts and Regional Devastation

Collisions with impactors of 2 to 3 km diameter (with energies in the 10^6 Mt range), producing impact craters from $\sim$ 30 to 60 km in diameter, should occur about every million years (Chapman and Morrison, 1994), or $\sim$ 600 such impacts during the 545 My of the Phanerozoic. About 30% of these impacts ($\sim$ 200) would have hit the continents, which means that North America, for example, (representing $\sim$ 15% of total continental area) would have suffered about 30 such impacts, or approximately one every 20 million years. Eleven such impact structures between about 30 and 60 km in diameter are known (Table 3). Calculations suggest that these energetic events should have left a significant imprint on regional geologic records.

Published scaling laws for radii of severe blast devastation around impacts give $R_s = 8E_k^{1/3}$, where R_s is in km, and E_k is the kinetic energy of the impactor in megatons (Adushkin and Nemtchinov, 1994). The estimated scaling for fire ignition by impacts is $R_f = 3E_r^{1/2}$, where R_f is in km and E_r is the thermal radiation energy in megatons (the ratio of thermal radiation to kinetic energy for an impact is $\sim$ 20%). In the case of a $\sim$ 10^6 Mt impact event, therefore, the radius of severe blast devastation could be $\sim$ 800 km, whereas the radius of fire ignition could be $>$ 1,000 km. The radius of severe seismic effects of a 10^6 Mt impact (and the resulting

TABLE 3. North American impact structures ($D = \sim 25$ to 60 km).

Impact Structure	Approx. Diameter (km)	Age (Ma)
Beaverhead	60	~ 600
Charlevoix	54	357
Slate Islands	30	< 350
Clearwater East	22	290
Clearwater West	32	290
Saint Martin	40	219.5
Carswell	39	115
Manson	35	73.8
Marquez	22	58 ± 2
Montagnais	45	50.5
Mistastin	28	38
Haughton	24	23.4 ± 1

magnitude 9 earthquake) is estimated at $\sim 1,000$ km.

A possible example of the regional environmental effects of such an impact involves the 35 km diameter Manson impact structure in northwest Iowa (73.8 ± 0.3 Ma) estimated to have resulted from an $\sim 10^6$ Mt impact. A search for Manson ejecta in Upper Cretaceous marine rocks to the west led to the discovery of shocked minerals in a widespread layer in southeastern South Dakota (the Crow Creek horizon), which has been interpreted as a probable tsunami deposit (Izett *et al.*, 1993; Steiner and Shoemaker, 1996).

There has been much discussion as to whether dinosaurs in western North America were declining prior to the end of the Cretaceous Period (Sloane *et al.*, 1986; Sheehan *et al.*, 1991). The confusion may be partly the result of a regional extinction caused by the Manson impact. Coincident with the timing of the Manson event, nine genera of dinosaurs disappeared in western North America, followed by the immigration and rapid evolution of other dinosaur taxa, and with major changes in marine vertebrates and mammals in the same region (Russell, 1993). Catastrophic dinosaur bone beds representing mass mortality (Varricchio and Horner, 1993), some indicating extremely violent conditions and containing herds of $> 10,000$ individuals, occur in rocks dated at ~ 74 Ma in Montana $\sim 1,400$ km from Manson (Rampino *et al.*, 1997).

The Azuara impact structure in Spain (Ernstson and Claudin, 1990; Ernstson and Fiebag, 1992) is a similar ~ 35 km diameter crater, tentatively dated as roughly 30 to 40 Ma. The aftermath of that impact should have been widespread environmental devastation in Europe. A drastic change did occur in the region's fauna and flora about 35 million years ago. More than 60% of the native western European species suddenly disappeared, to be subsequently replaced by immigrant species from eastern Europe and Asia, with a burst of evolution in surviving rodents and carnivores (see Rampino, 1997), an event so severe that it was named the "Grande Coupure" (the Great Cut).

5. Periodicity in Mass Extinction

A great deal has been written concerning periodicity in mass extinctions. Raup and Sepkoski (1984, 1986) reported a statistically significant 26.4 My periodicity in extinction time series for the last 250 Ma, with a secondary periodicity of 30 My. Periods of ~ 26 to 31 My have been derived using various subsets of extinction events (family and genus levels), different geologic time scales, and various methods of time-series analysis (see Rampino et al., 1997 for a review)

Extension of the record back to 515 Ma resulted in a spectrum with the highest peak at 27.3 My (Rampino and Haggerty, 1996a,b), and truncation analysis of the extinction time series revealed that all records longer than 144 My exhibited a major spectral peak at ~ 26 to 27 My. In the longer records, the second or third highest peak lies between 35 and 37 My (Fig. 3; Table 4).

Time series analyses has also been applied to the extinction data (using the 11 mass extinctions of the last 250 Ma) with allowances for the phase of the cycle to be either fixed (at the present time) or a free parameter. When the phase is allowed to vary, only one high spectral peak at 27 My is apparent. When the phase is fixed at the present, however, the highest peak shifts to 28 My. But also two somewhat smaller peaks at 32 and 35 My become considerably more prominent. Napier (1998) recently analyzed the Rampino/Caldeira (1993) compiled dataset of geologic events (which includes the extinctions) and found that the 25 to 27 Ma periodicity detected earlier (Rampino and Caldeira, 1993; see also Liritzis, 1993) was in agreement with periods expected from the galactic oscillation hypothesis.

The three most severe mass extinctions (Late Ordovician, ~ 435 Ma; Late Permian, ~ 250 Ma; and Late Cretaceous, 65 Ma) are separated by ~ 180 Ma. The Solar System undergoes a perigalactic revolution cycle with an estimated period of ~ 170 ± 10 My (Matese et al., 1995), and this cycle might also modulate the flux of Oort cloud comets (Bailey et al., 1990). Matese and Whitmire (1996) have found evidence for a perturbation of the present Oort comet cloud by the Galactic radial tide, due to the entire distribution of matter interior to the solar orbit (see also Matese et al., this volume).

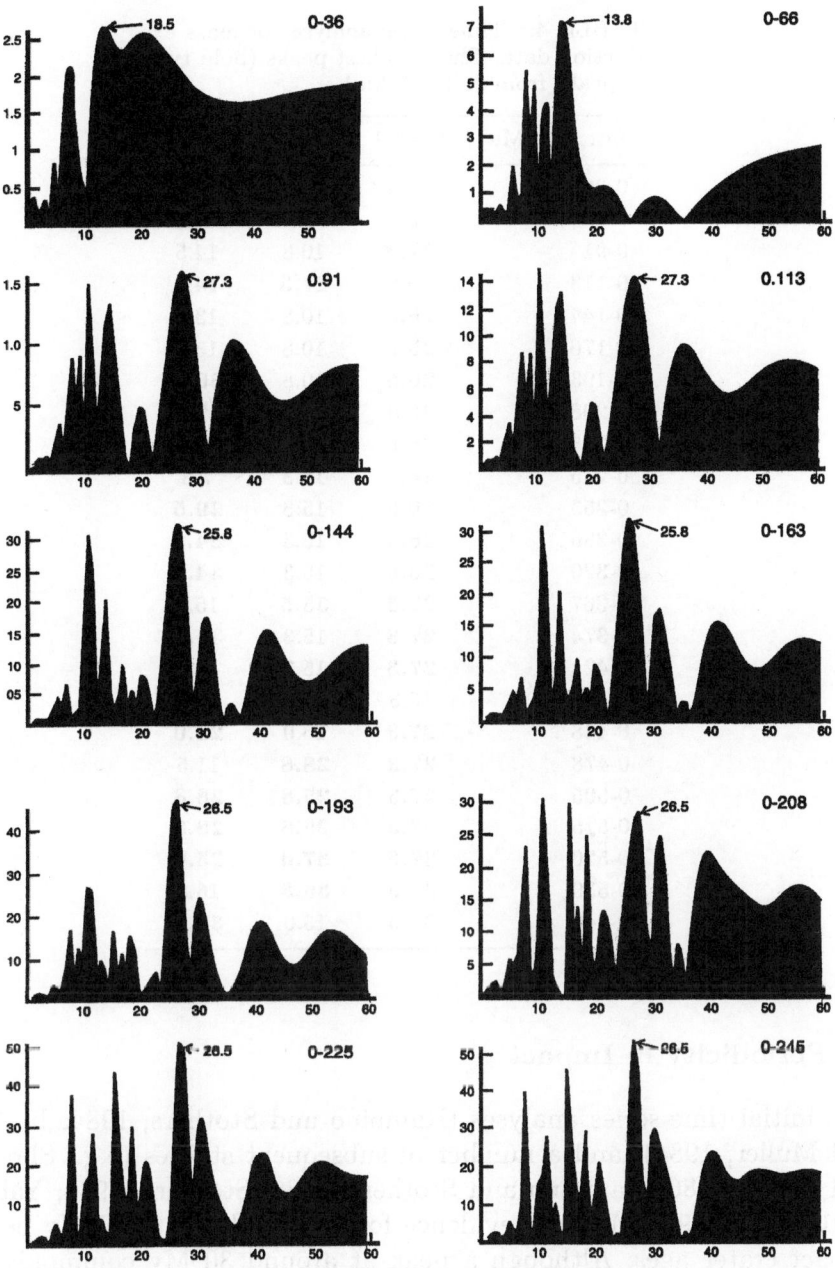

Figure 3.   Fourier power spectrum for extinction events from Figure 1 (subset from 0 to 250 Ma) computed as described in Rampino and Caldeira (1993). X-axis = period in My; y-axis = spectral power. The time series was truncated, one extinction at a time, from 250 to 36 Ma. The dominant stable peaks in the spectrum (aside from noise in the 0 to 20 My period range) are situated at $\sim 26 - 27$ My and $\sim 31 - 37$ My (see Table 4).

TABLE 4. Time-series analyses of mass extinction data: Three highest peaks (bold type = peaks from 25 to 37 Ma).

Interval (Ma)	Peak 1	Peak 2	Peak 3
0-36	13.8	20.5	7.8
0-66	13.8	7.8	9.3
0-91	**27.3**	10.8	14.5
0-113	10.8	**27.3**	14.4
0-144	**25.8**	10.8	13.8
0-176	**25.8**	10.8	13.8
0-193	**26.5**	10.8	**30.3**
0-208	10.8	15.3	**27.8**
0-225	**26.5**	15.3	7.8
0-245	**26.5**	15.3	7.8
0-253	**26.5**	15.3	**29.5**
0-286	**26.5**	15.3	**34.8**
0-320	**26.5**	15.3	**34.8**
0-367	**27.3**	**35.5**	15.3
0-374	**27.3**	15.3	**35.5**
0-421	**27.3**	15.3	11.5
0-438	15.3	**27.3**	11.5
0-448	**27.3**	16.0	**25.0**
0-478	**27.3**	**28.8**	11.5
0-505	**27.3**	**25.8**	**28.8**
0-515	**27.3**	**25.8**	**28.8**
0-520	**27.3**	**37.0**	**25.0**
0-530	**27.3**	**36.3**	16.8
0-540	**27.3**	13.0	**36.3**

6. Periodicity in Impact

The initial time-series analyses (Rampino and Stothers, 1984a,b; Alvarez and Muller, 1984), and a number of subsequent studies (e.g., Shoemaker and Wolfe, 1986; Rampino and Stothers, 1986; Stothers, 1988; Yabushita 1991, 1992, 1996a,b) found evidence for a possible 28 to 32 My period in impact crater ages. Although a peak at around 30 My commonly shows up in statistical analyses, the significance of the peak has been questioned (e.g., Grieve and Shoemaker, 1994; Grieve and Pesonen, 1996). Montanari et al. (1998) detected peaks of $\sim$ 33 My and 64 My (the harmonic?), but concluded that the peaks were artifacts of the statistical analysis.

The differences in the formal periods derived from analyses of extinctions and cratering might at first seem problematic. However, several studies have concluded that the observed differences in the formal periodicity are to be expected, taking into consideration problems in dating and the likelihood that both records would be mixtures of periodic and random events (Stothers, 1988, 1989; Trefil and Raup, 1987; Fogg, 1989).

Most earth-crossing asteroids fall in the $\lesssim 1$ km size range (Shoemaker et al., 1990), and thus we might expect that for impact craters on Earth $\lesssim 20$ km in diameter, the presumably random asteroid flux would dominate over any signal from periodic comets in that size range. In this case, the cratering period should show up preferentially in the largest craters. In a test of this idea, Matese et al. (1998) recently found a best-fitting period of 36 ± 2 My, using only the nine largest well-dated craters, when the phase was fixed to the time of most recent galactic plane crossing. Rampino and Stothers (1998) also performed linear time-series analyses on sets of various sized craters using the revised list of 34 impact craters shown in Table 5 (Grieve and Pesonen, 1996; with updates from Grieve, 1996, 1997 and other sources). As Table 6 shows, two peaks were detected, a narrow peak at 30 ± 0.5 My and a broader peak at 35 ± 2 My. In the cases where only the largest craters were utilized, the highest peak in the period spectrum is located at ~ 36 My, in agreement with the results of Matese et al. (1998).

Stothers (1988) had earlier shown that the dating errors alone were capable of shifting the dominant period between 30 and 35 million years. The periods detected in the mass extinction and cratering records, however, are statistically significant (at the 5% level) only when they are treated as periods (and phases) known a priori. Thus, it is difficult to say whether the "best" period in cratering is closer to 30 or 36 million years. Furthermore, the width of the comet flux pulses in which large comet impacts become most probable is ~ 8 My.

7. Conclusions

Geologic data on mass extinctions of life and evidence of large impacts on the Earth are thus far consistent with a quasi-periodic modulation of the flux of Oort cloud comets. A mean period of about 30 to 36 My is indicated, as predicted by galactic oscillation models utilizing the current range of estimates of the total mass of the galactic disk (including a dark-matter component). Discrepancies in the periodicities detected in the extinctions and cratering may be the result of a combination of dating errors, mixtures of periodic and non-periodic events, the small data sets, and real irregularities in the underlying cycle. Further astronomical and geological studies (including the discovery and accurate age-dating of large impact craters, and the better determination of the dark-matter component of galactic disk

TABLE 5. Impact craters ($D \gtrsim 5$ km, ages $\lesssim 250$ Ma with ± 10 Ma).

Crater	Location	Diameter (km)	Age (Ma)	Method
Zhamanshin	Kazakhstan	13.5	$0.90 \pm .10$	Ar-Ar
Bosumtwi	Ghana	10.5	1.03 ± 0.02	K-Ar
El'gygytgyn	Russia	18	3.5 ± 0.5	K-Ar
Bigach	Kazakhstan	7	63	Strat.
Karla	Russia	12	10 ± 10	Strat.
Ries	Germany	24	15.1 ± 1	Ar-Ar
Haughton	Canada	24	23.4 ± 1	Ar-Ar
Chesapeake	USA	90	35.2 ± 0.3	Biostrat.
Popigai	Russia	100	35.7 ± 0.8	Ar-Ar
Wanapitei	Canada	7.5	37 ± 2	Ar-Ar
Mistastin	Canada	28	38 ± 4	Ar-Ar
Logoisk	Belarus	17	40 ± 5	Strat.
Montagnais	Canada	45	50.5 ± 0.8	Ar-Ar
Ragozinka	Russia	9	55 ± 5	Strat.
Marquez	USA	22	58 ± 2	Strat.
Chicxulub	Mexico	180	64.98 ± 0.05	Ar-Ar
Kamensk	Russia	20	65 ± 2	Strat.
Kara, Ust-Kara	Russia	65	70.3 ± 2.2	Ar-Ar
Manson	USA	35	73.8 ± 0.3	Ar-Ar
Lappajärvi	Finland	23	77 ± 0.4	Ar-Ar
Boltysh	Ukraine	24	88 ± 3	K-Ar
Dellen	Sweden	15	89 ± 2.7	Rb-Sr
Steen River	Canada	25	95 ± 7	K-Ar
Avak	USA	12	100 ± 5	Strat.
Carswell	Canada	39	115 ± 10	Ar-Ar
Mien	Sweden	9	121 ± 2.3	Ar-Ar
Tookoonooka	Australia	55	128 ± 5	Strat.
Gosses Bluff	Australia	22	142.5 ± 0.5	Ar-Ar
Mjølnir	Barents Sea	40	142.6 ± 2.6	Strat.
Morokweng	South Africa	≥ 100	145 ± 2	Ar-Ar
Rochechouart	France	23	214 ± 8	Ar-Ar
Manicouagan	Canada	100	214 ± 1	U-Pb
Puchezh-Katunki	Russia	80	220 ± 10	Strat.
Araguainha	Brazil	40	247 ± 5.5	Rb-Sr

mass) should help to clarify and refine both the expected astronomical cycle times and the periodicities detectable in the geologic record. It would be of significant interest if the evolution of life on Earth can be reduced in part to so simple a relationship as $P_{1/2} = (\pi/4G\rho)^{1/2}$ for the half-period of

TABLE 6. Results of spectral analyses of impact crater ages.

Diam. (km)	No.	Phase	Highest Peak (My)	2nd highest Peak (My)
≥ 5	31	Free	30	35
		Fixed	30	35
≥ 35	11	Free	35	None
		Fixed	35	None
≥ 90	5	Free	36	29
		Fixed	36	30

vertical oscillation of the Solar System. The results could be a synthesis of astrophysics and earth science.

Acknowledgments This paper is dedicated to the memory of three pioneers in the impact and extinction field: Bob Dietz, Gene Shoemaker, and Jack Sepkoski. Thanks to K. Caldeira, B. Haggerty, J. Matese, A. Montanari, W. Napier, A. Prokoph, R. Stothers, M. Valtonen, and S. Yabushita for valuable discussions and criticism. J. Sepkoski kindly provided an updated version of his extinction data set.

References

Adushkin, V.V. & Nemtchinov, I.V. (1994) Consequences of impacts of cosmic bodies on the surface of the Earth, in *Hazards Due to Asteroids and Comets*, ed. Gehrels, T., University of Arizona Press, Tucson, pp. 721–778.

Alvarez, L.W., Alvarez, W., Asaro, F. & Michel, H.V. (1980) Extraterrestrial cause of the Cretaceous/Tertiary extinction, *Science* **208**, 1095–1108.

Alvarez, W. & Muller, R.A. (1984) Evidence from crater ages for periodic impacts on the Earth, *Nature* **308**, 718–720.

Bahcall, J.N., Flynn, C. & Gould, A. (1992) Local dark matter from a carefully selected sample, *Astrophys. J.* **389**, 234–250.

Bailey, M.E., Clube, S.V.M. & Napier, W.M. (1990) *The Origin of Comets*, Pergamon, Oxford, 577 p.

Chapman, C.R. & Morrison, D. (1994) Impacts on the Earth by asteroids and comets: Assessing the hazard, *Nature* **367**, 33–40.

Clube, S.V.M. & Napier, W.M. (1996) Galactic dark matter and terrestrial periodicities, *Q. J. R. Astron. Soc.* **37**, 617–642.

Crézé, M., Chereul, E., Bienaymé, O. & Pichon, C. (1998) The distribution of nearby stars in phase space mapped by Hipparcos: I. The potential well and local dynamical mass, *Astron. Astrophys.* **329**, 920–936.

Dypvik, H., Gudlaugsson, S.T., Tsikalas, F., Attrep, M., Jr., Ferrell, R.E., Jr., Krinsley, D.H., Mørk, A., Faleide, J.I. & Nagy, J. (1996) Mjølnir structure: An impact crater in the Barents Sea, *Geology* **24**, 779–782.

Ernstson, K. & Claudin, F. (1990) Pelarda Formation (Eastern Iberian Chains, NE Spain): Ejecta of the Azuara impact structure, *Neues Jahrbuch Geologie und Paläontologie* **10**, 581–599.

Ernstson, K. & Fiebag, J. (1992) The Azuara impact structure (Spain): New insights from geophysical and geological investigations, *Geol. Rundschau* **81**, 403–427.

Farley, K. A., Montanari, A., Shoemaker, E.M. & Shoemaker, C.S. (1998) Geochemical evidence for a comet shower in the Late Eocene, *Science* **280**, 1250–1253.

Fogg, M.J. (1989) The relevance of the background impact flux to cyclic impact/mass extinction hypotheses, *Icarus* **79**, 382–395.

Fowell, S.J., Cornet, B. & Olsen, P.E. (1994) Geologically rapid Late Triassic extinctions: Palynological evidence from the Newark Supergroup, in *Pangea: Paleoclimate, Tectonics, and Sedimentation During Accretion, Zenith, and Breakup of a Supercontinent*, ed. Klein, G.D. *et al.*, Boulder, Colorado, Geological Society of America Special Paper 288, pp. 197–206.

Gehrels, T. (1999) *A review of comet and asteroid statistics: Earth, Planets, and Space*, in press.

Gould, A., Bahcall, J.N. & Flynn, C. (1996) Disk M dwarf luminosity function from Hubble Space Telescope star counts, *Astrophys. J.* **465**, 759–768.

Gradstein, F.M. & Ogg, J.G. (1996) A Phanerozoic time scale, *Episodes* **19**, Nos. 1 & 2, insert.

Grieve, R.A.F. (1996) Chesapeake Bay and other terminal Eocene impacts, *Meteoritics and Planetary Science* **31**, 166–167.

Grieve, R.A.F. (1997) Extraterrestrial impact events: The record in the rocks and the stratigraphic column, *Palaeoclimatology, Palaeogeography, Palaeoecology* **132**, 5–23.

Grieve, R.A.F. & Pesonen, L.J. (1996) Terrestrial impact craters: Their spatial and temporal distribution and impacting bodies, *Earth, Moon & Planets* **72**, 357–376.

Grieve, R.A.F. & Shoemaker, E.M. (1994) The record of past impacts on Earth, in *Hazards Due to Comets & Asteroids*, ed. Gehrels, T., Univ. of Arizona Press, Tucson, pp. 417–462.

Hildebrand, A.R., Pilkington, M., Connors, M., Ortiz-Aleman, C. & Chavez, R.E. (1995) Size and structure of the Chicxulub Crater revealed by horizontal gravity gradients and cenotes, *Nature* **376**, 415–417.

Hut, P., Alvarez, W., Elder, W.P., Hansen, T., Kauffman, E.G., Keller, G., Shoemaker, E.M. & Weissman, P.R. (1987) Comet showers as a cause of mass extinctions, *Nature* **329**, 118–126.

Izett, G.A., Cobban, W.A., Obradovich, J.D. & Kunk, M.J. (1993) The Manson impact structure: ^{40}Ar/^{39}Ar age and its distal impact ejecta in the Pierre Shale in southeastern South Dakota, *Science* **262**, 729–732.

Liritzis, I. (1993) Cyclicity in terrestrial upheavals during the Phanerozoic Eon, *Q. J. R. Astron. Soc.* **34**, 251–260.

Matese, J.J., Whitman, P.G., Innanen, K.A. & Valtonen, M.J. (1995) Periodic modulation of the Oort Cloud comet flux by the adiabatically changing galactic tide, *Icarus* **116**, 255–268.

Matese, J.J., Whitman, P.G., Innanen, K.A. & Valtonen, M.J. (1996) Why we study the geological record for evidence of the solar oscillation about the galactic midplane, *Earth, Moon & Planets* **72**, 7–12.

Matese, J.J., Whitman, P.G., Innanen, K.A. & Valtonen, M.J. (1998) Variability of the Oort comet flux: Can it be manifest in the cratering record?, in *Highlights in Astronomy* **11A**, ed. Andersen, J., Kluwer, Dordrecht, pp. 252–256.

Matese, J. & Whitmire, D. (1996) Tidal imprint of distant galactic matter on the Oort comet cloud, *Astrophys. J.* **472**, L41–L43.

Matsumoto, M. & Kubotani, H. (1996) A statistical test for correlation between crater formation rate and mass extinctions, *Mon. Not. R. Astron. Soc.* **282**, 1407–1412.

McEwen, A.S., Moore, J.M. & Shoemaker, E.M. (1997) The Phanerozoic impact cratering rate: Evidence from the far side of the Moon, *J. Geophys. Res.* **102**, 9231–9242.

Melosh, H.J., Schneider, N.M., Zahnle, K.J. & Latham, D. (1990) Ignition of global wildfires at the Cretaceous/Tertiary boundary, *Nature* **343**, 251–254.

Montanari, A., Campo-Bagatin, A. & Farinella, P., (1998) Earth cratering record and impact energy in the last 150 Ma, *Planet. Space Sci.* **46**, 271–281.

Mossman, D.J., Grantham, R.G. & Langenhorst, F. (1998) A search for shocked quartz at the Triassic-Jurassic boundary in the Fundy and Newark basins of the Newark Supergroup, *Canadian Journal of the Earth Sciences* **35**, 101–109.

Napier, W.M. (1998) NEOs and impacts: The galactic connection, *Celest. Mech. & Dynam. Astron.* **69**, 59–75.

Poag, C.W. (1997) Roadblocks on the kill curve: Testing the Raup hypothesis, *Palaios* **12**, 582–590.

Pope, K.O., Baines, K.H., Ocampo, A.C. & Ivanov, B.A. (1997) Energy, volatile production, and climate effects of the Chicxulub Cretaceous/Tertiary impact, *J. Geophys. Res.* **102**, 21645–21664.

Pope, K.O., D'Hondt, S.L. & Marshall, C.R. (1998) Meteorite impact and the mass extinction of species at the Cretaceous/Tertiary boundary, *Proc. Natl Acad. Sci., USA* **95**, 11028–11029.

Rampino, M.R. (1995) Did a comet shower occur at the time of the Eocene/Oligocene transition?, *Eos, Transactions of the American Geophysical Union* **76**, S188.

Rampino, M.R. (1997) El impacto de Azuara y sus efectos, *Universo (Spain)* **No. 25**, 48–50.

Rampino, M.R. (1998a) The Shiva hypothesis: Impacts, mass extinctions, and the Galaxy, *The Planetary Report* **18**, 6–11.

Rampino, M.R. (1998b) The galactic theory of mass extinctions: An Update, *Celest. Mech. & Dynam. Astron.* **69**, 49–58.

Rampino, M.R. & Caldeira, K. (1993) Major episodes of geologic change: Correlations, time structure and possible causes, *Earth and Planetary Science Letters* **114**, 215–227.

Rampino, M.R. & Haggerty, B.M. (1996a) The "Shiva Hypothesis": Impacts, mass extinctions, and the Galaxy, *Earth, Moon & Planets* **72**, 441–460.

Rampino, M.R. & Haggerty, B.M. (1996b) Impact crises and mass extinctions: A working hypothesis, in *The Cretaceous-Tertiary Event and Other Catastrophes in Earth History*, ed. Ryder, G. *et al.*, Boulder, Colorado, Geological Society of America Special Paper 307, pp. 11–30.

Rampino, M.R., Haggerty, B.M. & Pagano, T.C. (1997) A unified theory of impact crises and mass extinctions: Quantitative tests, *Ann. New York Acad. Sci.* **822**, 403–431.

Rampino, M.R. & Schwindt, D.M. (1999) Comment on "The age of the Kara impact structure, Russia" by Trieloff *et al.*, *Meteoritics and Planetary Science* **34**, 1–2.

Rampino, M.R. & Stothers, R.B. (1984a) Terrestrial mass extinctions, cometary impacts and the Sun's motion perpendicular to the Galactic plane, *Nature* **308**, 709–712.

Rampino, M.R. & Stothers, R.B. (1984b) Geological rhythms and cometary impacts, *Science* **226**, 1427–1431.

Rampino, M.R. & Stothers, R.B. (1986) Geologic periodicities and the Galaxy, in *The Galaxy and the Solar System*, ed. Smoluchowski, R. *et al.*, Univ. of Arizona Press, Tucson, pp. 241–259.

Rampino, M.R. & Stothers, R.B. (1998) Mass extinctions, comet impacts, and the Galaxy, in *Highlights in Astronomy* **11A**, ed. Andersen, J., Kluwer, Dordrecht, pp. 246–251.

Raup, D.M. (1979) Size of the Permian-Triassic bottleneck and its evolutionary implications, *Science* **206**, 217–218.

Raup, D.M. (1992) Large-body impact and extinction in the Phanerozoic, *Paleobiology* **18**, 80–88.

Raup, D.M. & Sepkoski, J.J., Jr. (1984) Periodicity of extinctions in the geologic past, *Proc. Natl Acad. Sci. USA* **81**, 801–805.

Raup, D.M. & Sepkoski, J.J., Jr. (1986) Periodic extinctions of families and genera, *Science* **231**, 833–836.

Retallack, G.J., Seyedolali, A., Krull, E.S., Holser, W.T., Ambers, C.P. & Kyte, F.T. (1998) Search for evidence of impact at the Permian-Triassic boundary in Antarctica and Australia, *Geology* **26**, 979–982.

Sepkoski, J.J., Jr. (1995) Patterns of Phanerozoic extinction: A perspective from global data bases, in *Global Events and Event Stratigraphy in the Phanerozoic*, ed. Walliser, O.H., Springer, Berlin, pp. 35–51.

Sheehan, P.M., Fastovsky, D.E., Hoffmann, R.G., Berghaus, C.B. & Gabriel, D.L. (1991) Sudden extinction of the dinosaurs: latest Cretaceous, Upper Great Plains, U.S.A., *Science* **254**, 835–839.

Shoemaker, E.M. & Wolfe, R.F. (1986) Mass extinctions, crater ages, and comet showers, in *The Galaxy and the Solar System*, ed. Smoluchowski, R. *et al.*, Univ. of Arizona Press, Tucson, pp. 338–386.

Shoemaker, E.M., Wolfe, R.F. & Shoemaker, C.S. (1990) Asteroid and comet flux in the neighborhood of Earth, *Geological Society of America Special Paper* **247**, 155–170.

Sloan, R.E., Rigby, J.K., Van Valen, L.M. & Gabriel, D. (1986) Gradual dinosaur extinction and simultaneous ungulate radiation in the Hell Creek Formation, *Science* **232**, 629–632.

Steiner, M.B. & Shoemaker, E.M. (1996) A hypothesized Manson tsunami: Paleomagnetic and stratigraphic evidence in the Crow Creek Member, Pierre Shale, *Geological Society of America Special Paper* **302**, 419–432.

Stothers, R.B. (1985) Terrestrial record of the solar system's oscillation about the galactic plane, *Nature* **317**, 338–341.

Stothers, R.B. (1988) Structure of Oort's comet cloud inferred from terrestrial impact craters, *Observatory* **108**, 1–9.

Stothers, R.B. (1989) Structure and dating errors in the geologic time scale and periodicity in mass extinctions, *Geophysical Research Letters* **16**, 119–122.

Stothers, R.B. (1993) Impact cratering at geologic stage boundaries, *Geophysical Research Letters* **20**, 887–890.

Stothers, R.B. (1998) Galactic disc dark matter, terrestrial impact cratering and the law of large numbers, *Mon. Not. R. Astron. Soc.* **300**, 1098–1104.

Thaddeus, P. & Chanan, G. (1985) Cometary impacts, molecular clouds, and the motion of the Sun perpendicular to the galactic plane, *Nature* **314**, 73–75.

Toon, O.B., Zahnle, K., Morrison, D., Turco, R.P. & Covey, C. (1997) Environmental perturbations caused by the impacts of asteroids and comets, *Reviews of Geophysics* **35**, 41–78.

Trefil, J.S. & Raup, D.M. (1987) Numerical simulations and the problem of periodicity in the cratering record, *Earth & Planet. Sci. Lett.* **82**, 159–164.

Valtonen, M.J., Zheng, J.Q., Matese, J.J. & Whitman, P.G. (1995) Near-Earth populations of bodies coming from the Oort Cloud and their impacts with planets, *Earth, Moon & Planets* **71**, 219–223.

Varricchio, D.J. & Horner, J.R. (1993) Hadrosaurid and lambeosaurid bone beds from the Upper Cretaceous Two Medicine Formation of Montana: Taphonomic and biologic implications, *Canadian Journal of the Earth Sciences* **30**, 997–1006.

Yabushita, S. (1991) A statistical test for periodicity hypothesis in the crater formation rate, *Mon. Not. R. Astron. Soc.* **250**, 481–485.

Yabushita, S. (1992) Periodicity in the crater formation rate and implications for astronomical modeling, in *Dynamics and Evolution of Minor Bodies with Galactic and Geological Implications*, ed. Clube, S.V.M. *et al.*, Kluwer, Dordrecht, pp. 161–178.

Yabushita, S. (1996a) Are cratering and probably related geological records periodic?, *Earth, Moon & Planets* **72**, 343–356.

Yabushita, S. (1996b) Statistical tests of a periodicity hypothesis for crater formation rate – II, *Mon. Not. R. Astron. Soc.* **279**, 727–732.

Yabushita, S. (1998) A statistical test of correlations and periodicities in the geological records, *Celest. Mech. & Dynam. Astron.* **69**, 31–48.

Zakharov, V.A., Lapukhov, A.S. & Shenfil, O.V. (1993) Iridium anomaly at the Jurassic-Cretaceous boundary in northern Siberia, *Russian Journal of Geology and Geophysics* **34**, 83–90.

THE ROLE OF NON-GRAVITATIONAL FORCES IN DECOUPLING ORBITS FROM JUPITER

D.J. ASHER AND M.E. BAILEY
Armagh Observatory
College Hill, Armagh, BT61 9DG, U.K.

AND

D.I. STEEL
Joule Physics Laboratory, University of Salford
Salford, M5 4WT, U.K.

Abstract. There are many near-Earth objects (NEOs) with aphelia sufficiently sunward of 5 AU to be protected from close approaches to Jupiter.
Such NEOs may derive either from the main asteroid belt or from an intermediate cometary source such as the short-period family, and at least two
have exhibited comet-like physical activity. If a long-lived NEO is to arrive
through this cometary avenue, then a mechanism is required to decouple
its orbit from Jupiter. This may be achieved by gravitational perturbations
alone, but a long timescale is required. Attainment of such a cis-Jovian
orbit is feasible rather more rapidly, and with a higher efficiency, if non-
gravitational forces also are acting. Here we discuss the role of such forces
in producing NEOs from Jupiter-family comets.

1. Decoupled orbits

Earth-crossing bodies may be categorized into several dynamical classes,
which are not necessarily distinct. Although each is subjected to gravitational perturbations by all the planets, the orbital evolution is qualitatively
different if close approaches to Jupiter are possible. If an object's aphelion distance is well below 5 AU (a cis-Jovian orbit), it is decoupled from
Jupiter and dynamically stable over a greatly increased timescale: the effect
of Jupiter is secular, without large random perturbations occurring through
discrete approaches. Only close approaches to the terrestrial planets may

M. Ya. Marov and H. Rickman (eds.), Collisional Processes in the Solar System, 121–130.
© 2001 *Kluwer Academic Publishers. Printed in the Netherlands.*

then occur, and they have much lower masses with concomitant smaller effects.

Quantitatively, a decoupled state can be defined in one of the following ways (beginning with the simplest): by some maximum value of the aphelion distance Q; by some minimum value of $\Delta r = r_J - Q$ where r_J is Jupiter's heliocentric distance at the longitude of the particle's aphelion (this allowing for Jupiter's non-zero eccentricity); or by a minimum value of D, the minimum distance between the two ellipses (particle, Jupiter) in 3-dimensional space. A mean-motion resonance can prevent close approaches even if the orbits of the particle and Jupiter are not widely separated, i.e., even if D is small. Similarly, Kozai cycles (correlated variations in Q, inclination i and argument of perihelion ω at roughly invariant semi-major axis a) can result in D never falling to values as low as those attained by Δr (e.g., Milani et al., 1989). We are interested here in processes by which the particle's orbit does become separated from that of Jupiter and so shall not class orbits as decoupled merely on account of a resonance, of any description, which prohibits approaches; indeed arrival in some resonant states may require decoupling to occur in the earlier evolution.

The Hill radius of Jupiter is 0.35 AU although slow, nearly-tangent encounters producing large orbital changes are possible even at a distance of 0.6 AU (Valsecchi, 1992). The condition $Q < 4.35$ AU (as used by Wetherill, 1991) ensures that $D > 0.6$ AU, because the Jovian perihelion distance is $q \approx 4.95$ AU.

Decoupled objects either never were Jupiter approaching, originating instead in the asteroid belt, or have somehow separated from Jupiter. Physical data (e.g., Rabinowitz, 1998) may help to identify the source region. The injection rate from the asteroid belt may be adequate to supply half at least of the observed near-Earth asteroid (NEA) population (cf. Menichella et al., 1996; Zappalà & Cellino, this volume). The complementary consideration, with respect to determining the proportion of NEAs of cometary origin, is whether suitably efficient dynamical mechanisms exist whereby short-period or Jupiter-family (JF) comets can become decoupled. Although the dynamical trend of most JF objects is to be displaced by Jupiter on to larger orbits, a few may become decoupled through random changes in a and Q through approaches to Earth or Venus (Wetherill, 1991), and there are purely gravitational paths linking JF to Encke-like orbits that do not involve exceptional close approaches to the terrestrial planets (Bailey, 1995; Valsecchi, 1999).

However, cometary non-gravitational (NG) forces (i.e., the net force on the nucleus due to non-isotropic sublimation of volatiles) may also play a role (Yeomans, 1994), and that is the subject we investigate here. After attaining a decoupled cis-Jovian orbit, the objects can survive dynamically for $\sim 10^7$ yr (though some orbital types suffer secular perturbations producing rapid solar infall: Farinella *et al.*, 1994). Cometary outgassing persists for a timescale much shorter than this, so that there is a possibility that the short-term NG force may be responsible, at least in part, for delivering objects into decoupled orbits where they are stable for extended periods.

We need to quantify how often changes in a and Q of sufficient magnitude occur (*i.e.*, the decoupling efficiency). Because the minimum orbital distance between the particle and Jupiter varies systematically over a Kozai cycle (typically lasting a few millennia), we require D to remain larger than the nominal limit (0.6 AU) for at least one complete cycle. Secular perturbations related to the longitude of perihelion ϖ can cause large swings in Q over 10^5 yr timescales, but since we are interested in shorter-term behaviour we do not require orbits to be decoupled for that long to be of interest.

This work has wider implications for the origin of NEOs, but the problem can also be regarded as one of explaining the origin of Comet 2P/Encke. The age of the Taurid streams, related to 2P/Encke, is ~ 20 kyr, and the systematic action of the cometary NG force has been suggested as a mechanism for the aphelion distance of Encke, or its progenitor, decreasing from ~ 5 AU to its current 4.1 AU within that timescale (Sekanina, 1972). Such a systematic trend seems unlikely, based on observed comets. However, another decoupled, Earth-crossing active comet may have been D/1766 G1 Helfenzrieder ($q = 0.4$ AU, $Q \lesssim 5$ AU), while a third object, with $q = 1.0$ AU and $\Delta r \sim 1.1$ AU, namely Wilson-Harrington, classified both as comet 107P and minor planet (4015), is certainly a cometary NEO trapped in a decoupled orbit. Whether these objects originated in the main belt or elsewhere, however, is uncertain, as Elst-Pizarro ($q = 2.6$, $Q = 3.7$), which is both comet 133P and minor planet (7968), shows that the outer main belt also contains objects resembling comets. Here we consider whether these transitional objects may provide clues as to how decoupled NEOs are produced from JF comets, appearing later as NEAs when devolatilized.

2. Cumulative effect of random small changes to a

Steel & Asher (1996a,b) modelled the influence of NG forces on orbits like that of 2P/Encke by integrating various test particles whilst applying, at intervals t, impulses

$$\delta a = \sqrt{2}\, X \sin\theta \tag{1}$$

with θ uniformly distributed in $(0, 2\pi)$, and X used to parameterise the size of the NG force. They showed that, with $X = 0.0137$ AU and $t = 170$ yr, Jupiter-crossing objects could be produced from decoupled NEOs in $\sim 10^4$

yr without requiring that the NG force acts consistently in one sense. The overall effect is due to the NG force causing a scanning across the Jovian resonances which, for a few particles, results in a medium-term trend in a even though the values of δa imposed by the NG model are both positive and negative. The process could be referred to as "stochastic resonant transport". The parameter X is the r.m.s. NG-caused change in a after each t years and can be compared with the observed behaviour of comets, in particular 2P/Encke for which the evolution of the transverse NG parameter since 1786 has been evaluated (Marsden & Sekanina, 1974; Sitarski, 1988). The somewhat smaller value, $X=0.0037$ AU in $t=170$ yr, is *observed* for 2P/Encke (see further discussion in Steel & Asher, 1996a).

We follow much the same procedure here, except that we *begin* with Jupiter-approaching particles and look for decoupling. Using Eq. (1) we refer to model A. We also employ a slightly more sophisticated model we reference as B, as follows. As NG effects are dependent on q, we impose perturbations:

$$\delta a = (3q)^{-k} \sqrt{2}\, X \, \sin\theta \qquad (2)$$

The factor of 3 simply scales against 2P/Encke ($q \simeq 0.33$ AU since discovery). To avoid large single impulses, δa is calculated on the basis of q being set to 0.1 AU if the value is smaller. Integrations were performed with $k=1.5$; this is consistent with the widely used NG formulation of Marsden *et al.* (1973).

For six values of X, namely 0.000 (purely gravitational), 0.005, 0.010, 0.015, 0.020 and 0.030 AU, and using $t=100$ yr, sets of 360 particles evenly spaced in mean anomaly were integrated forwards from 50 kyr BP (before present). Initially, $a=3.83$ AU, eccentricity $e=0.37$ (thus $q=2.44$ AU), and $\varpi=187°$; this is consistent with typical JF orbits. However, the exact values are irrelevant as memory of these conditions is soon lost in JF comet evolution (Nakamura & Yoshikawa, 1991; Pittich & Rickman, 1994; Harris & Bailey, 1996), and these values were chosen in an arbitrary fashion based on earlier work. An initial inclination $i=25°$ was used alone in model A (Eq. 1), but both $i=10°$ and $i=25°$ with model B (Eq. 2). In all we integrated $6\times360=2160$ particles in each of these three cases, and searched for intervals where

$$q < 1.4 \text{ AU} \quad \text{and} \quad \Delta r > 0.5 \text{ AU} \qquad (3)$$

The conditions defining orbits that are both near-Earth and decoupled from Jupiter are somewhat arbitrary, but Eq. (3) suffices to identify particles whose evolution shows promise, at least from the perspective of our quest to demonstrate how decoupling may occur.

Although in most cases no decoupling was identified, in some cases it was, and the significance of this is quantified in Sect. 3 below. Fig. 1 repre-

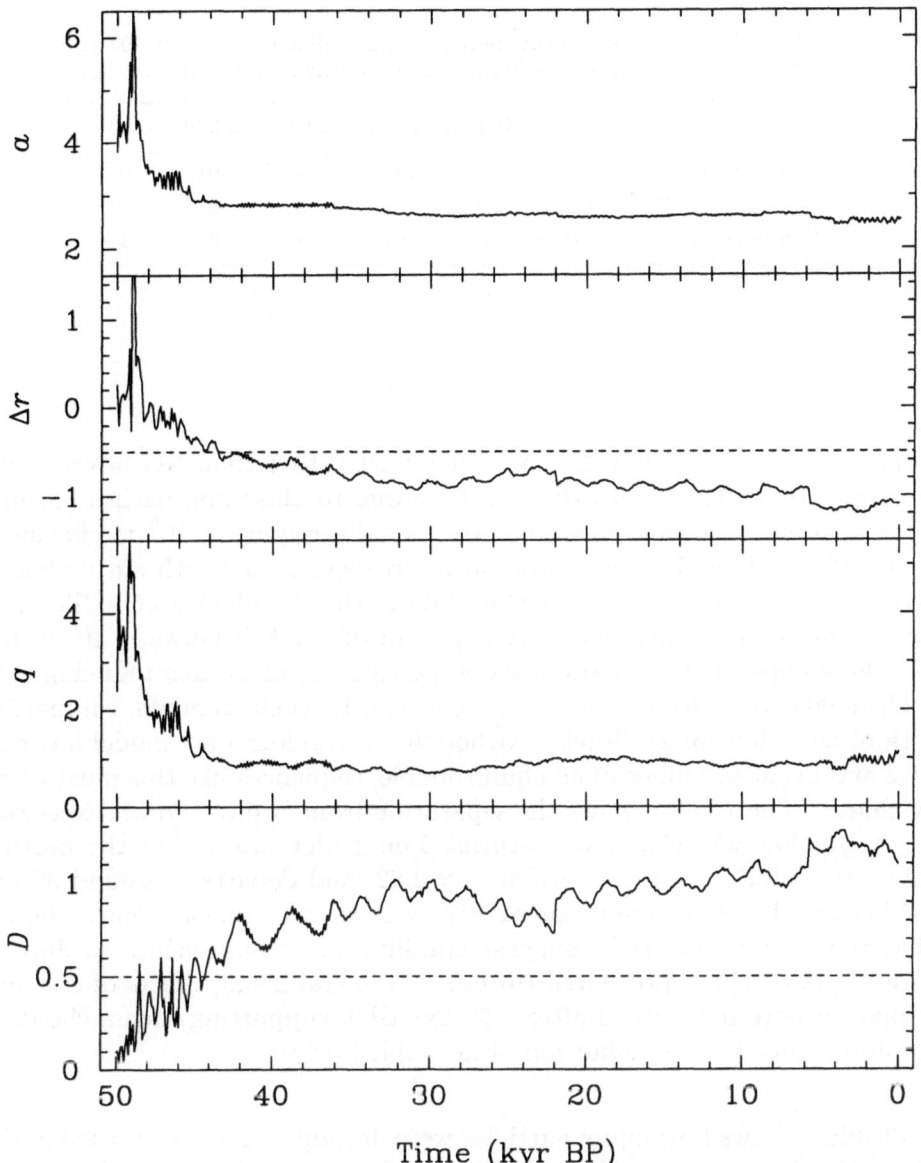

Figure 1.   Evolution of a, Δr, q, and D from 50 kyr BP until the present for one test particle which became decoupled. $X=0.015$ AU was used in Eq. (2), and impulses applied every $t=100$ yr.

sents an example of decoupling. This illustrates typical features of transitions to near-Earth, decoupled orbits. Firstly, it is clear that the reduction in q to become ~ 1 AU occurs during the Jupiter-approaching phase, being

TABLE 1. Numbers of orbits decoupled, as defined by (3), for 10^4 yr or more (not necessarily continuously), out of each set of 360 particles.

X:	0.000	0.005	0.010	0.015	0.020	0.030
Model A, $i_0=25°$	0	1	10	5	10	16
Model B, $i_0=25°$	1	2	4	3	2	7
Model B, $i_0=10°$	0	2	2	3	2	1

a consequence of the influence of that planet rather than NG forces. Over merely $\sim10^4$ yr timescales this must be due to close approaches (secular effects can cause wide swings in e but usually require $\gtrsim 10^5$ yr: Farinella et al., 1994). Thus in Fig. 1, the object reaches near-Earth status before decoupling (as shown by the dashed line in the Δr plot) occurs. The reason for decoupling was that over a period of ~2 kyr between 46–44 kyr BP, there happened to be upwards of a dozen impulses each reducing a by 0.002–0.004 AU, with only three increasing a by such amounts, compatible with observations of 2P/Encke. Although our random walk model has positive and negative values of δa equiprobable, sequences like this must occur by chance. The D plot shows the separation from Jupiter's orbit exceeding 0.5 AU, below which it never returns. The a plot shows that the particle enters the 5:2 Jovian resonance at $a \approx 2.82$, and departs it around 36 kyr BP because by chance four consecutive values $\delta a \simeq -0.006$ occur. The Δr plot, plus our other trials, suggest stability when approaches to Jupiter within $\lesssim 0.6$ AU are prevented (in Fig. 1, the stabilising effects of a strong resonance were not needed after ~36 kyr BP), supporting such a choice of cutoff distance for the definition of decoupled orbits.

Table 1 shows how many particles were decoupled for $>10^4$ yr using Eq. (3) as the discriminant. The strong dependence on heliocentric distance in the standard water sublimation model of the NG force (cf. the function $g(r)$ of Marsden et al., 1973) would imply that model A allows unrealistically large perturbations of that origin at asteroid belt distances, and unrealistic numbers of decouplings for the larger X values. It was for this reason that we introduced the more realistic model B, the results for which in Table 1 confirm that NG effects can facilitate decoupling. Next we perform a more quantitative assessment of whether decoupled comets can contribute significantly to the NEO population.

3. Decoupling efficiencies and the lifetime of the NG force

Harris & Bailey (1996, 1998) analysed intervals in their ~1 Myr integrations when particles had short period ($P < 20$ yr), non-decoupled ($Q > 4.2$ AU), near-Earth ($q < 1.4$ AU) orbits. Given the number of known, active JF comets with $q < 1.4$ AU, they derived the injection rate into the population of $q < 1.4$ AU JF comets, dependent on the assumed active lifetime. For an active lifetime of 1000 revolutions with $q < 1.4$ AU, such JF comets would be produced at a rate of about one per 90 yr. In addition a decoupling efficiency (*i.e.*, the fraction of such JF comets that decouple, other than very transiently) can be derived, this yielding an injection rate into the decoupled ($Q < 4.2$ AU) population. Decoupled cometary NEOs come from the subset of JF comets with $q < 1.4$ AU, since as discussed above deflection (as opposed to a secular trend over 10^5 yr) into a small q orbit is due to an approach to Jupiter and must happen before decoupling. Harris & Bailey (1998) derived an upper bound of 0.0023 for the decoupling efficiency if gravitational forces only are considered and noted that ~14 times even this upper bound would be needed to match the injection rate of km-sized NEAs from the main belt estimated by Menichella *et al.* (1996).

Considering a JF source population with $q < 1$ AU and defining decoupled as $Q < 4.35$ AU, Wetherill (1991) used a Monte Carlo method that modelled random effects from Earth and Venus to derive a decoupling efficiency of 0.0033. The typical decoupling timescale was 10^5–10^6 yr, but as short as 10^3–10^4 yr in ~0.5% of cases, consistent with the observation of one active comet (2P/Encke) co-existing with several hundred cometary NEOs.

These are purely gravitational decoupling efficiencies. We are interested in the effect of imposing realistic NG forces: do these significantly enhance the efficiencies? Can they cause a substantial decrease in a and Q within the available interval, the activity lifetimes of comets? Consider the transverse NG acceleration A_2 (Marsden *et al.*, 1973). For $q \approx 0.5$ AU and $Q \approx r_J$, A_2 in its conventional units of 10^{-8} AU day^{-2} is about the same as the change in a in AU kyr^{-1}. This relationship varies according to the orbit under consideration (*e.g.*, for smaller q the decrease in a is greater for the same A_2, although the loss of volatiles is faster, reducing the active lifetime; the opposite is true for larger q), but it conveniently allows NG forces, as measured by A_2, to be compared quantitatively with the change in a necessary for decoupling. For example, Harris & Bailey (1998), in full integrations using the Marsden *et al.* (1973) formulation, assumed an r.m.s. A_2 of 0.16 acting in a consistent direction for up to 1000 revolutions (a few kyr) at $q < 1.4$ AU. Thus several particles would be expected to suffer decreases in a by a few tenths of an AU, resulting in decoupling. Under

TABLE 2. Ratio of numbers of particles which became decoupled by the time of cessation of cometary activity, compared to the total number becoming inactive during the 50 kyr integrations. Decoupling was defined by (3) and the active lifetime as 1000 revolutions with $q < 1.4$ AU, In parentheses is the further number of particles that had $P < 20$ yr, $Q > 4.2$AU and $q < 1.4$ AU at some stage but $q < 1.4$ AU for less than 1000 revolutions.

Model	i_0	X=0.000	X=0.005	X=0.010	X=0.015	X=0.020	X=0.030
A	25°	0/40 (151)	0/38 (152)	4/61 (151)	5/57 (142)	10/46 (153)	12/59 (133)
B	25°	0/28 (173)	1/34 (169)	1/38 (177)	0/35 (157)	0/41 (165)	1/44 (153)
B	10°	0/15 (82)	0/23 (75)	3/10 (85)	2/12 (94)	1/17 (82)	1/16 (74)

that assumption a sufficient proportion of decoupled objects was found to match the NEA contribution from the main belt. The question is whether such a consistently acting large NG force is supported by observations of JF comets.

Many observed comets have values of $|A_2|$ of order 0.01 (in units of 10^{-8} AU day^{-2}), and a few as large as 0.1 (Marsden & Williams, 1996). However the magnitude of A_2 in many cases has changed significantly over a century, and the direction too, owing to such things as discrete regions becoming temporarily active (Yeomans, 1994). The average A_2 of 2P/Encke since discovery has been ~−0.02 (Sitarski, 1988) but the present value is less than 10% of that 200 yr ago. Sekanina (1991) discusses the nucleus precession which could be responsible. Such precession would result in the NG force not acting in a consistent sense over millennia.

The integrations described above with X=0 are over a shorter timescale than those of Harris & Bailey (1998) and so the latter better constrain the purely gravitational decoupling efficiency. However, integrations need only cover at least the expected active lifetime of the nucleus in order to reflect NG effects, which is our major aim here. We wish to estimate a decoupling efficiency if random small NG forces are included; it can then be compared to Harris & Bailey's gravitational upper bound of 0.0023. With this in mind, we search our integrations for epochs when $P < 20$ yr, $Q > 4.2$ AU and $q < 1.4$ AU, because these conditions defined the JF population for which Harris & Bailey derived an injection rate. The question is what fraction (compared to 0.0023) decouple, given various sizes of NG force (as parameterised by X).

Taking the active lifetime as limited to 1000 revolutions with $q < 1.4$ AU (cf. Harris & Bailey, 1998), the numbers of objects becoming inactive within 50 kyr are shown in Table 2. Only three particles (all model A) had $Q < 4.2$ AU (the Harris & Bailey definition of decoupled) at the time of becoming inactive; Table 2 lists the number decoupled at that time as defined by (3). In our model B, which is more realistic than model A, there were ten such decoupled objects (total of the last two lines of Table 2). To investigate whether (3) may be too weak a definition of decoupling, purely gravitational integrations were performed from the time of becoming inactive for a further 100 kyr (*i.e.*, we imagine that a decoupled comet has become inactive, and we see how it subsequently behaves as an NEA).

Of these ten particles, four were only marginally decoupled, *i.e.*, $\Delta r \approx$ 0.5 AU, and soon returned to being Jupiter-approaching; two appeared to owe their stability more to resonances than to distance from the Jovian orbit; and one fell into the Sun. Only three seemed to be good examples of stable, decoupled NEOs. The total number of particles in the last two lines and $X > 0$ columns of Table 2, that have completed 1000 revolutions with $q < 1.4$ AU and so by assumption can no longer be assisted by NG forces to become decoupled, is 270. These three out of 270 particles with $X > 0$ render a decoupling efficiency of ~0.011, a factor of about five higher than Harris & Bailey's gravitational upper bound, and may provide ~20% of the NEA population. On the other hand, comets which split after capture into decoupled orbits might produce many NEAs each (cf. Pittich & Rickman, 1994).

4. Conclusion

The results of this preliminary study are clear, despite the low number statistics. When NG forces are included in the integrations as impulsive effects, varying erratically in sense (accelerative or decelerative) but with magnitudes in accord with observed perturbations, the rate at which objects may be decoupled from Jupiter and attain orbits like NEOs is increased by a factor of four or five, producing ~20% of the observed NEA population. That factor may be somewhat higher if enhancements due to cometary splitting, producing several NEOs, are included. Similarly the yield may be higher if the stringent condition that $q < 1.4$ AU for at least 1000 orbits were relaxed, 30 to 40% of the test particles attaining such reduced values of q at least for part of the period studied (these correspond to NEOs which appear asteroidal, but are actually dormant comets, not yet extinct/devolatilized). We conclude that the NG force is an important feature of how JF comets may be fed into the NEO population, regardless of whether their NG forces are large or act in a consistent fashion for an extended period of time.

References

Bailey, M.E. (1995) Recent results in cometary astronomy: implications for the ancient sky, *Vistas Astron.* **39**, 647–671.

Farinella, P., Froeschlé, Ch., Froeschlé, Cl., Gonczi, R., Hahn, G., Morbidelli, A. & Valsecchi, G.B. (1994) Asteroids falling into the Sun, *Nature* **371**, 314–317.

Harris, N.W. & Bailey, M.E. (1996) The cometary component of the near-Earth object population, *Irish Astron. J.* **23**, 151–156.

Harris, N.W. & Bailey, M.E. (1998) Dynamical evolution of cometary asteroids, *Mon. Not. R. Astron. Soc.* **297**, 1227–1236.

Marsden, B.G. & Sekanina, Z. (1974) Comets and nongravitational forces. VI. Periodic comet Encke 1786–1971, *Astron. J.* **79**, 413–419.

Marsden, B.G. & Williams, G.V. (1996). *Catalogue of Cometary Orbits 1996*, MPC, Cambridge, Massachusetts.

Marsden, B.G., Sekanina, Z. & Yeomans, D.K. (1973) Comets and nongravitational forces. V, *Astron. J.* **78**, 211–225.

Menichella, M., Paolicchi, P. & Farinella, P. (1996) The main belt as a source of near-Earth asteroids, *Earth, Moon, and Planets* **72**, 133–149.

Milani, A., Carpino, M., Hahn, G. & Nobili, A.M. (1989) Dynamics of planet-crossing asteroids: classes of orbital behavior. Project Spaceguard, *Icarus* **78**, 212–269.

Nakamura, T. & Yoshikawa, M. (1991) Cosmo-dice: dynamical investigation of cometary evolution, *Publ. Nat. Astron. Obs. Japan* **2**, 293–383.

Pittich, E.M. & Rickman, H. (1994) Cometary splitting – a source for the Jupiter family?, *Astron. Astrophys.* **281**, 579–587.

Rabinowitz, D.L. (1998) Size and orbit dependent trends in the reflectance colors of Earth-approaching asteroids, *Icarus* **134**, 342–346.

Sekanina, Z. (1972) A model for the nucleus of Encke's Comet, in *The Motion, Evolution of Orbits, and Origin of Comets (IAU Symp. No. 45)*, eds Chebotarev, G.A., Kazimirchak-Polonskaya, E.I. & Marsden, B.G., Reidel, Dordrecht, pp. 301–307.

Sekanina, Z. (1991) Encke, the comet, *J. R. Astron. Soc. Can.* **85**, 324–376.

Sitarski, G. (1988) Long-term motion of Comet P/Encke, *Acta Astron.* **38**, 269–282.

Steel, D.I. & Asher, D.J. (1996a) The orbital dispersion of the macroscopic Taurid objects, *Mon. Not. R. Astron. Soc.* **280**, 806–822.

Steel, D.I. & Asher, D.J. (1996b) On the origin of Comet Encke, *Mon. Not. R. Astron. Soc.* **281**, 937–944.

Valsecchi, G.B. (1992) Close encounters, planetary masses and the evolution of cometary orbits, in *Periodic Comets*, eds Fernández, J.A. & Rickman, H., Universidad de la República, Montevideo, pp. 81–96.

Valsecchi, G.B. (1999) From Jupiter-family comets to objects in Encke-like orbits, in *Evolution and Source Regions of Asteroids and Comets (IAU Colloq. No. 173)*, eds Svoreň, J., Pittich, E.M. & Rickman, H., Slovak Acad. Sci., Tatranská Lomnica, pp. 353–364.

Wetherill, G.W. (1991) End products of cometary evolution: cometary origin of Earth-crossing bodies of asteroidal appearance, in *Comets in the Post-Halley Era (IAU Colloq. No. 116)*, eds Newburn, R.L., Neugebauer, M. & Rahe, J., Kluwer, Dordrecht, pp. 537–556.

Yeomans, D.K. (1994) A review of comets and nongravitational forces, in *Asteroids, Comets, Meteors 1993 (IAU Symp. No. 160)*, eds Milani, A., Di Martino, M. & Cellino, A., Kluwer, Dordrecht, pp. 241–254.

THE COMETARY CONTRIBUTION TO PLANETARY IMPACT RATES

H. RICKMAN
Astronomiska observatoriet, Box 515, S–75120 Uppsala, Sweden

J.A. FERNÁNDEZ, G. TANCREDI
Departamento de Astronomia, Facultad de Ciencias, Igua 4225, 11400 Montevideo, Uruguay

AND

J. LICANDRO
Centro Galileo Galilei, P.O. Box 565, E–38.700 Santa Cruz de La Palma, Tenerife, Spain

Abstract. We use statistics on the nuclear magnitudes of Jupiter Family comets to derive the number of km-sized members with perihelion distances less than 2 AU. When coupled with impact rates per comet taken from Levison *et al.* (2000), and with our estimate of the number of dormant comets in similar orbits yielding a dormant/active ratio of about 2, we find a terrestrial impact rate of about $1 \cdot 10^{-6}$ per year. This is already significant ($\sim 20 - 50\%$) compared with the total estimated impact rate by km-sized bodies from cratering statistics, and we estimate that further contributions by Halley-type and long-period comets are also substantial. Thus, in broad agreement with a number of earlier investigators, we find that comets yield a large, perhaps dominant, contribution to km-sized terrestrial impactors.

1. Introduction

Historically, comets were recognized as potential planetary impactors long before the first asteroid was discovered; in fact, already Isaac Newton had interesting thoughts about cometary impacts. During the last century, with the discovery of many Near-Earth Asteroids (NEAs), it became clear that for bodies of km-size the NEAs are far more numerous than the nuclei

M. Ya. Marov and H. Rickman (eds.), Collisional Processes in the Solar System, 131–142.
© 2001 *Kluwer Academic Publishers. Printed in the Netherlands.*

of active, Earth-crossing, short-period comets. Moreover, the frequency of passages of km-sized NEAs through the $r = 1$ AU sphere around the Sun was realized to be much larger than the corresponding passage rate of long-period comets. But even so, due to apparent problems to explain a sufficient supply of NEAs from the main belt by resonant transfer routes, it used to be a widely held opinion that a large fraction of the NEAs were really of cometary origin (Wetherill, 1979; see also Morbidelli, this volume).

The supply in the latter case involves problems that were poorly understood, i.e.: *(i)* the decoupling of typically cometary orbits from Jupiter's influence and ensuing dynamical stabilization; *(ii)* the quasi-permanent sealing off of cometary activity that would allow for very long physical lifetimes and our inability to easily recognize the respective NEAs as cometary by the development of comae.

For the first of these problems Asher *et al.* (this volume) conclude that current models indicate a rather minor cometary contribution to the NEA population (less than half). At the same time, the understanding of the second problem remains poor, and thus the cometary NEAs may be further limited by physical disintegration effects. And, very importantly, there is now a much improved understanding of resonant eccentricity pumping in the main belt that has shifted the general opinion into one where nearly all NEAs are seen as collisional fragments from the main belt (see Morbidelli, this volume).

However, there are several caveats to this simple picture. First, there is an albedo bias to flux-limited surveys of NEAs such that those dark spectral classes that might host the ex-comets are severely underrepresented (Rickman, 1991; see also Jewitt and Fernandez, this volume). The real contribution of these classes in a size-limited sample may still be considerable, and the main belt transfer efficiency for such classes may be limited by the paucity of large, dark parent bodies near the 3/1 and ν_6 resonances.

Secondly, the number of $\sim$ 10-km size NEAs is very small, and it is not clear that these can sustain the cratering rate on Earth for crater diameters exceeding 100 km during very long times. At present, the only 10-km size objects crossing Earth's orbit are active Halley-type comets (and some comets of long period), but not NEAs. Shoemaker *et al.* (1990) indeed suggested that the terrestrial cratering rate for such large scars should be dominated by a cometary component.

Thirdly, both the asteroidal and cometary contributions to the Near-Earth Object (NEO) population may be time variable (see Zappalà and Cellino; and Matese *et al.*, this volume), and we have no guarantee that the present ratio between the two sources is representative of the grand average over Gyr time scales. In this paper, we will touch upon some of the above points while primarily discussing the current population size of active comets in short-period orbits and possibilities for corresponding populations of dormant or permanently deactivated objects.

2. Impact rates by active Jupiter Family comets

2.1. IMPACT PROBABILITIES PER COMET

In this paper, like in our previous ones dealing with absolute magnitude estimates for short-period comets (Tancredi *et al.*, 2000; Fernández *et al.*. 1999) we define the Jupiter Family (JF) as the set of comets with orbital periods $P < 20$ yr and Tisserand parameters $T > 2$. We will concentrate in particular on such comets with perihelion distances $q < 2$ AU. With these constraints, the population under study corresponds almost exactly to the Ecliptic Comets (ECs) with $q < 2$ AU of Levison *et al.* (2000).

The latter paper derives planetary impact rates, updated with respect to those of Levison and Duncan (1997) but using the same set of long-term integrations of fictitious objects originating from the Edgeworth-Kuiper belt. Their fundamental result, which we will use in the following, is that the jovian and terrestrial impact rates for ECs, based on the number with $q < 2$ AU and absolute total magnitude $H_T < 9$, are $\simeq 6 \cdot 10^{-4}$ yr^{-1} and $\simeq 7 \cdot 10^{-8}$ yr^{-1}, respectively. These rates are for an estimated number of 40 such comets along with a number of "extinct" (*i.e.*, no longer active) comets that is assumed to be 3.5 times as large (Levison and Duncan, 1997), so taking those numbers for granted, we conclude that the rates per comet are 180 times less, i.e.: $\simeq 3 \cdot 10^{-6}$ yr^{-1} and $\simeq 4 \cdot 10^{-10}$ yr^{-1}, respectively. Note that these quantities depend only on the orbital parameters (essentially, T) but not on total magnitude, nucleus diameter, or activity level.

Nakamura and Kurahashi (1998) derived impact rates per object, based on 60 000 yr integrations for 228 known comets with $P < 1000$ yr. These rates are much smaller for terrestrial impacts in particular. However, this is due mainly to the fact that the perihelion distances of known JF comets are currently near their minima, so for most of the time spanned by these integrations the objects rather belong to an associated large-q population that the authors refer to as "unseen". Hence the corresponding impact rates are difficult to interpret, since the real number of such objects (of course, much larger than 40) is very difficult to estimate – the objects being, in fact, mostly unseen.

Weissman (1982, 1990) gave an average terrestrial impact rate by short-period ($P < 200$ yr) comets of $8.2 \cdot 10^{-10}$ yr^{-1} per comet, based on the known sample of 20 Earth-crossing such objects in 1981. Again, the interpretation of that number is a complicated task for the reason that the sample is a mixture of two separate dynamical classes of comets, the JF comets and the Halley-Types (HT). These have quite different impact probabilities, those of the JF being much larger due to the lower inclinations and shorter periods. And, in addition, the observed samples of the two classes are subject to very different amounts of magnitude bias (Rickman, 2000).

2.2. POPULATION SIZE AND TOTAL IMPACT RATE

For this reason, we stick to the Levison *et al.* (2000) impact rates and take up a thread left practically untouched by these authors. We have in mind the problem that the normalization to 40 comets with $q < 2$ AU and $H_T < 9$ refers to a sample limited by gas production rates rather than nuclear diameters. Levison *et al.* offer three crude ways to translate the above number to the one with $q < 2$ AU and diameter $d > 1$ km, arriving at a most likely factor $S = 5$ (*i.e.*, number of active, $q < 2$ AU, $d > 1$ km ECs $= 40S = 200$), though with a very large uncertainty.

As described in the papers already referred to, we have been involved since a long time with updating the information on nuclear magnitudes of JF comets. We have used published photometry of such comets, either at large heliocentric distances where their activity should be low or at smaller distances where coma correction methods have been applied. Part of the data, which is now essentially all CCD-based, comes from our own observations (Licandro *et al.*, 2000). We have thus derived a set of 105 absolute nuclear magnitudes of JF comets, of which 64 are believed to be accurate to better than 1^m, while 41 are of worse quality (Tancredi *et al.*, 2000).

These nuclear magnitudes were analyzed by Fernández *et al.* (1999) with, among others, the following results. For the entire sample of JF comets with $q < 2$ AU, a power law with index 0.53 ± 0.05 was fitted to the cumulative distribution of absolute nuclear magnitudes H_N. An analysis of the observed subpopulations in the ranges $0 \leq q < 0.5$ AU, $0.5 \leq q < 1$ AU, $1 \leq q < 1.5$ AU, and $1.5 \leq q < 2$ AU yielded the expected total numbers of JF comets with $H_N \leq 18.5$ listed in Table 1 for each of the q intervals. From this we derive the total number $N_{18.5} \simeq 350^{+250}_{-125}$ for JF comets with $q < 2$ AU.

Using a power-law index between 0.48 and 0.58, we can derive limits for the number with $H_N \leq 19$ (corresponding to $d \gtrsim 1$ km for geometric albedo $p_V = 0.04$) as: $N_{19} \simeq 650^{+550}_{-270}$. The most likely value yields an estimate of $S = 16^{+14}_{-6.5}$, about three times the result by Levison *et al.* (2000). From the above-quoted impact rates per comet we thus find, multiplying by $40S$:

$$(1.9^{+1.7}_{-0.8}) \cdot 10^{-3} \text{ yr}^{-1} \approx (1 - 4) \cdot 10^{-3} \text{ yr}^{-1}$$

for impacts onto Jupiter; and

TABLE 1.   Estimated number of JF comets
brighter than $H_N = 18.5$ in different q intervals

q (AU)	Number
0.0 – 0.5	2^{+1}_{-1}
0.5 – 1.0	12^{+4}_{-4}
1.0 – 1.5	80^{+40}_{-20}
1.5 – 2.0	250^{+200}_{-100}
2.0 – 2.5	450^{+300}_{-200}

$$(2.6^{+2.2}_{-1.1}) \cdot 10^{-7} \text{ yr}^{-1} \approx (1.5 - 5) \cdot 10^{-7} \text{ yr}^{-1}$$

for impacts onto the Earth. Note that there is some uncertainty of the latter
numbers due to the fact that the "observed" q distribution of Table 1 is
steeper than the dynamically calculated one of Levison *et al.*. If the one
that we have tabulated is closer to reality, the terrestrial impact rates per
comet may thus have been overestimated.

3.   Additional impacts by inactive Jupiter Family comets

3.1. DORMANT COMETS

We have to bear in mind that these numbers refer to impacts by active
comets such as those on record in the observed JF. However, some comets
are then likely to be missed because they are temporarily or permanently
deactivated. Various observational evidence has suggested since a long time
that short-period comets undergo periods of dormancy (*e.g.* Kresák, 1987),
but these data do not easily allow one to estimate how much of the time the
comets spend as active and dormant, respectively, or if this phenomenon is
common to all comets or only affects a minor fraction of them. Theoretical
models of dust mantling of cometary nuclei remain rather unconstrained as
to input parameters as well as observable output and are thus of little use
in predicting the number ratio of active/dormant comets.

Evidence on the physical lifetimes of JF comets also remains fairly in-
conclusive. Levison and Duncan (1997) estimated a 90% confidence interval
of $3\,000 - 30\,000$ yr for the active lifetime based on the statistics of observed
JF orbits, and this would correspond to a ratio between 2.0 and 6.7 of the
numbers of "extinct" and active JF comets. Since physical death (or "extinction") may mean either deactivation or disintegration, this estimate
leaves room for a dormant/active ratio R_{da} between 0 and 6.7.

TABLE 2.   Near-Earth asteroids in JF orbits, as of 1 January 2000

Name or designation	q (AU)	Q (AU)	Absolute magnitude
(3552) Don Quijote	1.21	7.25	13.0
(5370) Taranis	1.23	5.46	15.7
1982 YA	1.10	6.22	16.1
1992 UB	1.28	4.86	16.1
1994 LW	1.21	5.12	16.8
1995 QN$_3$	1.18	5.43	17.1
1997 SE$_5$	1.24	6.21	14.9
1997 YM$_3$	1.06	5.43	16.9
1998 GL$_{10}$	1.06	5.31	18.2
1998 HN$_3$	1.21	5.06	18.5
1999 LT$_1$	1.02	4.93	17.4
1999 VX$_{15}$	1.21	4.82	18.9

In order to place direct constraints on the value of R_{da}, we have scrutinized the statistics of discovered Near-Earth asteroids in JF orbits, *i.e.*, with $2 < T < 3$. Table 2 lists all such objects as of 1 January 2000, with the restriction of $Q > 4.8$ AU applied in order to make the distinction clear against the "usual" NEAs with $T > 3$ (there is actually a continuum field of objects – see Fig. 1, where cometary and asteroidal objects partially overlap). Even with this rather conservative choice, the number of entries in Table 2 amounts to 12.

3.2. ASTEROIDS IN JF ORBITS

First, then, let us estimate how many of these could be *bona fide* asteroids. The underlying scenario would be twofold. First, inner main belt objects are fed into Earth-crossing orbits, and a certain fraction of these are expelled by the Earth or Venus into Jupiter-crossing orbits (like those of the JF comets), from which Jupiter rather quickly ejects them into interstellar space. How many such objects could there be in a steady state?

Recent estimates of the population size for $d > 1$ km NEAs (supposed to be in typical orbits with aphelia well inside 4.8 AU) are in the range of $500 - 1000$ including likely errors (Bottke *et al.*, 2000; Rabinowitz *et al.*, 2000) or somewhat larger (Harris, this volume). The dynamical fates of such objects have been studied by Gladman *et al.* (1997), and it thus appears that the fraction experiencing the jovian ejection process is $\sim 20\%$.

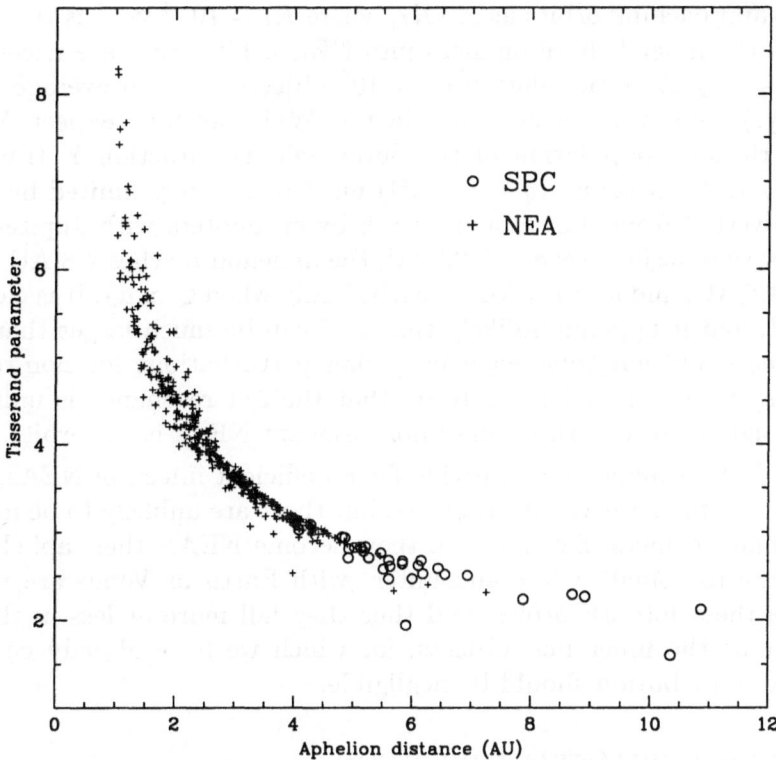

Figure 1. Tisserand parameter vs aphelion distance for two populations of Earth-approaching bodies: NEAs and $q < 1.3$ AU JF comets.

Most of the rest end up in the Sun.

The steady-state number of asteroidal NEAs in JF-type orbits is then: $N_{SS} \simeq (500-1000) \cdot 0.2 \cdot T_{JF}/T_{NEA}$, where T_{JF} and T_{NEA} are the dynamical residence time scales for Earth-approaching JF orbits and typical NEA orbits, respectively. The former is $\sim 10^3 - 10^4$ yr (Lindgren, 1992) and the latter is $\sim 10^6 - 10^7$ yr (Gladman *et al.*, 1997). The expectance for N_{SS} is $<< 1$, so asteroids emanating from the inner belt appear to contribute negligibly.

The other scenario may be much more efficient. In this case the asteroids emanate from the outer belt resonances, like the 2/1 and 5/2 mean motion resonances. It is very difficult to estimate their number, because the number of km-sized main belt asteroids near these resonances and their infeed rate into the eccentricity pumping zones are essentially unknown.

Some remarks are nonetheless relevant. Consider first the 2/1 resonance, and write the unknown transfer rate of objects from the main belt into Jupiter-approaching orbits as $X \cdot R_I$, where $R_I \sim 10^{-4}$ yr^{-1} is the transfer rate via the inner belt resonances into NEA orbits (the rate necessary to keep a steady-state population of $\sim 10^3$ objects with an average lifetime $\sim 10^7$ yr), and X is an unknown factor. While we may expect $X >> 1$ due to the rich population of the outer belt, the fraction Y transferred further into NEA orbits ($q < 1.3$ AU) must be severely limited by objects being diverted from the resonant track by encounters with Jupiter. For a constant semimajor axis $a = 3.28$ AU, the aphelion reaches 4.8 AU already at $q \simeq 1.8$ AU, and $q = 1.3$ AU is reached only when $Q \simeq a_J$. It is clear that $Y << 1$, and it appears unlikely that XY can be much larger than unity. Since the expulsion time scale by jovian perturbations for non-resonant orbits is $\lesssim 10^4$ yr, it thus appears that the 2/1 resonance is unlikely to yield a significant contribution of non-resonant NEAs in JF orbits.

The 5/2 resonance may provide for an efficient infeed of NEAs, as evidenced by some observed NEA orbits, but those are unlikely to be mistaken for dormant comets. By the time they become NEAs, their aphelion distances are too small. Close encounters with Earth or Venus are required to place them into JF orbits, and thus they fall more or less in the same category as the inner belt objects, for which we have already concluded that the contribution should be negligible.

3.3. THE JF DORMANT/ACTIVE RATIO

From the previous discussion, we have good reasons to conclude that a significant fraction or even most objects of Table 2 are dormant comets. These objects can be separated into different ranges of absolute magnitude, e.g.: 3 objects with $H \leq 16.0$; 4 with $16.1 \leq H \leq 17.0$; 2 with $17.1 \leq H \leq 18.0$; and 3 with $18.1 \leq H \leq 19.0$. But this is just a subsample of the real population, which should have nearly the same absolute magnitude distribution as that of nuclear magnitudes of JF comets. The latter is exponential with an index of 0.53 ± 0.05 (Fernández et al., 1999), so it is obvious that the fainter objects are severely underrepresented in Table 2. Therefore, we will concentrate on the brighter ones for a comparison with the number of active JF comets with $q < 1.3$ AU.

That number is found by interpolation from Fig. 7 and Table 2 of Fernández et al. (1999), and we estimate about 35 comets brighter than $H_N = 18.5$. Using the above distribution index, the number brighter than $H_N = 16.0$ should be about 1.7. Those in the ranges $16.1 - 17.0$, $17.1 - 18.0$, and $18.1 - 19.0$ should similarly be 4.0, 13.5 and 45.8, respectively. We see that for $H \leq 16$, the asteroids dominate over the comets by a factor 1.7. For fainter magnitudes the ratio decreases rapidly, but most of these small comets in fact remain undiscovered, and the corresponding asteroids must be even more difficult to detect, lacking as they are in cometary activity.

The best estimate of R_{da} would thus seem to be 1.7, but this is actually just a lower limit, because probably some more asteroidal objects with $H \leq 16$ remain to be discovered. The discovery years for the three known objects are 1983, 1986 and 1997 for (3552) Don Quixote, (5370) Taranis, and 1997 SE_5, respectively. Even if only one more exists, then R_{da} would be about $2 - 3$. This is actually in good agreement with the maximum-likelihood extinct/active ratio $R_{ea} = 3.5$ by Levison and Duncan (1997; see Sect. 2) for ECs with $q < 2.5$ AU. Recall that "extinct comets" means comets that have ceased to be active, whether they be dormant or disintegrated. Further comparison is difficult, because (1) R_{ea} must be larger than R_{da} by an unknown amount; (2) both R_{ea} and R_{da} should depend on q in an unknown fashion.

Our estimate of $R_{da} \simeq 2-3$ holds for comets with $H_N < 16$ (*i.e.*, nuclear radii $\gtrsim 4$ km) and with $q < 1.3$ AU. Transferring this into a corresponding estimate for $H_N < 19$ and $q < 2$ AU involves some degree of guesswork, but according to the modelling results for dust mantling by Rickman *et al.* (1990) there should be two opposing trends. Reducing the nuclear radius limit by a factor four should involve a decrease of R_{da}, but going to larger perihelion distances should rather increase it. We will use an estimate of $R_{da} = 2$ for km-sized nuclei with $q < 2$ AU, which we believe to be rather conservative. The JF cometary impact rates are thus found, multiplying the values at the end of Sect. 2 by $(1 + R_{da})$, to be:

$$(3 - 11) \cdot 10^{-3} \text{ yr}^{-1}$$

for Jupiter, and

$$(5 - 15) \cdot 10^{-7} \text{ yr}^{-1}$$

for the Earth.

4. Impacts by Halley-type and long period comets

It is unfortunately much more difficult to assess the contributions by other types of comets to the planetary impact rates. The reason is that the observational statistics of active, and in particular, dormant members of those groups is much poorer. Moreover, determinations of nuclear magnitudes are nearly completely missing.

Nonetheless, some rough estimates can be made. Based on statistics of total absolute magnitudes, Rickman (2000) concluded, in agreement with Shoemaker *et al.* (1994), that the number of Earth-crossing Halley types in a given total absolute magnitude range is ~ 10 times as large as the corresponding number of Jupiter family comets. On the other hand, their orbital periods are also $10 - 20$ times longer, so the impression is that the impact rates due to active comets within a given size range are similar for the two groups. Note, however, that Fernández and Gallardo (1994) found a theoretical, steady-state Halley-type population size for $q < 2$ AU that is about equal to the corresponding number of JF comets. This underscores the fact that the true Halley-type population size remains very uncertain.

For a given impactor size, however, the energy of impact is much larger for Halley types than it is for the Jupiter family. Hence, counting craters down to a given size limit, the statistics might be dominated by Halley-type impactors. However, we are explicitly referring to active comets, and while we estimated a dormant/active ratio $R_{da} \sim 2$ for JF comets, nothing is known about that ratio for the Earth-crossing Halley types.

Recent years have seen an increasing number of asteroids discovered in Halley-type orbits, but these have in general perihelion distances larger than any of the known Halley-type comets. It may thus be that R_{da} is extremely large for large-q Halley-type comets, but for low-q comets we do not know. Suffice it to recall that even if R_{da} were zero for Earth-crossing Halley types, the Halley-type impact rate should not be insignificant compared to the Jupiter family one.

For long-period (including Oort Cloud) comets, we may take Bailey and Stagg's (1988) estimate of the Earth-crossing flux as about 1 yr^{-1} with total absolute magnitude $H_{10} \leq 7$ and extrapolate that to about 3.5 yr^{-1} with $H_{10} \leq 9$ using the magnitude distribution slope by Everhart (1967) and Sekanina and Yeomans (1984). If we naively use our own estimate of $S \sim 16$ for the translation factor from $H_{10} \leq 9$ to $d \geq 1$ km in the Jupiter Family, and combine this with Marsden and Steel's (1994) terrestrial impact probability of $(2-3) \cdot 10^{-9}$ per orbit, we arrive at an expected impact rate of $\sim (1 - 1.5) \cdot 10^{-7}$ yr^{-1} by long-period comets onto the Earth.

It is true that $S \sim 16$ may be an overestimate for long-period comets, if these are more active than Jupiter Family ones, but we also must bear in mind that the estimated impact rate is only for active comets, and very little is known about the dormant/active ratio for long-period comets. Fitting their orbital energy distribution by physico-dynamical evolutionary models works best, if dormancy with possible rejuvenation is assumed to be common (Emel'yanenko and Bailey, 1996). So there is some reason to consider a large value of R_{da} for long-period comets (and, by implication, perhaps also a smaller level of activity for the active ones than indicated above). The recent discoveries of a couple of long-period, low-q asteroids lends support to this picture. However, the actual value of R_{da} is as yet impossible to tell.

In spite of all the uncertainties, there are clear indications that the total cometary impact rate onto the Earth is not completely dominated by the Jupiter Family. The total rate may thus significantly exceed the JF one, estimated in the previous Section.

5. Discussion

The result of our analysis is that km-sized comets probably yield a terrestrial impact rate of at least $1 \cdot 10^{-6}$ yr^{-1}. This basic conclusion of ours is enough to guarantee the comets an important and perhaps even dominant role among all km-sized impactors. On the other hand, it would be unwise to think that the upper part of our range, extending to at least $3 \cdot 10^{-6}$ yr^{-1}, may mean that asteroids could contribute negligibly. That is certainly not the case, but it is anyway important to remember that the main belt fragments should be far from dominant.

Conclusions similar to ours have been reached before (e.g., Shoemaker et al., 1994; see also Marov and Ipatov, this volume) but contrast with those of other investigators (e.g., Bailey, 1991; Neukum and Ivanov, 1994; see also Ivanov et al., this volume). Let us only mention a final caveat regarding the cometary impact rate. Rickman and Greenberg (1998) argued that cometary nuclei with $d \gtrsim 2$ km stand a much larger chance of being tidally split than of hitting the Earth upon a random close encounter. Thus it may be that this process causes an important shift of the cometary impactor population toward smaller sizes, but better physical models for tidal splitting (see Davidsson, 1999) need to be used to clarify the issue.

References

Bailey, M.E. (1991) Comet Craters Versus Asteroid Craters, *Adv. Space Res.* **11**, (6)43-(6)60.

Bailey, M.E., and Stagg, C.R. (1988) Cratering constraints on the inner Oort cloud: steady-state models, *Mon. Not. R. Astron. Soc.* **235**, 1-32.

Bottke, W.F., Jedicke, R., Morbidelli, A., Gladman, B., and Petit, J.-M. (1999) Understanding the Distribution of Near-Earth Asteroids, *Science* **288**, 2190-2194.

Davidsson, B.J.R. (1999) Tidal Splitting and Rotational Breakup of Solid Spheres, *Icarus* **142**, 525-535.

Emel'yanenko, V.V., and Bailey, M.E. (1996) Dynamical Evolution of Comets and the Problem of Cometary Fading, in *Worlds in Interaction: Small Bodies and Planets of the Solar System*, eds. H. Rickman and M.J. Valtonen, Kluwer Acad. Publ., Dordrecht, pp. 35-40.

Everhart, E. (1967) Intrinsic distributions of cometary perihelia and magnitudes, *Astron. J.* **72**, 1002-1011.

Fernández, J.A., and Gallardo, T. (1994) The transfer of comets from parabolic orbits to short-period orbits: numerical studies, *Astron. Astrophys.* **281**, 911-922.

Fernández, J.A., Tancredi, G., Rickman, H., and Licandro, J. (1999) The population, magnitudes, and sizes of Jupiter family comets, *Astron. Astrophys.* **352**, 327-340.

Gladman, B., Migliorini, F., Morbidelli, A., Zappalà, V., Michel, P., Cellino, A., Froeschlé, Ch., Levison, H.F., Bailey, M., and Duncan, M. (1997) Dynamical Lifetimes of Objects Injected into Asteroid Belt Resonances, *Science* **277**, 197-201.

Kresák, L'. (1987) Dormant phases in the aging of periodic comets, *Astron. Astrophys.* **187**, 906-908.

Levison, H., and Duncan, M. (1997) From the Kuiper belt to Jupiter-family comets: The spatial distribution of ecliptic comets, *Icarus* **127**, 13-32.

Levison, H.F., Duncan, M.J., Zahnle, K., Holman, M., and Dones, L. (2000) Note: Planetary Impact Rates from Ecliptic Comets, *Icarus* **143**, 415-420.

Licandro, J., Tancredi, G., Lindgren, M., Rickman, H., and Gil Hutton, R. (2000) CCD Photometry of Cometary Nuclei I: Observations from 1990–1995, *Icarus*, in press.

Lindgren, M. (1992) Dynamical Timescales in the Jupiter Family, in *Asteroids, Comets, Meteors 1991*, eds. A.W. Harris and E. Bowell, Lunar & Planetary Institute, Houston, pp. 371-374.

Marsden, B.G., and Steel, D.I. (1994) Warning Times and Impact Probabilities for Long-Period Comets, in *Hazards Due to Comets & Asteroids*, ed. T. Gehrels, Univ. Arizona Press, Tucson, pp. 221-239.

Nakamura, T., and Kurahashi, H. (1998) Collisional probability of periodic comets with the terrestrial planets – An invalid case of analytic formulation, *Astron. J.* **115**, 848-854.

Neukum, G., and Ivanov, B.A. (1994) Crater Size Distributions and Impact Probabilities on Earth from Lunar, Terrestrial-Planet, and Asteroid Cratering Data, in *Hazards Due to Comets & Asteroids*, ed. T. Gehrels, Univ. Arizona Press, Tucson, pp. 359-416.

Rabinowitz, D., Helin, E., Lawrence, K., and Pravdo, S. (2000) A reduced estimate of the number of kilometre-sized near-Earth asteroids, *Nature* **403**, 165-166.

Rickman, H. (1991) On the Properties of Comets, Asteroids, and Terrestrial Planet Impactors, *Adv. Space Res.* **11**, (6)7-(6)18.

Rickman, H. (2000) Cometary Populations, in *Asteroids, Comets, Meteors 1996*, in press.

Rickman, H., Fernández, J.A., and Gustafson, B.Å.S. (1990) Formation of stable dust mantles on short-period comet nuclei, *Astron. Astrophys.* **237**, 524-535.

Rickman, H., and Greenberg, J.M. (1998) Tidal Splitting of Comets in Earth's Vicinity, in *Highlights of Astronomy*, ed. J. Andersen, Vol. **11A**, pp. 262-265.

Sekanina, Z., and Yeomans, D.K. (1984) Close encounters and collisions of comets with the Earth, *Astron. J.* **89**, 154-161.

Shoemaker, E. M., Wolfe, R.F., and Shoemaker, C.S. (1990) Asteroid and Comet Flux in the Neighborhood of the Earth, *Geological Society of America Special Paper 247*, 155-170.

Shoemaker, E.M., Weissman, P.R., and Shoemaker, C.S. (1994) The Flux of Periodic Comets Near Earth, in *Hazards Due to Comets & Asteroids*, ed. T. Gehrels, Univ. Arizona Press, Tucson, pp. 313-335.

Tancredi, G., Fernández, J.A., Rickman, H., and Licandro, J. (2000) Catalog of Nuclear Magnitudes of Jupiter Family Comets, *Astron. Astrophys. Suppl.*, in press.

Weissman, P.R. (1982) Terrestrial impact rates for long and short-period comets, *Geological Society of America Special Paper 190*, 15-24.

Weissman, P.R. (1990) The cometary impactor flux at the Earth, *Geological Society of America Special Paper 247*, 171-180.

Wetherill, G.W. (1979) Steady–state populations of Apollo–Amor objects, *Icarus* **37**, 96-112.

PHYSICAL PROPERTIES OF PLANET CROSSING OBJECTS

DAVID JEWITT AND YANGA FERNANDEZ

Institute for Astronomy, University of Hawaii
2680 Woodlawn Drive, Honolulu, HI 96822 USA

Abstract. The physical properties of planet crossing objects are reviewed, with special focus on the Near Earth Objects, the nuclei of the short-period comets, and the Centaurs. These objects share in common dynamical lifetimes that are short compared to the age of the Solar System, requiring continual replenishment from more stable source regions to maintain steady-state populations. The planet-crossers therefore convey information about their source populations, including the Kuiper Belt (the source of the Centaurs and short-period comets). Many of the objects in these superficially different groups may in fact be inter-related. Observations are urgently required to accurately assess the populations in the various planet-crossing groups, to determine their physical properties with greater confidence, and so to elucidate the underlying connections.

1. Introduction

The planet-crossing objects of our Solar System include a diverse set of bodies classified (somewhat arbitrarily) according to different dynamical and observational criteria. They include Near-Earth Objects (NEOs), Short- and Long-Period Comets (SPCs and LPCs, respectively, distinguished loosely by whether their orbital periods are smaller or larger than 200 years), Centaurs, Resonant Kuiper Belt Objects (KBOs) and, potentially, Vulcanoids (asteroids with semi-major axes interior to the orbit of Mercury; Evans and Tabachnik, 1999). Some of these objects are shown in Fig. 1, distributed by perihelion distance, q, and aphelion distance, Q.

Most planet-crossing objects share in common the property that their dynamical lifetimes are short compared to the age of the Solar System. Gravitational interactions with the planets produce chaotic motions, leading to catastrophic collisions with the planets themselves, or with the Sun (Bailey, Chambers and Hahn, 1992; Beust and Morbidelli, 1996), or to

143

M. Ya. Marov and H. Rickman (eds.), Collisional Processes in the Solar System, 143–161.
© *2001 Kluwer Academic Publishers. Printed in the Netherlands.*

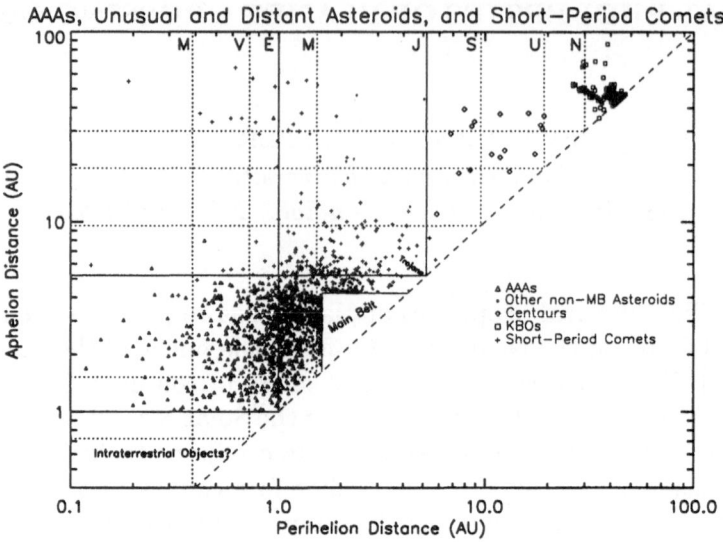

Figure 1. Distribution of planet-crossing objects in the q, Q (perihelion, aphelion) distance plane. Symbols distinguish the NEOs, Centaurs, KBOs and SPCs. The sun-grazing comets ($q \sim 0.005$ AU, $Q \sim 1000$ AU) are omitted from the plot. Orbits of the major planets are marked. Note the overlap between the cometary and asteroidal regions.

ejection from the Solar System on timescales of only $\sim 10^6$ yr. The short dynamical lifetimes imply that the planet-crossing objects must be continually resupplied from well-stocked and more stable reservoirs in order to maintain steady state.

The physical properties of planet crossing objects are of interest primarily in connection with locating their source regions. In the current paradigm, most SPCs (specifically, the Jupiter Family comets) can be traced through the Centaurs back to an origin in the Kuiper Belt (Fig. 2). The NEOs are linked both to source regions in the main asteroid belt and, less certainly, the short period comets. Planet crossing objects have been reviewed by a number of authors in recent years (McFadden *et al.*, 1989; Weissman *et al.*, 1989; McFadden, 1994; Luu, 1994; Jewitt, 1996; Morbidelli, 1999). In this short review, we focus attention on physical observations reported – and new understanding gleaned – in the last decade.

2. Near Earth Objects

2.1. OVERVIEW

The main-belt asteroids are located primarily between the orbits of Mars and Jupiter and number about 10^6 objects larger than 1 km. They consist of refractory materials (mostly rock with a few percent metals). Orbits of

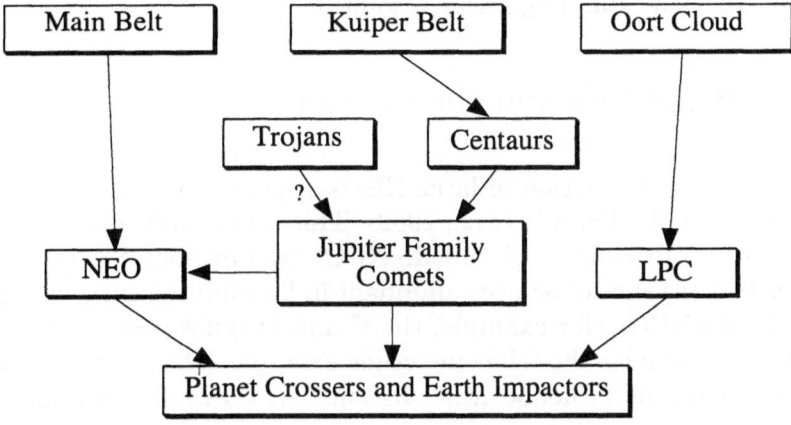

Figure 2. Inter-relationships between small bodies in the Solar System.

some main-belt asteroids are unstable to strong perturbations by nearby Jupiter, leading to eccentricity pumping and the excitation of orbits that cross those of the terrestrial planets, giving rise to NEOs. Specific Jupiter resonances have been explored as likely sources of the NEOs (*e.g.*, the 3:1 mean motion resonance at 2.5 AU, the ν_6 secular resonance at 2.1 AU; Wisdom, 1983, Gladman *et al.*, 1997) but these resonances are themselves fed in ways that are not yet fully understood (for further discussion of this problem, see Morbidelli, this volume).

About 900 NEOs are known as of January 2000. They are divided into classes based on their orbital parameters: Those with $a \leq 1$ AU and $Q \geq 0.983$ AU are "Atens", those with $q < 1.017$ AU and $a > 1$ AU are "Apollos" and those with $1.017 \leq q \leq 1.3$ AU are "Amors" (the perihelion and aphelion distances of Earth are 0.983 AU and 1.107 AU, see Shoemaker *et al.*, 1979). Their known populations are in the order Apollo:Amor:Aten = 400:400:60. Together, these objects are sometimes referred to as AAA objects. An additional NEO class with $Q < 0.983$ AU probably exists but, with perpetually small solar elongations, they cannot be detected by conventional night-time surveys. A single tentative example, 1998 DK36 ($a, e, i = 0.6943, 0.4115, 2.0, Q = 0.98$ AU), has been reported by R. Whiteley and D. Tholen (private communication).

The measurable physical properties of the NEOs include the surface colors and reflection spectra (constraints on the surface composition), the rotational lightcurves (constraints on the body shape and rotation state) and the optical and thermal emission (from which the size, albedo and thermal diffusivity can be estimated). In addition, ground-based radar is playing a larger role in the study of near-Earth objects. It has been used to image NEOs at startling linear resolution.

2.2. SPECTRAL TYPES AND COMPOSITIONS

A disproportionate fraction of large NEOs appear to be of asteroidal spectral type S (*e.g.*, McFadden *et al.*, 1989). This is probably a result of albedo bias, according to which high albedo NEOs (the average S-type geometric albedo is ≈ 0.16) should be over-abundant in flux-limited surveys compared to low albedo NEOs (for example, the C and D types, which have albedos $0.04 - 0.06$). Consider the following crude example. The scattered flux from an object of radius r, heliocentric distance R, geocentric distance Δ and albedo p varies as

$$f \propto pr^2/R^2\Delta^2.$$

For objects of a given size ($r = $ constant) in near-Earth space ($R = 1$ AU), $f \propto p/\Delta^2$. In a flux-limited survey, the maximum distance of visibility of an object varies as $\Delta \propto (p/f)^{1/2}$, and the number of objects within distance Δ is $N \propto \Delta^3 \propto (p/f)^{3/2}$. For example, dark NEOs ($p = 0.04$) of a given size would be under-represented relative to bright NEOs ($p = 0.16$) in a flux-limited survey by a factor $(0.16/0.04)^{3/2} = 8$, about an order of magnitude. Measurements of the albedos of NEOs (by comparison of the reflected and thermally emitted fluxes) indeed show that high albedo objects dominate the sample (Fig. 3). The albedo bias has bearing on the fraction of NEOs which might be dead comets.

The colors of NEOs show a trend with diameter, with the small objects being more nearly neutral than large ones: in spectral parlance, S-types dominate at large sizes while Q-types dominate at diameters $d \leq 2$ km km (Binzel *et al.*, 1996; Rabinowitz, 1998; Hicks *et al.*, 1998). If small NEOs are produced from larger main-belt asteroids by collisional shattering, their mean age should be a function of their size. This suggests to some that the color-diameter trend among NEOs may be an artifact of "space weathering" – damage done to the surface minerals of asteroids by prolonged exposure to radiation and particle bombardment in space. Alternatively, the excess fraction of nearly neutral small-sized NEOs could be due to differently friable materials in main-belt asteroids, with the neutral NEOs being over-abundant simply because they are more easily fractured and comminuted by impacts. Whatever the cause, the color-diameter trend brings the smaller NEOs closer to the reflected colors shown by ordinary chondrites, strengthening the link between the chondrites and their parent asteroids.

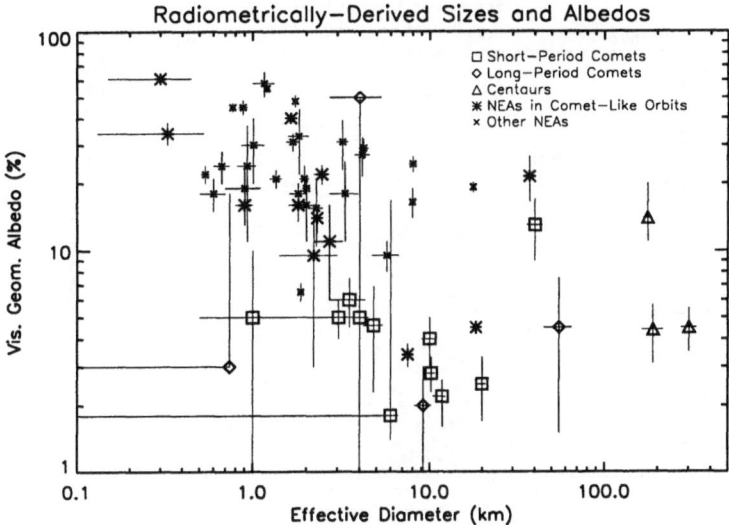

Figure 3. Radiometrically derived diameters and albedos of planet crossing objects. The NEO data are primarily from Veeder *et al.*(1989). Figure from Fernandez (1999).

2.3. SHAPES AND ROTATIONS

The projected axial ratios and rotations of a considerable number of NEOs are now known (Fig. 4). Most NEOs (and most main-belt asteroids) rotate with periods longer than 2 to 3 hours. This may be compared with the critical period at which centripetal and gravitational accelerations become equal. For a spherical body of density ρ [kg m^{-3}], this critical period is $T_C = (3\pi/G\rho)^{1/2}$. For example, for $\rho = 1000$ kg m^{-3}, the critical period is $T_C = 3.3$ hrs, quite close to the minimum periods of most asteroids. For more realistic prolate spheroidal objects, the implied critical period for a given density can be longer by a factor ≈ 2 (cf. Jewitt and Meech, 1988; Fig. 4). Nevertheless, the observation that no large asteroids are spinning with periods $<< T_C$ suggests the importance of centripetal effects. In fact, many NEOs (and, indeed, main-belt asteroids) may possess little internal tensile strength, presumably because they consist of gravitationally bound rubble piles (Benz and Asphaug, 1999). Rubble piles would be natural products of extensive collisional processing, as is thought to have occurred in the main belt. Strengthless asteroids spinning at high angular velocities will deform into a series of elongated figures of revolution known as MacLaurin spheroids. While it is clear from Fig. 4 that asteroids do not adopt the exact shapes of MacLaurin spheroids, rubble pile models provide at least a viable explanation of the observed lower limit to the rotation periods in Figure 4.

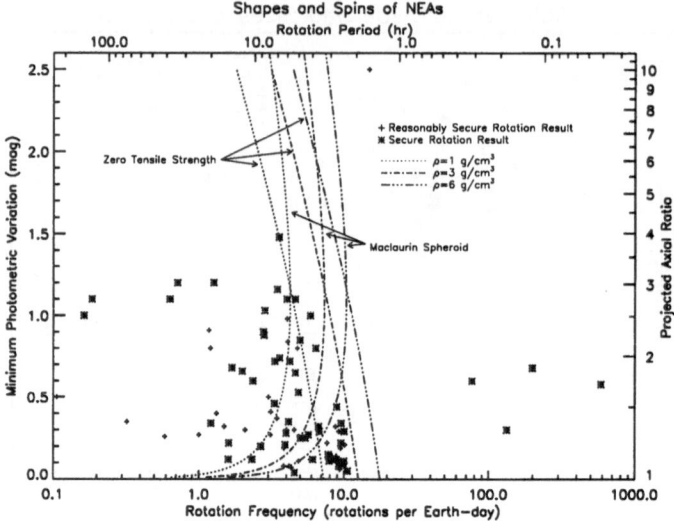

Figure 4.   Rotational frequency of NEOs is plotted against the observed maximum photometric variation. Curves show the trajectories of MacLaurin spheroids of three different densities, as indicated. Objects with rotational frequencies > 10 day^{-1} are rare, but there is otherwise no clear relation with the MacLaurin curves. Data extracted from the Planetary Data System node at `pdssbn.astro.umd.edu`.

Of course, materially strong objects can rotate faster than the critical period. Recently, five small NEOs have been found spinning with periods $\tau < T_C$, implying that these bodies possess internal tensile strength sufficient to overcome centripetal acceleration. The characteristic sizes of these (monolithic?) objects fall in the 10-m to 100-m range. They are presumably coherent splinters produced by impact into parent bodies (Pravec *et al.*, 2000). Numerical simulations (albeit of a single assumed impact geometry) show that impact fragments should spin with periods distributed over more than two orders of magnitude, with the shortest periods confined to the smallest fragments (Asphaug and Scheeres, 1999). The fragments should also initially be in excited (non-principal axis) rotation. So far, no small slow rotators have been identified, and the objects in Table 1 exhibit simple periodic lightcurves that suggest principal axis rotation. The damping timescale, t_d [My], due to internal frictional energy dissipation, is of order

$$\tau_d \approx T^3/(4.9d^2),$$

where T [hr] is the rotation period and d [km] is the diameter (Harris, 1994). The damping times for the majority of the rapid rotators (*i.e.*, all except 1995 HM) are small compared to the few My dynamical lifetimes of the NEOs (Gladman *et al.*, 1997; Table 1), suggesting that these objects should already be rotationally damped and consistent with principal axis rotation.

TABLE 1. Rapidly Rotating Asteroids[a]

Name	T^b	Δm^c	H^d	d^e	τ_d^f
1995 HM	97.2	2	22.5	130	5.1×10^7
1998 KY$_2$6	10.7015 ± 0.0004	0.30	25.5 ± 0.3	30	1.3×10^6
1998 WB2	18.8 ± 0.3	0.6	22.1 ± 0.2	120	4.4×10^5
1999 SF$_1$0	2.4663 ± 0.0005	0.58	24.0 ± 0.5	60	3.9×10^3
1999 TY2	7.2807 ± 0.0003	0.68	23.1 ± 0.3	80	5.7×10^4

[a] Modified from Pravec et al.(2000).   [b] Rotation period (min).
[c] Photometric range (mag).   [d] Absolute magnitude.
[e] Estimated diameter (m).   [f] Damping time (yr).

A further consequence of rubble pile structure among the NEOs has been noted by Solem and Hills (1996). They considered the effects of tidal stresses due to planetary encounters on NEO shapes and rotations. Tidal deformation models provide a convincing match to both the shape and rotation of 1620 Geographos (Bottke et al., 1999; see Fig. 5) . However, strong tidal effects are restricted to the closest and most rare planetary approaches. For this reason, only $\approx 2\%$ of the NEOs are likely to be tidally deformed even if all such objects are rubble piles (Bottke et al., 1999).

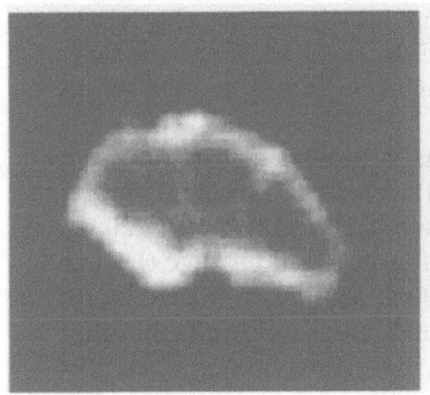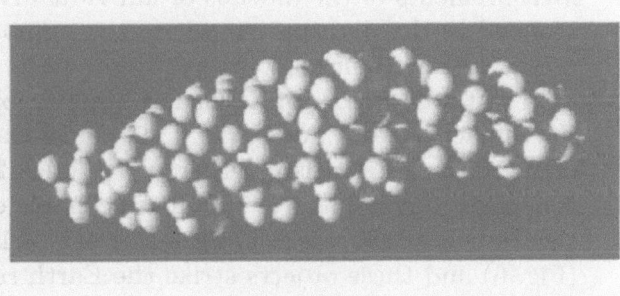

Figure 5.   (Left) The silhouette of Geographos from radar observations (Ostro et al., 1996). (Right) A rubble-pile model asteroid that has been tidally distorted by a close pass to the Earth (Bottke et al., 1999). The similarity between the asteroid and model shapes is obvious.

Figure 5 demonstrates how well radar can provide detailed rotational, shape and topographic information on NEOs. The main limitation to the technique is the strong (inverse 4th power) dependence of the signal on geocentric distance. Nevertheless, 10 NEO radar detections were secured in 1999 and 9 of those were of newly discovered objects. If this rate is sustained, we will soon possess a substantial number of radar measurements of NEOs that will rival or surpass what could plausibly be achieved using spacecraft.

2.4. POPULATION

As noted above, our knowledge of the population of NEOs is incomplete and subject to important observational biases. A list of productive survey programs may be found on the World Wide Web at URL

http://cfa-www.harvard.edu/iau/NEO/TheNEOPage.html.

We show in Fig. 6 the cumulative distribution of NEOs as a function of absolute visual magnitude and year of observation. Note that the total number has almost doubled in just the last two years as a result of aggressive survey observations (see Table 2). The sample of objects with absolute magnitude brighter than 15.5 or 16 (corresponding roughly to a diameter of 8 km) is evidently nearly complete. The dashed-dotted lines in Fig. 6 represent possible differential power law size distributions of the objects; q is the exponent. A value $q = 3.5$ corresponds to a collisionally relaxed system (Dohnanyi, 1969); the current best fit to the objects brighter than $V = 15.0$ gives $q \approx 3.75$, consistent with the Dohnanyi index within the uncertainties of measurement. The accuracy of the determination of q is limited in part by the assumption of constant albedo among the NEOs. Measurements of the albedos of individual NEOs, particularly at the large size end of the distribution where the number of objects is comparatively small, are sorely needed.

The total number of NEOs has implications for the current impact rate, the impact hazard, and small body collisional evolution. The number with diameter $d > 1$ km is of order $N = 700 \pm 250$ (Rabinowitz et al., 2000). Asteroids as small as $d = 100$ m are capable of unimpeded passage through the Earth's atmosphere. The number of $d > 100$ m NEOs is $\approx (0.5 \text{to} 2) \times 10^5$ (Fig. 6) and these objects strike the Earth roughly once per millennium.

The number of NEOs with $d < 100$ m may be higher than predicted by power law extrapolations from larger objects, with an excess of two orders of magnitude reported at $d \approx 10$ m (Rabinowitz, 1993). These small NEOs are so numerous that it is unlikely that they can be fragments of "split" comets. Instead, a source might be found near resonances in the main belt, where the Yarkovsky effect (a non-radial force caused by thermal emission from a rotating, anisothermal body) could drive small asteroidal fragments into the resonance (Farinella et al., 1998).

TABLE 2. Discoveries of NEOs[a]

Type	Apollos	Amors	Atens
Total	411	403	58
1999	106	104	12
1998	95	90	19
1997	27	23	4
1996	24	20	2
1995	14	17	1
1994	20	21	4
1993	18	20	2
1992	10	19	2
1991	25	13	2
1990	13	11	1
≤ 1989	59	65	9

[a] From Minor Planet Center's WWW site (URL
http://cfa-www.harvard.edu/iau/mpc.html)

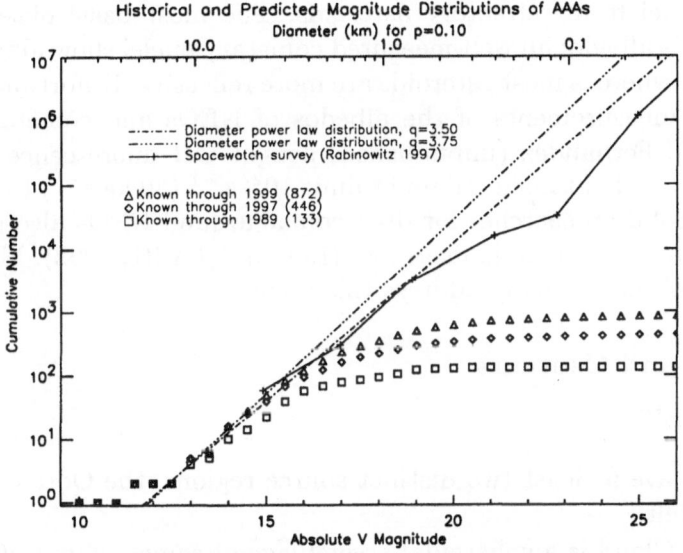

Figure 6. Cumulative numbers of NEOs as a function of absolute V magnitude measured at different epochs. The flattened curves suggest that the sample is complete to about $V(1,1,0) = 15$, corresponding to diameter $D \approx 8$ km (albedo 0.10 assumed).

The fraction of the NEOs (of all sizes) that might be dead comets is uncertain. The standard dynamical discriminant is the Tisserand invariant, T, with most asteroids having $T > 3$ and most comets $T < 3$. However, as shown in Fig. 1, there are members of each class that overlap the dynamical region of the other. Moreover, there is no reliable observational test to uniquely discriminate dead comets from ordinary asteroids. The D type NEO 3552 Don Quixote ($T = 2.3$), with a presumed cometary origin, could be literally the tip of the iceberg of a substantial population of NEOs derived from comets. The limitation in applying simple dynamical discriminants is that the known NEO and comet populations are highly biased (notably against dark NEOs and weakly or inactive comet nuclei). Recent attempts to correct observational surveys for bias are reportedly compatible with a wholly asteroidal source for the NEOs (A. Morbidelli, private communication, January 2000). Harris and Bailey (1998) have independently tried to quantify the fraction of the NEOs that might be dead comets by integrating the equations of motion over long timescales. Over time non-gravitational forces due to anisotropic outgassing can significantly affect the orbit and bring comets into NEO-like orbits. They find the fractional comet population to be somewhere between 10 and 50 percent, depending on the efficacy of non-gravitational forces (cf. Asher et al., this volume).

The short sublimation lifetimes of comets imply a large population of dead nuclei (Sect. 3.2). Observationally, the problem is to discriminate dead cometary nuclei from refractory asteroids. The most basic observational discriminant is albedo: all well-measured cometary nuclei show albedos near a few percent whereas most asteroids are more reflective. Unfortunately, few high quality measurements of the albedos of NEOs and cometary nuclei exist (Fig. 3). Fernandez (unpublished) sought OH fluorescence emission from 1984 BC and obtained an upper limit 10^{27} s^{-1} (30 kg s^{-1}) to the mass loss. High resolution searches for dust comae around NEOs also produced only upper limits to possible mass loss (Luu and Jewitt, 1992). More work is needed to observationally address this issue.

3. Comets and Centaurs

3.1. OVERVIEW

The comets have at least two distinct source regions, the Oort Cloud and the Kuiper Belt.

The Oort Cloud is a spheroidal assemblage of comets about 50,000 AU (0.2 pc) in radius loosely bound to the Sun. Orbits of comets in the Oort Cloud evolve in response to random perturbations from passing stars and from the gradient in the Galactic gravitational potential. Comets whose perihelia diffuse to $q < 35$ AU are strongly scattered, first by Neptune then by other planets, eventually to be ejected to interstellar space or injected into the terrestrial planet region, where they are classified as Long Period Comets (LPCs). Roughly 10^{12} comets larger than about 1 km diameter are thought to be contained in the Oort Cloud.

The Kuiper Belt is a disk-shaped collection of comets located beyond Neptune (heliocentric distance 30 AU). Kuiper Belt objects (KBOs) follow a variety of orbits, some of which are chaotic on timescales comparable to the age of the Solar System. Objects dislodged from the Kuiper Belt diffuse first through the region of the gas-giant planets and then, if they survive ejection to the interstellar medium, into the region of the terrestrial planets (where they are known as Short-Period Comets; SPCs). Planet-crossing objects with both perihelia and semi-major axes located between Jupiter's 5 AU orbit and Neptune's 30 AU orbit are referred to as Centaurs. About 20 Centaurs have been observed as of January 2000, and about 250 SPCs possess reliable orbital elements. The location of the source of the SPCs within the Kuiper Belt is not precisely known (Morbidelli, 1997; Duncan and Levison, 1997). The number of KBOs larger than about 1 km diameter is estimated to be about 5×10^9. The flux of SPCs arriving in the inner Solar System is of order 10^{-2} yr^{-1}, but this number is only as reliable as our knowledge of the SPC population. There is no systematic or well controlled search program for faint comets and, therefore, every reason to expect that the number of SPCs is poorly known.

The ensemble properties of cometary nuclei are poorly known. The main practical problem is that cometary nuclei are small in size and faint. When a comet is close to Earth (and Sun) outgassing produces a bright coma that conceals the nucleus. Conversely, when the comet is far from the Sun and the coma is minimal the nucleus is typically too faint to study in detail. In general, attempts to subtract the coma are suspect because of uncertainties in the extrapolation of the coma surface brightness in the near-nucleus environment. In this regard, note that 1 arcsec corresponds to 730 km at 1 AU, while the characteristic sizes of the nuclei are in the 1 to 10 km range. The most notable success with coma subtraction has been obtained by capitalizing on the high angular resolution afforded by Hubble Space Telescope (e.g., Lamy et al., 1999) and, to a lesser extent, on near-Earth comets like C/1996 B2 Hyakutake (Lisse et al., 1999). J.A. Fernández et al.(1999) have recently reported a study of the size distribution of cometary nuclei obtained by their own and historical optical magnitudes. They attempt to correct for coma contamination and extract a power law for the nuclei whose slope is close to the ≈ 3.5 value found for the NEOs. However, it is difficult to assess the systematic errors of their determination. Sub-kilometer nuclei appear to be under-abundant (relative to a $q = 3.5$ power law), presumably because these bodies quickly lose their volatiles by sublimation when near the Sun.

3.2. MANTLES

Albedos of most nuclei fall in the range $0.02 < p < 0.06$. Such low albedos
are thought to be caused by surface mantles consisting of carbon rich, refractory substances. There are two basic ideas for the formation of cometary
mantles (but little direct evidence relating to either). So-called irradiation
mantles might form by prolonged bombardment of cometary ices by solar
and Galactic cosmic rays, leading to the dissociation of simple molecules
and their subsequent reassembly into chemically complex forms (e.g.,, Johnson, 1991). Hydrogen is liberated in this process and, because of its small
size, can leak from the irradiated layers into space. The resulting mantle extends to a column density of order 10^3 kg m^{-2} (physical thickness of order 1
meter), is free of ice, depleted of hydrogen and dark because of high molecular weight carbon compounds (Moroz et al., 1998). Alternatively, rubble
mantles might form as a by-product of cometary sublimation, and consist
of rocks and other debris from the body of the nucleus too large to be
ejected against nuclear gravity by gas drag (Rickman et al., 1990). Perhaps
both kinds of mantles are present on comets at different stages of their evolution. Irradiation mantles could grow while the nuclei are "stored" in the
deep freeze of the Kuiper Belt or Oort Cloud (the timescales for irradiation
mantle growth are uncertain, but perhaps near $10^{8\pm1}$ yr). Once deflected
towards the Sun, solar heating would lead to gas pressure build up, followed by mantle fracturing and outgassing from isolated vents or "active
areas" on the nucleus. In this picture, the ultimate fate of the irradiation
mantle is to be either ejected by gas drag from sublimated volatiles located
beneath the mantle, or buried by sub-orbital debris ejected from the active
regions. We should note that Kührt and Keller (1994) have observed that
cohesive forces between grains in porous mantles are likely to be orders of
magnitude stronger than gravity on the nuclei of comets. They suggest that
sticky, cohesive mantles are more probable structures on comets than pure
rubble mantles held in place by nuclear gravity alone.

Some clues about these processes are provided by recent physical observations of the Centaurs. These objects are dynamically intermediate between the Kuiper Belt and the short-period comets. They are susceptible
to strong scattering by the gas giants, and have dynamical lifetimes limited to about 10^7 yr. Their relevant characteristic is that they orbit the
Sun beyond Jupiter, and so maintain surface temperatures that are too
low to permit the sublimation of water ice (although supervolatiles such as
CO can be active to much larger heliocentric distances). Some Centaurs,
notably 2060 Chiron and 29P/Schwassmann-Wachmann 1, display comae
indicative of mass loss. In 29P, the mass loss is firmly established as being
due to the sublimation of CO (Senay and Jewitt, 1994) at a rate of order

1 tonne s^{-1}. Outgassing of CO has also been reported in 2060 Chiron by Womack and Stern (1999). However, their detection appears to us to be of questionable significance.

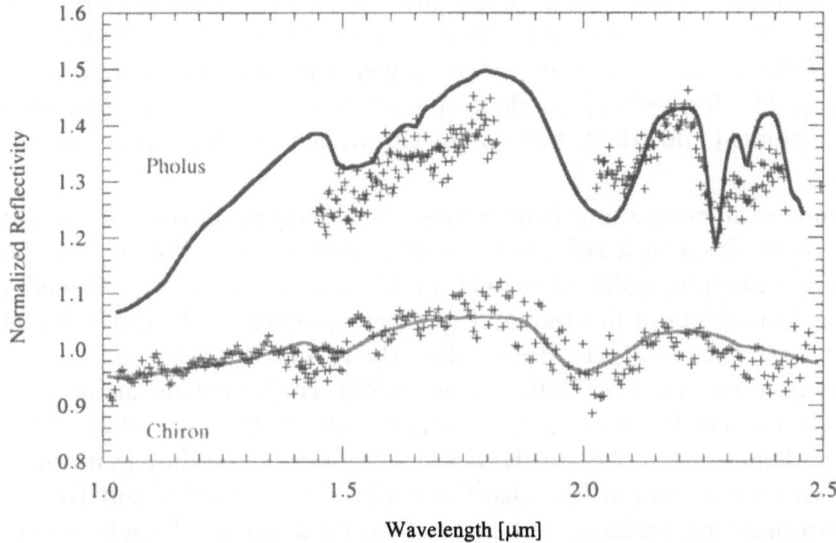

Figure 7. Near infrared reflection spectra of Centaurs 5145 Pholus (Luu *et al.*, 1994) and 2060 Chiron (Luu *et al.*, 2000). The spectrum of Pholus has been shifted vertically for clarity. The lines show model fits to the Pholus (Cruikshank *et al.*, 1998) and Chiron (Luu *et al.*, 2000) spectra. The prominent feature near 2 μm in both objects is due to water ice. Pholus contains an additional absorption near 2.25 μm that may be due to a hydrocarbon.

Spectral evidence from the Centaurs is limited and poorly understood but nevertheless suggestive. There is a wide diversity in the optical and near-infrared reflection spectra of these bodies (Davies *et al.*, 1998) as in their Kuiper Belt parents (Luu and Jewitt, 1996; Jewitt and Luu, 1998). 5145 Pholus is (optically) very red and shows absorption features near 2.0 and 2.25 μm. The 2.0 μm feature is presumably caused by OH stretch in water ice. The 2.25 μm band has been identified with methanol or a similar hydrocarbon (Cruikshank *et al.*, 1998). 2060 Chiron is known to be an outgassing body complete with a prominent but highly variable dust coma. The nuclear spectrum has become clear with the decay of the surrounding dust coma, and shows only water ice (Foster *et al.*, 1999; Luu *et al.*, 2000). A plausible, but non-unique, interpretation of these differences suggests itself. Inactive Pholus could be a pristine body from the Kuiper Belt, complete with an irradiation mantle (hence the hydrocarbon absorption). Active Chiron's surface, on the other hand, has been blanketed by fall-back debris (including water ice grains responsible for the 2.0 μm feature) to a depth sufficient to hide the irradiation mantle from view. It is natural, in this context, to infer that Pholus is a more recent and less thermally evolved escapee from the Kuiper Belt than is Chiron. Unfortunately, this cannot be tested using dynamical models of the evolution of Centaurs since their motions are intrinsically chaotic in nature. One testable prediction of this hypothesis is that the nuclei of active comets should be spectrally more similar to Chiron than to Pholus, since they should all be blanketed by fall-back debris. The few reliable nucleus spectra show evidence for differences among the nuclei (Luu, 1993) but do not yet prove that Pholus type spectra are absent.

Evidence that outgassing from comets proceeds primarily through active areas embedded in a refractory mantle, rather than globally, is compelling. For example, collimated jets in the coma are best modelled as emanating from vents of limited spatial extent (Keller et al., 1994). In-situ images of the nucleus of 1P/Halley directly reveal the localised roots of jets in the inactive mantle (Keller et al., 1987). High nucleus surface temperatures in comets C/1983 XX IRAS-Araki-Alcock (Hanner et al., 1985), 49P/Arend-Rigaux (Brooke and Knacke, 1986) and 1P/Halley (Emerich et al., 1987) appear incompatible with the cooling that would result from an actively sublimating surface, and instead imply a surface largely covered by refractory mantle. Less directly, the total sublimation rate from many comets is much less than expected if the nucleus were coated uniformly by volatile ice. Among the SPCs, the sublimating fraction of the nucleus surface is $0.1 < f < 10\%$ (Jewitt, 1996). It is clear that mantles play a role in limiting the escape of volatiles from cometary nuclei. It is even possible that mantles sometimes completely stifle the flow of gas, leading to comets which are asteroidal in appearance or "dead" (Luu, 1994; Jewitt, 1996). Such objects would be leading candidates for cometary NEOs (Sect. 2.4). A few objects are transitional in nature between comets and asteroids including the (briefly reactivated?) comet 107P/Wilson-Harrington (Fernandez et al., 1997) and the (now inactive) parent of the Geminid meteors, 3200 Phaethon. It is likely that comets are repeatedly activated and deactivated in response to changes in the total insolation and mantling caused by progressive orbital evolution among the planets (Rickman et al., 1990). The number of dead comets can be estimated by comparing the dynamical lifetime of the SPCs ($\tau_d \approx 4 \times 10^5$ yr; Levison and Duncan, 1994) with the timescale for the loss of volatiles by sublimation. An unmantled, perfectly absorbing water ice nucleus of radius r would completely sublimate on timescale

$$\tau_V \approx \rho r L R^2 / S f$$

where $S = 1380$ W m^{-2} is the Solar constant, L [J kg^{-1}] is the latent heat of vaporization ($L = 2 \times 10^6$ J kg^{-1} for water ice), $\rho \approx 10^3$ [kg m^{-3}] is the nucleus bulk density, f is the mantling fraction and R [AU] is the heliocentric distance. In deriving this relation, $R \leq 1$ AU is assumed so that the absorbed energy lost by radiation is small compared to that consumed in sublimation. With $r = 5$ km and $f = 0.01$ (typical of a weakly active comet) we find $\tau_V \approx 2.5 \times 10^4$ yr. The expected ratio of dead to active comets is

$$N_D/N_A \approx \tau_d/\tau_V \approx 10,$$

meaning that dead comets should substantially outnumber their living counterparts. It is sometimes asserted that comets completely disintegrate into dust and do not leave behind a monolithic, inactive remnant. In this case, our estimate of N_D/N_A should be regarded as an upper limit. However, while comets sometimes throw out small secondaries in a process misleadingly known as "cometary splitting" (Chen and Jewitt, 1994) there is little compelling evidence for the complete disintegration of any nucleus. Indeed, weakly active comets (*e.g.*, 49P/Arend-Rigaux, 28P/Neujmin 1, 6P/d'Arrest) certainly appear to be nearly devolatilised bodies which retain very substantial mass, and quite distinct from clouds of dust! These objects appear completely asteroidal except when near perihelion and it is easy to imagine that even a small additional decline in the outgassing rate would render them asteroidal to all appearances. Observationally, the value of N_D/N_A is not well established, since we possess few unambiguous discriminants between defunct nuclei and the near Earth asteroids they would presumably resemble. The albedos of 2 or 3 of the 40 NEOs plotted in Fig. 3 fall in the realm otherwise occupied by the nuclei of SPCs. Given the large albedo bias against finding dark objects in NEO surveys, we conclude that there is room for a substantial number of dead comets among the NEOs. Clearly, this is an area deserving of much more observational attention.

Our thoughts about the physical nature of the cometary nucleus are in the process of revision. The classical view is that the cometary nucleus is a primordial relic from the formation of the Solar System, unchanged but for the loss of volatiles from the surface layers heated by the Sun. The realization that the SPCs are derived from the collisionally active Kuiper Belt, rather than from the much more extended and collision-free Oort Cloud, raises the possibility that the nuclei are not primordial (see discussion by Davis and Farinella, this volume). Specifically, with a mean impact velocity near 1.5 km s^{-1} (Jewitt and Luu, 2000), collisions between KBOs would produce shock heating and volatile loss. It seems likely that the larger KBOs are internally differentiated. If so, the material eroded from the outer regions will be depleted in supervolatile ices (CO, N_2) compared to the bulk composition, and the SPCs cannot be regarded as compositionally representative.

4. Some Questions

We end with a brief list of (mostly inter-related) observational questions about the physical properties of planet crossing objects.

1. What fraction of NEOs are of cometary, as opposed to asteroidal, origin? What is the ratio of numbers of dead to active comets? How do comets die? How can we best discriminate dead comets from asteroids? Most observational work on this question has been limited to the measurement of spectral types of NEOs in a data set that is known to be highly biassed against low albedo objects, such as the nuclei of comets. Only a few observers have targeted specifically cometary properties such as mass loss driven by outgassing (so far with negative results). Larger telescopes and better detectors should permit a meaningful observational search for outgassing from NEOs.

2. What is the form of the size distribution of the sub-kilometer NEOs and what does this imply for the impact timescale on Earth? Are there any differences in the size distributions of different taxonomic types?

3. How many monoliths are there in near-Earth space? How many binary asteroids?

4. What is the population and size distribution of the Centaurs? Only 20 large Centaurs are presently known but the sample is severely biassed against faint objects. Deeper surveys with large aperture telescopes should be used to determine the Centaur population to larger distances and smaller sizes.

5. What is the cause of the spectral diversity seen among the Centaurs? Can we observationally discriminate irradiation mantles on Centaurs from rubble mantles in the way described in 3.2, or are the physical processes more complicated?

6. What fraction of the Centaurs are measurably outgassing and which volatiles are responsible? At present, 2060 Chiron and P/SW1 both exhibit outgassing. It is likely that other Centaurs, especially those of small perihelion distance, lose mass due to the sublimation of embedded volatiles, but no additional examples have been discovered.

7. What is the population and size distribution of the short-period comets? The available sample is biased against comets that are faint (*i.e.*, small and/or far away, or both). Recent all-sky observations in search of NEOs (*e.g.*, LINEAR) have identified a large number of faint, weakly outgassing comets. What are their physical properties? How large are they? What is their total number and flux into the inner planetary region from the Kuiper Belt? Systematic investigation of these objects using simultaneous optical and thermal infrared photometry is required.

5. Acknowledgements

Work on this review was supported by an NSF Planetary Astronomy grant to DCJ.

References

Asphaug, E., and Scheeres, D. (1999). Deconstructing Castalia: Evaluating a Post impact State, *Icarus* **139**, 383-386.

Bailey, M., Chambers, J., and Hahn, G. (1992). Origin of Sungrazers: A Frequent Cometary End-State. *Astron. Ap.* **257**, 315-322.

Benz, W., and Asphaug, E. (1999). Catastrophic Disruptions Revisited. *Icarus* **142**, 5-20.

Beust, H., and Morbidelli, A. (1996). Mean-Motion Resonances as a Source for Infalling Comets Toward Beta Pictoris. *Icarus* **120**, 358-370.

Binzel, R., Bus, S., Burbine, T., and Sunshine, J. (1996). Spectral Properties of Near-Earth Asteroids: Evidence for Sources of Ordinary Chondrite Meteorites. *Science* **273**, 946-948.

Bottke, W., Richardson, D., Michel, P., and Love, S. 1999. 1620 Geographos and 433 Eros: Shaped by Planetary Tides? *Astron. J.* **117**, 1921-1928.

Brooke, T., and Knacke, R. (1986). The Nucleus of Comet P/Arend-Rigaux. *Icarus* **67**, 80-87.

Chen, J., and Jewitt, D. (1994). On the Rate at Which Comets Split. *Icarus* **108**, 265-271.

Cruikshank, D., et al. (1998). The Composition of Centaur 5145 Pholus. *Icarus* **135**, 389-407.

Davies, J.K., McBride, N., Ellison, S., Green, S.F., and Ballantyne, D. R. 1998. Visible and infrared photometry of six Centaurs. *Icarus* **134**, 213-227.

Dohnanyi, J. 1969. Collisional Models of Asteroids and their Debris. *J. Geophys. Res.* **74**, 2531-2554.

Duncan M, Levison H. 1998. A disk of scattered icy objects and the origin of Jupiter-family comets. *Science* **276** 1670-1672.

Emerich, C., et al. (1987). Temperature and Size of the Nucleus of Comet P/Halley from IKS Infrared Vega 1 Measurements. *Astron. Ap.* **187**, 839-842.

Evans, W. and Tabachnik, S. 1999. Possible long-lived asteroid belts in the inner Solar System. *Nature* **399**, 41.

Farinella, P., Vokrouhlicky, D. and Hartmann, W. (1998). Meteorite Delivery via Yarkovsky Orbital Drift. *Icarus* **132**, 378-387.

Fernández J.A., Tancredi G., Rickman H., Licandro J. (1999). The population, physical properties, and sizes of Jupiter family comets. *Astron. Ap.* **352**, 327-340.

Fernandez, Y. (1999). Physical Properties of Cometary Nuclei. PhD Thesis, University of Maryland.

Fernandez, Y., McFadden, L., Lisse, C., Helin, E., and Chamberlin, A. (1997). Analysis of POSS Images of Comet-Asteroid Transition Object 107P/1949 W1 (Wilson-Harrington). *Icarus* **128**, 114-126.

Foster, M., Green, S., McBride, N., and Davies, J. (1999). Detection of Water Ice on 2060 Chiron. *Icarus* **141**, 408-410.

Gladman, B. et al. (1997). Dynamical Lifetimes of Objects Injected into Asteroid Belt Resonances. *Science* **277**, 197-201.

Hanner, M., et al. (1985). Infrared Spectrophotometry of Comet IRAS-Araki-Alcock (1983d). *Icarus* **62**, 97-109.

Harris, A. (1994). Tumbling Asteroids. *Icarus* **107**, 209.

Harris, N., and Bailey, M. (1998). Dynamical Evolution of Cometary Asteroids. *MNRAS* **297**, 1227-1236.

Hicks, M., Fink, U., and Grundy, W. (1998). The Unusual Spectra of 15 Near-Earth Asteroids and Extinct Comet Candidates. *Icarus* **133**, 69-78.

Jewitt, D. and Meech, K. (1988). Optical Properties of Cometary Nuclei and a Preliminary Comparison with Asteroids, *Ap. J.* **328**, 974-986.

Jewitt, D. (1996). From Comets to Asteroids: When Hairy Stars Go Bald. *Earth, Moon and Planets* **72**, 185-201.

Jewitt, D. and Luu J. 1998. Optical-infrared spectral diversity in the Kuiper belt. *Astron. J.* **115**, 1667-70.

Jewitt, D. and Luu J. 2000. Physical Nature of the Kuiper Belt. In *Protostars and Planets IV*, eds. V. Mannings, A. Boss and S. Russell, Univ. Az. Press, Tucson, pp. 1201-1230.

Johnson, R. (1991). Irradiation Effects in a Comets Outer Layers. *J. Geophys. Res.* **96**, 17553-17557.

Keller, H.U. et al. (1987). Comet Halley's Nucleus and its Activity, *Astron. Ap.* **187**, 807-823.

Keller, H.U., Knollenberg, J., and Markiewicz, W.J. (1994). Collimation of Cometary Dust Jets and Filaments. *Planet. Space Sci.* **42**, 367-382

Kührt, E., and Keller, H.U. (1994). The Formation of Cometary Surface Crusts. *Icarus* **109**, 121-132.

Lamy, Ph., Toth, I., A'Hearn, M. and Weaver, H. (1999). Hubble Space Telescope Observations of the Nucleus of Comet 45P/Honda-Mrkos-Pajdušáková and Its Inner Coma. *Icarus* **140**, 424-438.

Levison, H.F., and Duncan, M. (1994). The Long-term Dynamical Behavior of Short-period Comets. *Icarus* **108**, 18-36.

Levison, H.F., Shoemaker, E.M. and Shoemaker, C.S. 1997. The Dispersal of the Trojan Asteroid Swarm. *Nature* **385**, 42-44.

Lisse, C., et al. (1999). The Nucleus of Comet Hyakutake (C/1996 B2). *Icarus* **140**, 189-204

Luu, J. and Jewitt, D. (1992). High Resolution Surface Brightness Profiles of Near-Earth Asteroids. *Icarus* **97**, 276-287.

Luu, J. (1993). Spectral Diversity Among the Nuclei of Comets. *Icarus* **104**, 138-148.

Luu, J. (1994). Comets Disguised as Asteroids. *P.A.S.P.* **106**, 425-435.

Luu, J. and Jewitt, D. (1996). Color Diversity Among the Centaurs and Kuiper Belt Objects, *Astron. J.* **112**, 2310-2318

Luu, J., Jewitt, D., and Cloutis, E. (1994). Near-infrared Spectroscopy of Primitive Solar System Objects. *Icarus* **109**, 133-144.

Luu, J., Jewitt, D., and Trujillo, C. (2000). Water Ice in 2060 Chiron and Its Implications for Centaurs and Kuiper-Belt Objects. *Ap. J. Lett.* **531**, L151.

Marzari, F., Farinella, P., Davis, D.R., Scholl, H. and Campo Bagatin, A. 1997. Collisional Evolution of Trojan Asteroids. *Icarus* **125**, 39-49.

McFadden, L., Tholen, D., and Veeder, G. (1989). Physical Properties of Aten, Apollo and Amor Asteroids. In *Asteroids II*, eds. R. Binzel, T. Gehrels and M. Matthews, Univ. Az. Press, Tucson. pp. 442-467.

McFadden, L. (1994). The Comet-Asteroid Transition: Recent Telescopic Observations. In *Asteroids, Comets, Meteors 1993*, IAU Symp. 160, eds. A, Milani, M Di Martino, and A. Cellino, Kluwer Academic Publ., Dordrecht, pp. 95-110.

Morbidelli A. (1997). Chaotic diffusion, and the origin of comets from the 3/2 resonance in the Kuiper belt. *Icarus* **127**, 1-12.

Morbidelli, A., and Gladman, B. (1998). Orbital and Temporal Distributions of Meteorites Originating in the Asteroid Belt. *Meteoritics and Planetary Science* **33**, 999-1016.

Morbidelli, A. (1999). Origin and Evolution of Near Earth Asteroids. *Celestial Mechanics and Dynamical Astronomy* **73**, 39-50.

Moroz, L, Arnold, G, Korochantsev, A, Wäsch, R. 1998. Natural solid bitumens as possible analogs for cometary and asteroid organics: 1. Reflectance spectroscopy of pure Bitumens. *Icarus* **134**, 253-268.

Ostro, S. et al. (1996). Radar Observations of Asteroid 1620 Geographos. *Icarus* **121**, 46

Pravec, P., Hergenrother, C., Whiteley, R., Sarounova, L., Kusnirak, P., and Wolf. M.

(2000). Fast Rotating Asteroids 1999 TY2, 1999 SF10 and 1998 WB2. *Icarus*, submitted.

Rabinowitz, D. (1993). The Size Distribution of the Earth-Approaching Asteroids. *Ap. J.* **407**, 412-427.

Rabinowitz, D. (1998). Size & Orbit Dependent Trends in the Reflectance Colors of Earth-Approaching Asteroids. *Icarus* **134**, 342-346.

Rabinowitz, D., Helin, E., Lawrence, K., and Pravdo, S. 2000. A Reduced Estimate of the Number of Kilometer-Sized Near-Earth Asteroids. *Nature* **403**, 165-166.

Rickman, H., Fernández, J.A., and Gustafson, B.Å.S. (1990). Formation of Stable Dust Mantles on Short Period Comet Nuclei, *Astron. Ap.* **237**, 524-535.

Senay, M., and Jewitt, D. (1994). Activity in a Distant Comet: First Detection of Carbon Monoxide, *Nature* **371**, 229-231.

Shoemaker, G., Williams, J., Helin, E., and Wolfe, R. (1979). Earth-Crossing Asteroids. In *Asteroids*, ed. T. Gehrels, Univ. Arizona Press, Tucson. pp. 253-282.

Solem, J., and Hills, J. 1996. Shaping of Earth-Crossing Asteroids by Tidal Forces. *Astron. J.* **111**, 1382-1387.

Tegler, S.C., and Romanishin, W. 1998. Two distinct populations of Kuiper Belt objects. *Nature* **392**, 49-51.

Veeder, G., et al. (1989). Radiometry of Near-Earth Asteroids. *Astron. J.* **97**, 1211-1219.

Weissman, P., A'Hearn, M., McFadden, L., and Rickman, H. (1989). Evolution of Comets into Asteroids. In *Asteroids II*, eds. R. Binzel, T. Gehrels and M. Matthews, Univ. Az. Press, Tucson. pp. 880-920.

Wisdom, J. (1983). Chaotic Behavior and the Origin of the 3:1 Kirkwood Gap. *Icarus* **56**, 51-74.

Womack, M., and Stern, S.A. (1999). The Detection of Carbon Monoxide Gas Emission in (2060) Chiron. *Astronomicheskii Vestnik* **33**, 187.

Zellner, B., Tholen, D.J., and Tedesco, E.F. (1985). The Eight-Color Asteroid Survey – Results for 589 Minor Planets. *Icarus* **61**, 355-416.

DISTANT COMET OBSERVATIONS

K.J. MEECH
Institute for Astronomy
2680 Woodlawn Drive, Honolulu, HI 96822 USA

AND

O.R. HAINAUT
European Southern Observatory
Casilla 19001, Santiago 19, Chile

Abstract. This paper summarizes the motivation for observing distant comets: they can provide information about the dynamics, collisions, physical and chemical conditions in the early solar nebula. The challenges and techniques of observing faint moving objects at large distances and searching for activity will be discussed. Our state of knowledge about distant comet size distributions, evidence for activity and spectral diversity is presented.

1. Introduction

The observation of distant comets is extremely challenging, but of fundamental importance to the understanding of the formation and evolution of the Solar System. A large body of comet knowledge has been amassed from observing short- and long-period comets at perihelion, using imaging and spectroscopic techniques to get information on production rates of daughter species, dust-to-gas ratios and information on the comae and tails of comets. More recently, observations in the sub-mm wavelength regime have opened up a new understanding about parent molecules.

New sophisticated techniques are emerging to infer nucleus properties and primordial composition for comets at perihelion when they are surrounded by a coma partially hiding the nucleus. However, few of these techniques have been verified by direct observations. There is fundamental information about comets and their relation to the early Solar System

163

M. Ya. Marov and H. Rickman (eds.), Collisional Processes in the Solar System, 163–172.
© 2001 *Kluwer Academic Publishers. Printed in the Netherlands.*

which can only be obtained through observations of distant comets. This chapter will highlight these observations, discuss the observational challenges of distant comet observations and discuss some of the observational results.

2. Importance of Observing Distant Comets

2.1. PLANETESIMAL FORMATION REGION

The short-period (SP) comets have short dynamical lifetimes. Therefore, their population must be replenished in order to maintain a steady-state. Recent dynamical models suggest that the SP comets have low-inclination sources in the Edgeworth-Kuiper region beyond Neptune (Duncan *et al.*, 1988), but that the long-period (LP) and dynamically new (DN) comets formed in the Uranus-Neptune zone and were perturbed outwards (Fernández & Ip, 1981), where they are stored in a vast reservoir, the Oort cloud, before being perturbed toward the inner Solar-System as LP/DN comets. The planetesimals grew by collisional coagulation until they were big enough (10–100m) to decouple from the turbulence between the particle layer and the gas (Weidenschilling, 1997). Many of the collisions were caused by gas drag-induced differential radial velocities, thus a single comet nucleus may have incorporated planetesimals from different heliocentric distances. The dynamical models give information about the formation locations of the different comet classes, and observations of different comet dynamical classes will provide information about the physical and chemical conditions in the solar nebula at different heliocentric distances.

2.2. COLLISIONAL ACCRETION AND SIZE DISTRIBUTION

Collisions in the early trans-Neptunian region probably altered the size distributions of the planetesimals (Farinella & Davis, 1996; Davis & Farinella, 1997), and helped transport the SP comets into the inner Solar System. The Centaurs probably represent the transition objects between the Edgeworth-Kuiper belt objects and SP comet populations (Stern & Campins, 1996). While we cannot directly observe this era of Solar System formation, we can observe nucleus size distributions and comet nuclei sub-structure size scales (through splitting, and compositional inhomogeneities).

2.3. FORMATION CONDITIONS AND DISTANT ACTIVITY

As the solar nebula collapsed, interstellar grains settled to the midplane. They may have undergone processing of their icy mantles as they fell (Lunine *et al.*, 1991), for instance shock-induced sublimation and volatile recondensation. In the outer part of the nebula temperatures were below 100K, thus the H_2O-ice would have condensed as amorphous ice. Amorphous ice has the ability to trap gases as high as 3.3 times the amount of the water-ice (Laufer *et al.*, 1987). Between 50–125K, these gases are released in distinct temperature regimes in response to the restructuring of the water-ice (annealing). Beginning near 120K and peaking at 137K, the ice undergoes an exothermic amorphous to crystalline phase transition. For specific aspects on this gas trapping, see Owen and Bar-Nun (this volume).

The amount of gas which can be trapped is a very strong function of the condensation temperature (see Fig. 1). The trapped gases in the planetesimals can thus serve as an extremely sensitive cosmic thermometer for their formation locations. The "thermometer" is read as the temperature (*i.e.* heliocentric distance, r) at which activity is present.

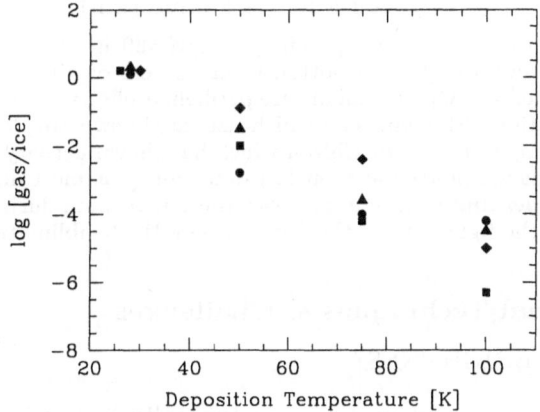

Figure 1. Total amounts of trapped gases in water ice versus deposition for $\Diamond$ CH_4; $\triangle$ CO; $\bigcirc$ N_2; and $\square$ Ar. Figure after (Bar-Nun & Kleinfeld, 1989).

2.4. ACTIVITY & EVOLUTION

Water-ice sublimation begins near 180K. Low albedo (*e.g.* a few percent) objects can reach this equilibrium temperature near $r=5$–6AU, thus most comet observations have concentrated on nuclei when they were inside this distance, and active. However, the activity driven by trapped gases in the amorphous ice, by the presence of highly volatile ices, or caused by the amorphous-to-crystalline ice phase transition will occur at much lower temperatures, and much larger distances. Furthermore, it is expected that comets will age (evolve physically and chemically) with repeated close perihelion passages as they lose volatiles. Thus, understanding and comparing primordial comet chemistry requires observations at over a wide range of distances, and observations at large r.

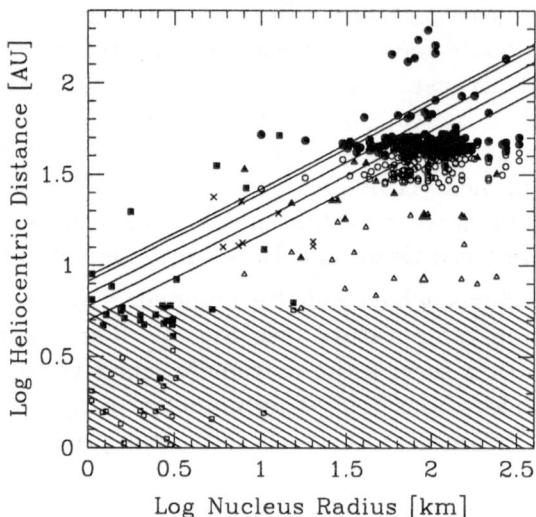

Figure 2. Plot of the curves corresponding to S/N=30 in 1h for telescope apertures 10m, 8m, 4m, 2m, and 1m (top to bottom), for comets of given radii and distances, assuming an albedo of 4%. Open symbols are perihelion observations, and filled symbols are aphelion observations. SP comets for which sizes are known are plotted as □, Centaurs as △ (the larger triangles represent Chiron which has shown activity), and EKO objects as ○. The LP comets are plotted at their last detection (*e.g.* most distant) as × and the nucleus radii are upper limits since in all cases there was coma during the observations. The shaded area of the figure shows the region where H_2O-sublimation is possible.

3. Observational Techniques & Challenges

3.1. EXPECTED BRIGHTNESS

The known SP comet nucleus sizes are typically in the few-km range, and measured albedos suggest that comets are very dark, near 4% (see Meech, 2000 for a review). The expected brightness of the nucleus depends on the solar flux scattered from the nucleus, which is a function of its heliocentric and geocentric (Δ) distances (in AU), the phase angle, α, the nucleus size and its scattering properties (Russell, 1916):

$$p_\lambda R_N^2 \phi(\alpha) = 2.24 \times 10^{22} r^2 \Delta^2 10^{+0.4(m_\odot - m_\lambda)} \tag{1}$$

where p_λ is the geometric albedo, $\phi(\alpha)$ is the phase function, and R_N the radius. Here m_λ is the observed magnitude, and $m_\odot$ is the solar magnitude (in the V filter $m_{\odot V} = -26.74$). The phase function is given by Eqn. 2, where $\Delta m = 0.035\,\alpha$ (Meech & Jewitt, 1987).

$$\phi(\alpha) = 10^{-0.4\Delta m} \tag{2}$$

By computing the magnitude for which a telescope can achieve a good detection (S/N ≈ 30, needed for physical studies) in 1 hour, the above equations can be used to determine the maximum r at which comet nuclei of various sizes can be detected. Figure 2 shows the result of such a calculation. It is clear from the figure that in order to observe comet nuclei as small as a few km in size in the outer Solar System requires long integration times with large apertures (see discussion in §4). Some of the known objects (top of the graph) are so faint that they cannot be observed with S/N~30, even on the largest telescopes.

3.2. OBSERVATIONS & IMAGE PROCESSING

When the comets are faint and total integration times are long, data are taken in an optimal manner if the exposure time of individual images is kept short enough that the stars will not trail by more than a small fraction of the seeing disk. This allows for high signal-to-noise detection of the nucleus after the images are combined in a composite, as well as a sensitive search for any faint coma. The telescope should be dithered between each exposure (*e.g.* offset by up to $20''-30''$ in both R.A. and Dec.) so that field stars and galaxies do not fall on the same part of the CCD in each frame. It is therefore possible to combine the images using a median filter to create a dark-sky flat to apply as a correction after the standard flat-fielding has been done using bright twilight sky flats or dome flats. The latter have very high S/N, but do not accurately represent the response of the CCD to the dark sky. When searching for very faint coma, it is critical to get the CCD flat to better than 0.01%, across the chip and this is only possible using auto-flat fielding techniques. This technique is discussed in some detail in Tyson (1990) and Hainaut *et al.* (1994). The flattening can be further improved using an optimized combination of dome, twighlight sky and night sky flats (Hainaut *et al.*, 1998).

3.3. PHOTOMETRIC PROFILES & COMETARY ACTIVITY

To the first order, the frames are then combined (*i*) using the relative offsets between the frames caused by the dithering and guiding errors – creating a composite image with well-guided stars, and (*ii*) in addition using the comet rates to create a composite image guided at the rate of the comet's motion. The latter requires the image plate scale, orientation, and knowledge of the comet ephemeris rates.

In an uncrowded field, this technique produces two composite images from which surface brightness profiles of both the stars and comet nucleus are computed to search for very faint coma when no activity is directly visible. A comparison of the normalized surface brightness profile of the comet and the averaged profile of several field stars in the untrailed image can be used to place extremely sensitive limits on the presence of any coma or activity (Meech & Weaver, 1996). The surface brightness profile difference between the comet and comparison stars gives an upper limit to the flux, F (Eqn. 3), contributed by a dust coma, where $S_\odot$ is the solar flux through the bandpass [W m^{-2}], ϕ the projected size of the aperture [m], a_{gr} [m] the grain radius, p_v the grain albedo, Q [kg s^{-1}] is the dust production rate, v_{gr} [m s^{-1}] the grain velocity, and r is in AU and Δ in m.

$$F = \left[\frac{S_\odot \pi \phi}{2r^2\Delta^2}\right]\left[\frac{a_{gr}^2 p_v}{v_{gr}}\right] Q \qquad (3)$$

However, usually in long sequences of exposures, the comet will have moved over background objects. Techniques of point spread function fitting may be used to effectively remove these objects from the individual frames before combining the images – but only for point sources. A large fraction of the very faint background objects will be galaxies which cannot be easily removed in this manner. Hainaut et al. (1994) discuss removal of non-stellar background objects by making artificially trailed comet and "anti-comet" frames. This paper also rigorously discusses the increase in noise expected from the object removal process.

4. Results

4.1. SIZE DISTRIBUTIONS

There are only a handful of sizes which are "directly" measured for any of the comets using the traditional techniques of radiometry. However, even this technique is model dependent, because it relies on assumptions about the optical and thermal phase function of the nucleus (Meech et al., 2000a). Most nucleus and EKO size estimates are now made on the basis of the total brightness and by assuming an albedo of 4%. All available size estimates are shown in Fig. 3. There appears to be a clear difference in size distributions between the different dynamical classes, however there are many unaccounted for observational biases. All of the observations are affected by selection effects, but this is particularly severe for the EKOs.

The cumulative luminosity function, or number of EKOs per unit area brighter than a limiting magnitude is measured for the magnitude range 20–26 (Jewitt, 1999), and provides important information about the comet size distribution and total number of EKOs. The luminosity function may be fit by a power law size distribution, which has an index between 4.0 $< \beta < 4.8$, which is probably steeper than the $\beta = 3.5$ expected from a collisionally evolved distribution for objects with radii greater than 50km. With better knowledge of the SP comet size distribution, it will be possible to obtain dynamical models for injection of SP comets into the Solar System

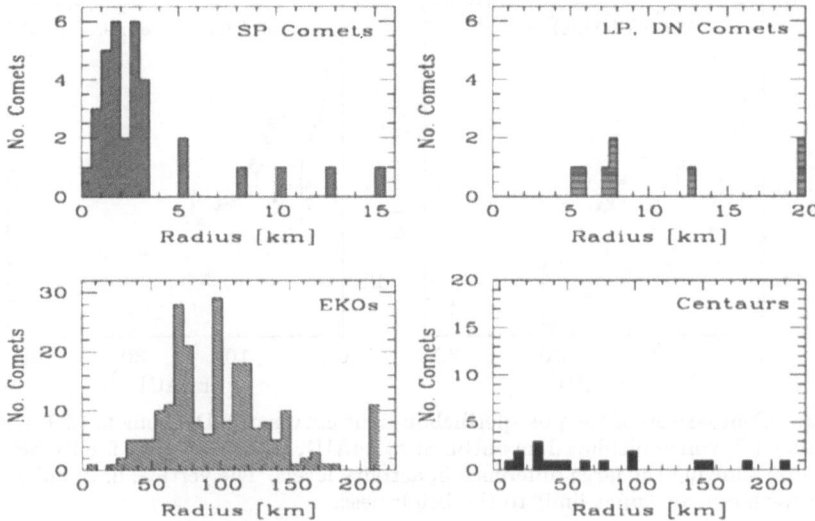

Figure 3.   Observed size distributions of the 4 main classes of outer Solar System objects

with the estimates of the numbers of small EKOs based on the EKO size distributions (Jewitt, 1999).

4.2. ACTIVITY

Observations over the last decade have shown that cometary activity in LP comets extends well beyond 10 AU (Meech, 1999), and that at least one Centaur, Chiron, has extensive activity at large r (Meech & Belton, 1990). It is worth noting that, in some cases, cometary activity was detected at extreme distance, such as in comet 1987H1 (Shoemaker) between 18-20 AU (Meech *et al.* 1996; Meech 2000).

Unsuccessful searches for activity in other Centaurs have been made at different sensitivity levels. There have been intriguing reports of possible activity in EKO objects. At the 1998 ESO workshop on Minor Bodies in the Outer Solar System, Fletcher *et al.* (1998) presented HST observations of EKO 1994TB which showed a 2-σ difference in the surface brightness profile of the EKO and a PSF-star between 0.2″–0.5″ at the 26 mag arcsec^{-2} level. They interpreted the observation as a possible dust coma. The results were preliminary since telescope jitter had not been accounted for. This would have been more important for the EKO which had longer exposure time (by a factor of seven). Subsequent observations of this object from Keck have not revealed any coma in 0.4″ seeing (Meech *et al.*, 2000b).

Observations of the rotational light curve of EKO 1996TO$_{66}$ showed a change in the shape of the lightcurve between Aug.-Oct. 1997 and Sep. 1998 (Hainaut *et al.*, 2000). A proposed cause of the change is that 1996TO$_{66}$

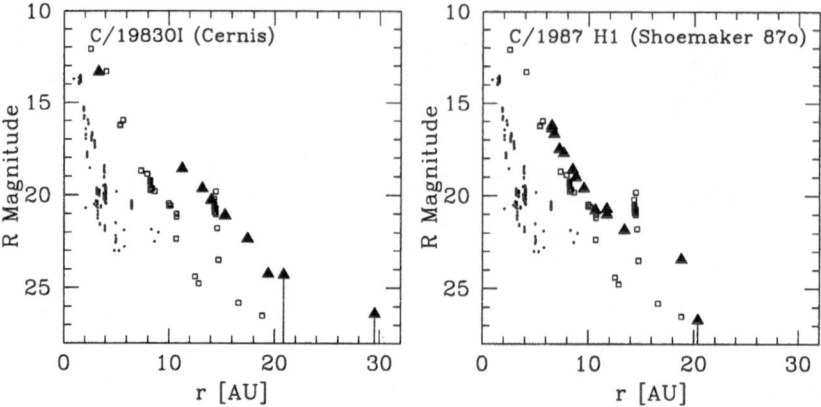

Figure 4. Comparison of the post-perihelion light curves of 2 DN comets (△), with that of P/Halley (□, which displayed an outburst at 14AU), and the Jupiter family SP comets (small dots) showing the large difference in activity levels. The vertical lines indicate that the observation is an upper limit to the brightness.

experienced a phase of cometary activity during the interval between observations. Other possible causes of the change due to complex rotation, or a collision were examined and determined to be highly unlikely. Instead, the lightcurve was easily reproduced by introducing a high albedo spot (caused by deposition of fresh material) on the surface of a bi-axial ellipsoid.

4.3. EVOLUTION & AGING

Long-term observations of distant comets over a range of r from near perihelion out to $r=30$ AU shows a striking difference in the activity level between LP, DN and the SP comets. Figure 4 shows a comparison of the fading or decrease in activity post-perihelion for two DN comets, P/Halley and a group of SP comets. The DN and LP comets fade much more slowly than do the SP comets and comet P/Halley, which implies a different underlying cause of the activity (Meech, 1999).

4.4. COLORS & SPECTRA

Although limited in number, multi-wavelength photometric measurements of comet nuclei in the optical and the near-infrared, for both the SP comets and the EKO objects show a wide range in color which suggests a diversity of surface types. Tegler and Romanishin (1998) have claimed that there is a division of the EKO objects by optical color into 2 distinct groups. However, other observers do not see a color separation, and this observation awaits confirmation (Barucci *et al.*, 1999). Diversity in colors and surface composition may arise from irradiation and subsequent collisional resurfacing or activity (*e.g.* aging effects), or may result from compositional differences related to different formation locations. Distinguishing between these possible mechanisms will require a larger statistical sample of accurate colors so that trends may be searched for as a function of dynamical age, r, and probable formation location.

Three EKOs have been observed spectrometrically: 1993SC, in the visible (Luu and Jewitt, 1996) and near-IR (Brown *et al.*, 1997); 1996 TL_{66} in the visible and near-IR (Luu and Jewitt, 1998) and 1996 TO_{66} in the near-IR showing the signature of H_2O-ice (Brown *et al.*, 1999). With the growing availability of large telescopes, it is expected that this technique will soon be able to constrain the surface composition of these minor bodies.

5. Conclusions

With the discovery of the Edgeworth-Kuiper Belt, there has been a tremendous push in the planetary community for large telescope time in order to make physical observations of minor bodies in the distant outer Solar System. The observations are challenging owing to the small sizes and low albedos of the objects, as well as to the extremely low surface brightnesses of the comae of comets at large r. While critical fundamental information about the chemistry and physics of comets continues to be obtained while they are at perihelion, there is an urgent need for observation of comets at large r in order to understand their primordial composition, the evolutionary effects they have undergone since formation (both physical and dynamical), and to understand the early Solar System formation processes and the interrelation between the formation of the cometesimals and the giant planets. Some distant observations are necessary because the objects do not venture into the inner Solar System, while other observations at large r are necessary because of the information they can give about volatile processes in the nucleus, and because it is a more direct means of getting information about the nucleus.

Acknowledgements We would like to acknowledge support from NASA Grant No. NAGW 5015 for this work.

References

Bar-Nun, A and I. Kleinfeld (1989) On the Temperature and Gas Composition in the Region of Comet Formation, *Icarus* **80**, 243-253

Barucci, M. A., J. Romon, A. Le Bras, M. Fulchignoni and D. Tholen (1999) Broad Band Optical Colors of Trans-Neptunian Objects, *BAAS* **31**, 23.04

Brown, H., P. D. Cruikshank, Y. Pendleton and G. J Weeder (1997) Surface composition of Kuiper Belt Object 1993SC, *Science* **276**, 397

Brown, R. H., D. P. Cruikshank and Y. Pendleton (1999) Water Ice on Kuiper Belt Object 1996 TO_{66}, *ApJ* **519**, L101-L104

Duncan, M. F., T. Quinn and S. Tremaine (1988) The Origin of Short-Period Comets, *ApJ* **328**, L69-L73

Davis, D. R. and P. Farinella (1997) Collisional Evolution of Edgeworth-Kuiper Belt Objects, *Icarus* **125**, 50-60

Farinella, P. and D. R. Davis (1996) Short-Period Comets: Primordial Bodies, or Collisional Fragments?, *Science* **273**, 938-941

Fernández, J. A. and W.-H. Ip (1981) Dynamical Evolution of a Cometary Swarm in the Outer Planetary Region, *Icarus* **47**, 470-479

Fletcher, E. *et al.* (1998) HST Observations of the Kuiper Belt Presented at the *ESO Workshop on Minor Bodies in the Outer Solar System*

Hainaut, O. R., C. E. Delahodde, H. Boehnhardt, E. Dotto, M. A. Barucci, K. J. Meech, J. M. Bauer, R. M. West and A. Doressoundiram (2000) Physical Properties of TNO 1996TO$_{66}$, *Astron. Astrophys.*, in press.

Hainaut, O., R. M. West, A. Smette, B. G. Marsden (1994) Imaging of Very Distant Comets: Current and Future Limits, *Astron. Astrophys.* **289**, 311-324

Hainaut, O. R., K. J. Meech, H. Boehnhardt and R. M. West (1998) Early Recovery of Comet 55P/Tempel-Tuttle, *Astron. Astrophys.* **333**, 746-752

Jewitt, D. (1999) Kuiper Belt Objects, *Ann. Rev. Earth, Planet. Sci.* **27**, 278-312

Laufer, D., E. Kochavi and A. Bar-Nun (1987) Structure and Dynamics of Amorphous Water Ice, *Phys. Rev B* **36**, 9219-9227

Lunine, J. I, S. Engel, B. Rizk and M. Horanyi (1991) Sublimation and Reformation of Icy Grains in the Primative Solar Nebula, *Icarus* **94**, 333-344

Luu, J. X. and D. J. Jewitt (1996), Reflection Spectrum Spectrum of Kuiper Belt Object 1993SC, *AJ* **111**, 499-503

Luu, J. X and D. J. Jewitt (1998),Optical and Infrared Reflectance Spectrum of Kuiper Belt Object 1996 TL$_{66}$, *ApJ* **494**, L117-L129

Meech, K. J. (1999) Chemical and Physical Aging of Comets, in Proc. of *IAU Colloq. 173*, pp. 195-210

Meech, K. J. (2000) Physical Properties of Cometary Nuclei, in Proc. of *ACM 1996*, in press

Meech, K. J. and M. J. S. Belton (1990) The Atmosphere of 2060 Chiron, *AJ* **100**, 1323-1338

Meech, K. J., O. R. Hainaut, J. Bauer (1996) Distant Comet Imaging With the Keck and the HST, *BAAS* **28**, 8.09

Meech, K. J., O. R. Hainaut, and B. G. Marsden (2000a) Comet Size Distributions and Distant Activity, in Proc. of *Minor Bodies in the Outer Solar System Meeting*, ESO Nov 1988; in press

Meech, K. J., O. R. Hainaut, and B. G. Marsden (2000b) Search for Coma Around EKO 1994TB, in preparation

Meech, K. J. and D. C. Jewitt (1987) Observations of Comet P/Halley at Minimum Phase Angle, *Astron. Astrophys.* **187**, 585-593

Meech, K. J. and H. A. Weaver (1996) Unusual Comets (?) as Observed from the Hubble Space Telescope, *Earth Moon Plan.* **72**, 119-132

Russell, H. N. (1916) On the Albedo of the Planets and Their Satellites, *ApJ* **43**, 173-195

Stern, A. and H. Campins (1996) Chiron and the Centaurs: Escapees from the Kuiper Belt, *Nature* **382**, 507-510

Tegler, S. C. and W. Romanishin (1998) Two Distinct Populations of Kuiper-Belt Objects, *Nature* **392**, 49-51

Tyson, J. A. (1990) The Shift-and-Stare Technique and a Large Area CCD Mosaic, *ASP Conf. Ser.* **8**, pp. 1-10

Weidenschilling, S. J. (1997) The Origin of Comets in the Solar Nebula, A Unified Model, *Icarus* **127**, 290-306

LONG-TERM ORBITAL EVOLUTION OF PROTOPLANETS

KEIKO YOSHINAGA

Department of Astronomy, School of Science, University of Tokyo

AND

EIICHIRO KOKUBO AND JUNICHIRO MAKINO

Department of General Systems Studies, College of Arts and Sciences, University of Tokyo,
3-8-1 Komaba, Meguro-ku, Tokyo, 153-8902 Japan

Abstract. We investigated the stability of systems of ten protoplanets using three-dimensional N-body simulations. We found that the timescale of instability T depends strongly on the initial random velocities v (eccentricities e and inclinations i) and orbital separations Δa of protoplanets. For $v = 0$, we confirmed the result of Chambers et al. (1996) that T is proportional to $\exp(\Delta a)$. For $v > 0$, we found that T depends strongly on the initial random velocities. The relation between T and Δa is still expressed as $\log T = b\Delta a + c$. However, both b and c depend on initial random velocities and the slope, b, decreases with v. Even for relatively small initial eccentricities such as $e \sim 2r_{\mathrm{H}}/a$, where r_{H} is the Hill radius, the timescale can be reduced by a factor of 10 compared with the case of the zero random velocity. Therefore, the timescale of the formation of inner planets might be much shorter than implied by Chambers et al..

1. Introduction

It is now widely accepted that terrestrial planets and cores of Jovian planets are formed through the accretion of many small bodies known as planetesimals (Safronov, 1969, Hayashi et al., 1985). The growth mode of planetesimals is called "runaway growth" (Greenberg et al., 1978; Wetherill and Stewart, 1989; Kokubo and Ida, 1996).

M. Ya. Marov and H. Rickman (eds.), Collisional Processes in the Solar System, 173–180.
© 2001 *Kluwer Academic Publishers. Printed in the Netherlands.*

There are very few studies of the planetary growth after the runaway stage. After the formation of protoplanets, the growth slows down because the velocity dispersion of the planetesimals around the protoplanets is increased due to scattering by the protoplanets (Ida and Makino, 1993). When runaway growth is halted in the terrestrial planet region, a typical protoplanet has a mass about $1/10$ of the present planets (Kokubo and Ida, 1998). At present, it is not at all understood how these protoplanets evolve into the present terrestrial planets. Chambers et $al.$ (1996) investigated the stability of systems of three or more protoplanets by numerical integration. The time of the first close encounter T is expressed approximately as $\log T = b\Delta a + c$, where b and c are constants, and Δa is the initial separation of the protoplanetary orbits in units of their Hill radius r_H. If we extrapolate their result to $\Delta a \sim 10 r_H$, which is a typical orbital separation of protoplanets after runaway growth, the instability timescale would be well beyond 10^8 years. In other words, it would have taken a very long time for terrestrial planets to be actually formed.

In this paper, we investigate the effect of the initial random velocities of protoplanets on the stability of a protoplanetary system by performing three-dimensional N-body simulations. We performed many simulations in which the orbits of protoplanets have initial eccentricities e and inclinations i. For coplanar, circular cases, our result agrees well with that of Chambers et $al.$ (1996). For cases with finite eccentricities and inclinations, we found that the timescale becomes shorter as the initial eccentricities and inclinations become larger. Even for relatively small initial eccentricities such as $e \sim 2 r_H/a$, the timescale can be reduced by a factor of 10. In other words, the timescale of the formation of inner planets might be much shorter than implied by Chambers et $al.$ (1996).

We describe the numerical method and initial conditions in Section 2. In Section 3 we show the results for systems of 10 protoplanets. Section 4 gives a summary. More details can be found in Yoshinaga et $al.$ (1999).

2. Method of Calculation

2.1. NUMERICAL METHOD AND INITIAL CONDITIONS

We used the 4th-order $P(EC)^n$ Hermite scheme (Makino and Aarseth, 1992; Kokubo et $al.$, 1998) with hierarchical timestep (Makino, 1991). We adopted a the maximum stepsize of $\Delta t_{max} = 2^{-6}$, where the initial orbital period of the innermost protoplanet is 2π. This scheme is effectively time-symmetric when applied to a nearly circular orbit, and therefore conserves important orbital elements, in particular, the semi-major axis and the eccentricity, up to the round off error. For details, see Kokubo et $al.$ (1998) and Yoshinaga et $al.$ (1999).

TABLE 1. Initial Condition Parameters

$\Delta a[r_{\rm H}]$	4	4.5	5	5.5	6	6.5	7	7.5	8	9	10
$\langle \tilde{e}^2 \rangle^{1/2}$	0	1	2	3	4						

We investigate the evolution of a system of 10 protoplanets. We set the initial masses of the protoplanets all equal ($m = 10^{-7}M_\odot$). This value is about 1/10 of the Earth mass. Table 1 lists the values of parameters we adopted for initial conditions, where Δa is the initial separation between neighbering protoplanets in unit of the mutual Hill radius $r_{\rm H}$,

$$r_{\rm H} = \left(\frac{2m}{3M_\odot} \right)^{\frac{1}{3}} a \qquad (1)$$

and $\langle \tilde{e}^2 \rangle^{1/2}$ is the RMS value of the reduced eccentricity $\tilde{e} = e/h$. Here, h is the reduced Hill radius defined as $h = r_{\rm H}/a$ (When $m = 10^{-7}M_\odot$, $h \simeq 4.055 \times 10^{-3}$). The Hill radius for outer pair of protoplanets is larger than that for an inner pair, so the orbital separations are not equal. Note that the random velocity v is related to the eccentricity and inclination as $v \sim \sqrt{(e^2 + i^2)}v_K$, where v_K is the Kepler velocity.

We assigned initial eccentricities and inclinations so that they obey the Rayleigh distribution with dispersions $\langle \tilde{e}^2 \rangle^{1/2}_{\rm ini} = 2\langle \tilde{i}^2 \rangle^{1/2}_{\rm ini}$ (Ida and Makino, 1992). Protoplanets are distributed at random phases. We calculated 40 cases with different random numbers of the Rayleigh distribution and with different starting random phases for each set of parameters.

2.2. CRITERION FOR INSTABILITY

We used the following criterion to determine the instability timescale. Let us define a_j as the semi-major axis of the protoplanet initially at the jth location counted from the innermost protoplanet. Initially, the condition $a_j < a_{j+1}$ ($j = 0, 1, \cdots 8$) is satisfied. We regard the system as unstable when the condition

$$a_j > a_{j+1} \qquad (2)$$

is satisfied for at least one value of j. We call $T_{\rm cross}$ the time at which this condition is satisfied. In addition, we define $T_{\rm collision}$ as the time when two protoplanets physically collide, in other words, when the distance of protoplanets i and j, r_{ij}, becomes smaller than the sum of the physical radii of two planets, $r_i + r_j$, where r_i, r_j are the radii of protoplanets i and j, respectively. We define the the instability timescale T as,

$$T = \min(T_{\text{cross}}, T_{\text{collision}}) \tag{3}$$

We compared our criterion with that of Chambers *et al.* (1996) and found that our criterion gave more reliable estimate for runs with non-circular initial orbits (Yoshinaga *et al.*, 1999).

3. Results

3.1. TIME EVOLUTION OF SEMI-MAJOR AXIS AND ECCENTRICITY

The time evolutions of the semi-major axes of all protoplanets are shown in Figure 1 for the cases, $\Delta a_{\text{ini}} = 8r_{\text{H}}$, $\langle \tilde{e}^2 \rangle_{\text{ini}}^{1/2} = 0$ or 2. Solid curves show the average values of a for 1000 years and broken lines show the values of $a(1 \pm e)$. The system with a finite $\langle \tilde{e}^2 \rangle_{\text{ini}}^{1/2}$ became unstable faster than the system with $\langle \tilde{e}^2 \rangle_{\text{ini}}^{1/2} = 0$.

3.2. TIMESCALE OF INSTABILITY

Figure 2 shows the instability timescale T defined in Sect. 2.2 against initial orbital separations. Since the results are sensitive to the initial conditions, we calculated 40 cases for the same set of initial condition parameters, Δa_{ini}, $\langle \tilde{e}^2 \rangle_{\text{ini}}^{1/2}$ and $\langle \tilde{i}^2 \rangle_{\text{ini}}^{1/2}$, and used the median of the results of 40 runs as the value of T. The thick straight line is the result of Chambers *et al.* (1996). Figure 2 suggests that the relation between T and Δa_{ini} is approximated pretty well by the relation,

$$\log T = b\Delta a_{\text{ini}} + c \tag{4}$$

Figure 3 shows the distribution of the resulting instability times for the cases with $\Delta a_{\text{ini}} = 8r_{\text{H}}$, $\langle \tilde{e}^2 \rangle_{\text{ini}}^{1/2} = 0, 2, 4$. The spread of instability times is about a factor 10^2. The spread is larger for larger $\langle \tilde{e}^2 \rangle_{\text{ini}}^{1/2}$.

Finally, we determined the values of the coefficients b and c, by a least-square-fit. We used the data for $\Delta a_{\text{ini}}/r_{\text{H}} = 4 - 7$ in the case of the zero random velocity, and that for $\Delta a_{\text{ini}}/r_{\text{H}} = 4 - 8$ in the non-zero random velocity cases. The results are given in Fig. 4. In the case of the zero initial random velocity, $\langle \tilde{e}^2 \rangle_{\text{ini}}^{1/2} = 0$, our result ($b = 0.84 \pm 0.058$, $c = -0.79 \pm 0.33$) is in good agreement with that of Chambers *et al.* (1996) ($b = 0.756 \pm 0.027$, $c = -0.358 \pm 0.176$). In the cases of non-zero initial random velocity, T seems to decrease linearly with $\langle \tilde{e}^2 \rangle_{\text{ini}}^{1/2}$ within the studied range of $\langle \tilde{e}^2 \rangle_{\text{ini}}^{1/2}$.

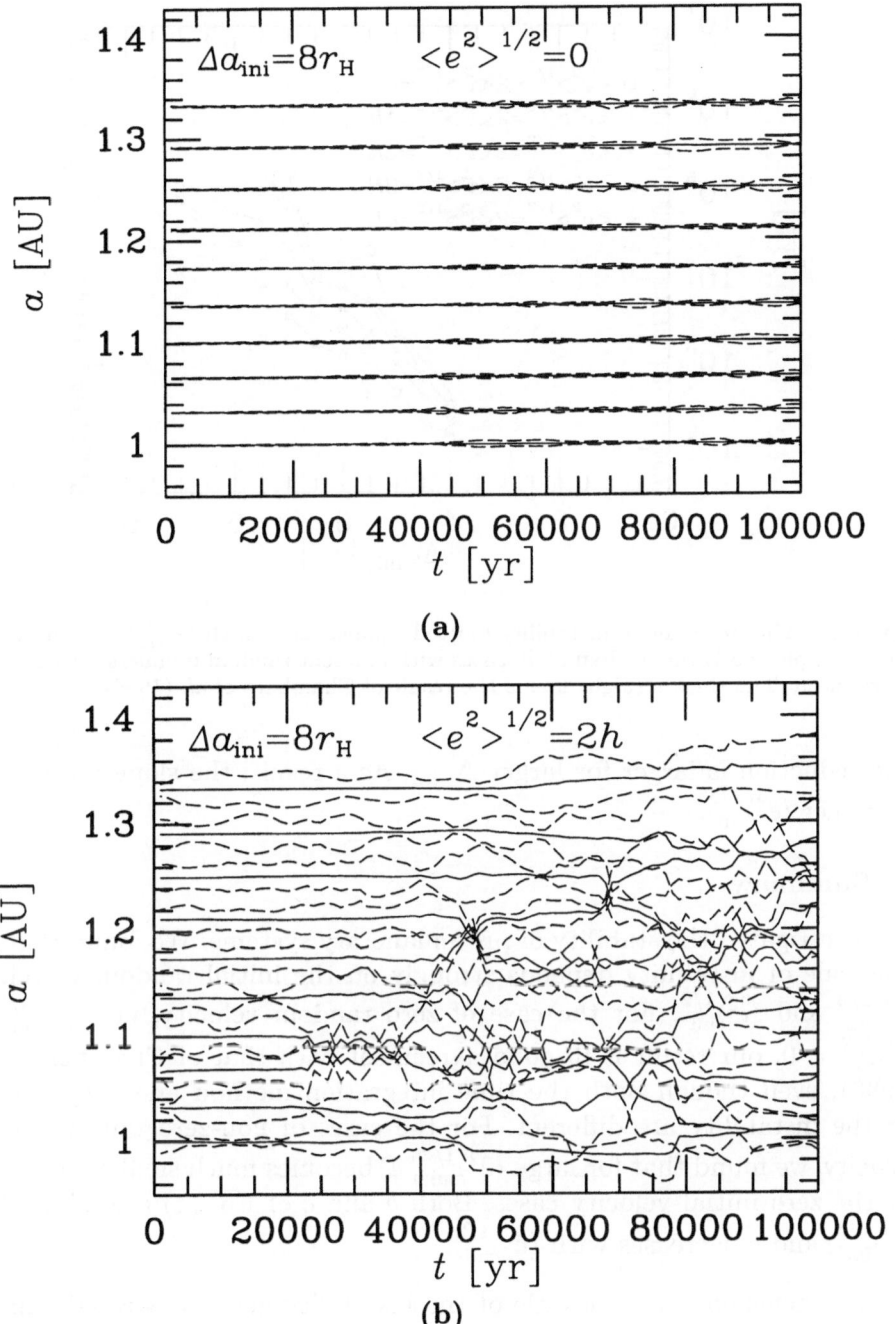

Figure 1.   The time evolution of semi-major axes of 10 protoplanets for the case $\Delta a_{\mathrm{ini}} = 8 r_{\mathrm{H}}$ and (a) $\langle \tilde{e}^2 \rangle_{\mathrm{ini}}^{1/2} = 0$; (b) $\langle \tilde{e}^2 \rangle_{\mathrm{ini}}^{1/2} = 2$. Solid curves show the average values a for 1000 years, and broken lines show the values of $a(1 \pm e)$.

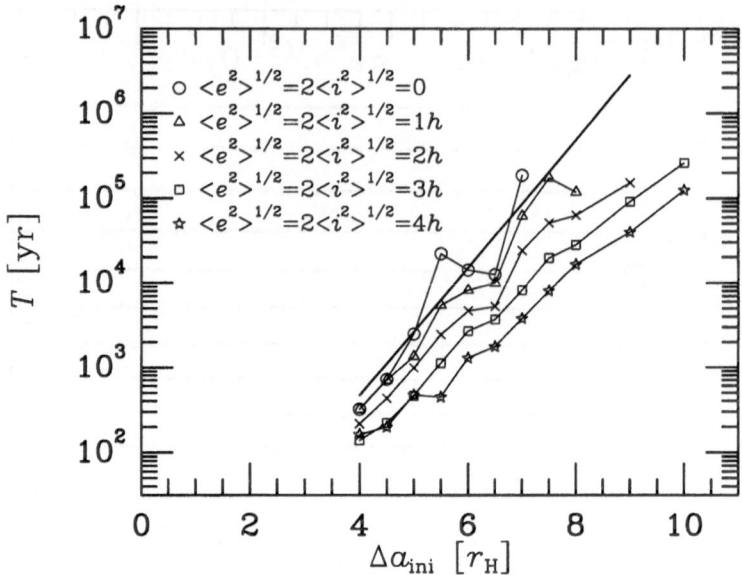

Figure 2.   The timescale of instability plotted against Δa_{ini} with $\langle \tilde{e}^2 \rangle_{\mathrm{ini}}^{1/2} = 0$ through 4. The value plotted is the median of 40 cases with different random numbers for the initial distribution. The thick straight line is the result of Chambers *et al.* (1996).

The reduction is larger for larger Δa_{ini}. As a result, the slope b is smaller for larger $\langle \tilde{e}^2 \rangle_{\mathrm{ini}}^{1/2}$.

4. Summary

We investigated the stability of protoplanetary systems. We found that the timescale of instability depends strongly on the initial random velocities, $\langle \tilde{e}^2 \rangle_{\mathrm{ini}}^{1/2}$ and $\langle \tilde{i}^2 \rangle_{\mathrm{ini}}^{1/2}$. For the case of zero random velocity, i.e., $\langle \tilde{e}^2 \rangle_{\mathrm{ini}}^{1/2} = \langle \tilde{i}^2 \rangle_{\mathrm{ini}}^{1/2} = 0$, our result is in good agreement with that of Chambers *et al.* (1996), even though both the time integration method and the criterion for the instability are different. For the cases of non-zero initial random velocity, we found that for large $\langle \tilde{e}^2 \rangle_{\mathrm{ini}}^{1/2}$, T becomes much smaller than that for the zero initial velocity cases. Both b and c of Eq. (4) are affected by $\langle \tilde{e}^2 \rangle_{\mathrm{ini}}^{1/2}$, and b decreases with $\langle \tilde{e}^2 \rangle_{\mathrm{ini}}^{1/2}$.

In conclusion, the timescale of the instability depends strongly on the initial random velocities and the orbital separation of protoplanets. Recent works have shown that the typical separation of protoplanets that are formed through runaway growth is $10 r_{\mathrm{H}}$ (Kokubo and Ida, 1998). Therefore, if we consider only the case with $\langle \tilde{e}^2 \rangle_{\mathrm{ini}}^{1/2} = 0$, it would take very long time

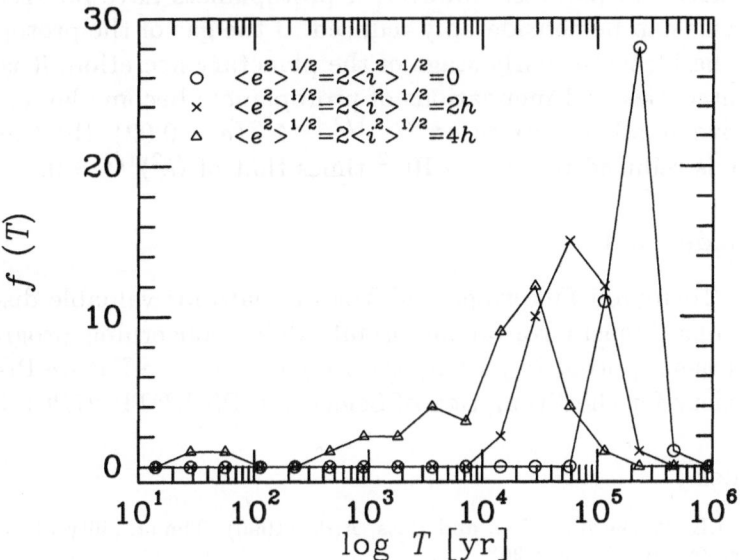

Figure 3. The distribution of the results of the timescale of instability plotted for the cases with $\Delta a_{\mathrm{ini}} = 8r_{\mathrm{H}}$ and $\langle \tilde{e}^2 \rangle_{\mathrm{ini}}^{1/2} = 0, 2, 4$. The vertical axis is the number of cases in which the results are located in area $[T_i, T_{i+1}]$, where $T_{i+1}/T_i = 2$.

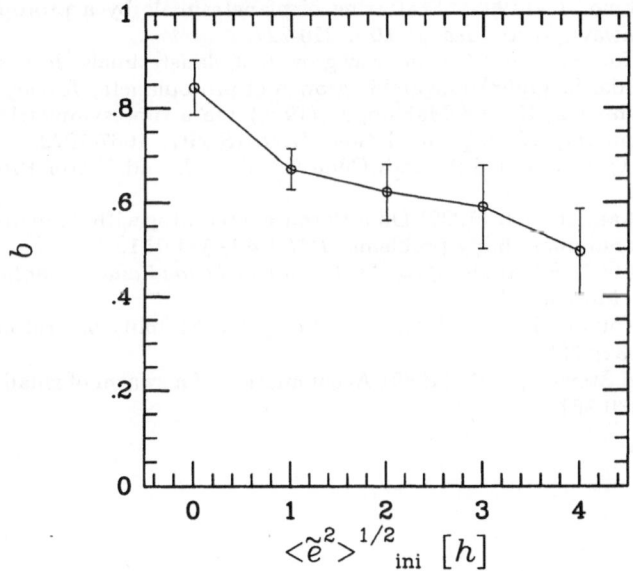

Figure 4. The value of b, the coefficient in the fitting formula $\log T = b\Delta a_{\mathrm{ini}} + c$, is plotted as a function of $\langle \tilde{e}^2 \rangle_{\mathrm{ini}}^{1/2}$. Error bars are 1σ values.

to form terrestrial planets. However, if protoplanets have non-zero e or i, the timescale can be considerably reduced. If the gas of the protoplanetary nebula is dissipated at early stage of the planetary accretion, it is possible that eccentricities and inclinations of protoplanets become large. If protoplanets have small eccentricities, $\langle \tilde{e}^2 \rangle^{1/2} = 4$ $(e \sim 0.02)$, the timescale of instability is reduced to $10^{-1} \sim 10^{-2}$ times that of $\langle \tilde{e}^2 \rangle^{1/2} = 0$.

Acknowledgments

We thank Toshiyuki Fukushige and Yoko Funato for valuable discussions. K. Y. thanks Atsushi Kawai for useful advice concerning programming. This work is supported in part by the Research for the Future Program of Japan Society for the Promotion of Science, JSPS-RFTP 97P01102.

References

Chambers, J.E. Wetherill, G.W., and Boss, A.P. (1996) The stability of multi-planet systems, *Icarus* **119**, 261-268.

Greenberg, R., Wacker, J., Chapman, C.R. and Hartmann, W.K. (1978) Planetesimals to planets: Numerical simulation of collisional evolution, *Icarus* **35**, 1-26.

Hayashi, C., Nakazawa, K. and Nakagawa, Y. (1985) in *Protostars and Planets II*, eds. D.C. Black and M.S. Matthews, Tucson: Univ. Arizona Press, pp. 1100-1153.

Ida, S. and Makino, J. (1992) *N*-body simulation of gravitational interaction between planetesimals and a protoplanet I. Velocity distribution of planetesimals, *Icarus* **96**, 107-120.

Ida, S. and Makino, J. (1993) Scattering of planetesimals by a protoplanet: Slowing down of runaway growth, *Icarus* **106**, 210-227.

Kokubo, E. and Ida, S. (1996) On runaway growth of planetesimals, *Icarus* **123**, 180-191.

Kokubo,E. and Ida, S. (1998) Oligarchic growth of protoplanets, *Icarus,* **131**, 171-178.

Kokubo, E., Yoshinaga, K. and Makino, J. (1998) On a time-symmetric Hermite integrator for planetary *N*-body simulation, *MNRAS* **297**, 1067-1072.

Makino, J. (1991) A Modified Aarseth Code for GRAPE and Vector Processors, *PASJ* **43**, 859-876.

Makino, J. and Aarseth, S.J. (1992) On a Hermite integrator with Ahmad-Cohen scheme for gravitational many-body problems, *PASJ* **44**, 141-151.

Safronov, V.S. (1969) *Evolution of the Protoplanetary Cloud and Formation of the Earth and Planets*, Moscow: Nauka.

Yoshinaga, K., Kokubo, E. and Makino, J. (1999) The Stability of Protoplanet systems, *Icarus* **139**, 328-335.

Wetherill, G. and Stewart, G.R. (1989) Accumulation of a swarm of small planetesimals, *Icarus* **77**, 330-367.

ACCRETIONAL ORIGIN OF THE GIANT PLANETS AND ITS CONSEQUENCES

WING-HUEN IP

Institute of Astronomy and Institute of Space Science
National Central University, Chung-Li, Taiwan 320, ROC

AND

JULIO A. FERNÁNDEZ

Departamento Astronomía, Facultad de Ciencias
Igua 4225, 11400 Montevideo, URUGUAY

Abstract. The basic mass distribution and orbital structure of the Solar System are determined by the original mass and angular momentum of the solar nebula and planetary accretion effect. Dust-gas aerodynamic interaction in the gaseous solar nebula could lead to the formation of an outer boundary or edge of the condensed matter in the outer planetary region. Such inward drift of the small solid bodies was subsequently reversed by the angular momentum transfer effect associated with the collisional accretion process. The orbits of Neptune and Uranus were found to expand outward during the accretional phase in numerical simulations. Such orbital migration mechanism might have played a key role at the same time in capturing Pluto into its 3:2 resonance with Neptune and the trapping of a significant number of Kuiper belt objects in this resonance. Gravitational scattering of icy planetesimals by the growing proto-Uranus and, most importantly, proto-Neptune was very effective in planting comets in the distant Oort cloud reservoir and driving the heavy bombardment event of the terrestrial planets and the Kuiper belt objects. To some extent the quasi-resonant relation of the major planets might have also been influenced by the orbital migration process. Besides gravitational scattering and accumulation of small planetesimals there is evidence that the protoplanets had collided with large planetoids of Mars- to Earth-size. The formation of an extended gaseous envelope/disk after the megaimpact event could be important in reprocessing the isotope (*i.e.*, D/H) ratio of the cometary material. This might provide a possible explanation to the unexpected difference between the D/H ratio in the water ice of comets (with an enrichment factor of f

181

M. Ya. Marov and H. Rickman (eds.), Collisional Processes in the Solar System, 181–201.

≈ 10 in comparison with the interstellar value) and those in the cores of Uranus and Neptune (with $f \approx 3$).

1. Introduction

How did the Solar System form is a fundamental question since the beginning of scientific thoughts. In addition to the fact that the formation and evolution of planets and planetary satellites must have been the result of a combination of many complicated processes, we are hampered by the lack of direct evidence on physical and chemical events that took place 4.5 billion years ago. The reconstruction of the early history and key processes during planetary formation therefore depends critically on testing theoretical models against observed properties of our Solar System and other solar systems. Because of the rapid advent in space technology in the past four decades, our knowledge of the structures and compositions of planets and small bodies like comets and asteroids has increased in a most dramatic manner. For example, the early phase of spacecraft reconnaissance of planets and satellites is now followed by orbiter/rendezvous missions like Galileo, Cassini, NEAR, and Rosetta. For very important scientific reasons, there will be many spacecraft visiting asteroids, comets and near-Earth objects (NEOs) for in-situ measurements and sample-return missions (*e.g.*, Deep Space 1, Contour, and Deep Impact) in the next decade. To a certain extent, comets and NEOs may be considered as carriers of the genetic codes of the Solar System. Only after proper understanding of their origins and physicochemical natures can we fully unlock the secret of the Solar System history. By implication, this also means that only with a very clear and precise understanding of our Solar System would we have the hope to extrapolate our knowledge base to other extra-solar systems. From this point of view, our paper touches upon a still relatively unexplored (and hence critically important) topic, namely, the accretion of the outer planets and its dynamical consequences.

Another very active related field in recent time has to do with the study of the Kuiper belt objects. Since the early suggestions by Edgeworth (1949) and Kuiper (1951), not much attention was given to the possible existence of a belt or disk of objects outside the orbit of Neptune (see Whipple, 1964) until the assessment by Fernández (1980) on the trans-Neptunian belt acting as a source of the short-period comets. This view was supported by the numerical work of Duncan *et al.* (1988) who showed that the flat inclination distribution of the short-period comets can not be accounted for by the isotropic injection of new comets from the cometary Oort cloud, but is in good agreement with the continuous injection of stray bodies from the trans-Neptunian belt or Kuiper belt. The presence of Kuiper belt objects proved to be elusive until the discovery of 1992 QB_1 of about 280-km diameter by Jewitt and Luu (1993). Since then, many more (more than 250 in June 2000) have been detected making inventory of the outer Solar System the most fast-moving topic in the last and future decade (Jewitt and Luu, 1995; Jewitt, 1999). One of the peculiarities of the structure of the Kuiper belt is that a large fraction of the detected objects (Kuiper belt objects or KBOs from now on) are found to be in 3:2 orbital resonance with Neptune (*i.e.*, the ratio of the orbital period of a resonant KBO to that of Neptune is nearly 3 to 2). Such resonant orbital configurations are very similar to that of Pluto. Note that Pluto's perihelion distance is actually inside Neptune's orbit. It is the libration (oscillation) of the position of Pluto relative to Neptune as a result of the orbital resonance which keeps Pluto from making close encounters to Neptune. Similarly, the orbital stability of the KBOs in 3:2 resonance is also maintained in this manner. Otherwise, they would have relatively short lifetimes against planetary perturbations (Duncan *et al.*, 1995). A view developed by Malhotra (1995) suggested that the resonant strucuture of the Kuiper belt is a natural consequence of the accretional formation of the outer planets. Our present theme therefore threads through the deep past and far future of our Solar System. To set the scene, we will discuss the basic scheme and assumptions (and uncertainties) involved in the current generation of theoretical models in Section 2, the orbital migration and the sweeping-resonant capture of Pluto and other Kuiper belt objects (KBOs) in Section 3, the megaimpact effects of planets and the Solar System-wide heavy bombardment event in Section 4, and the interrelation between the cometary Oort cloud and the Kuiper belt in Section 5.

2. The Basic Scheme

A determining factor of the Solar System formation has to do with the dynamical and chemical evolution of the solar nebula which are coupled to the activity of the proto-Sun. It is generally assumed that the gaseous accretion disc formed by the collapsing interstellar cloud is cleared away by a strong stellar wind at the end of the T-Tauri phase. Radio observations at 115 GHz and 230 GHz detected the presence of extended CO lobes as a result of collimated bipolar outflows near the young stellar objects (YSOs) (Uchida and Shibata, 1985; Edwards *et al.*, 1993; Fukui *et al.*, 1993). There are also indications that the inner circumstellar disc (radius $\lesssim 100$ AU) can be a significant source region of the energetic outflow, presumably because of magnetic dynamo activity at the central accretion disc (Koenigl and

Ruden, 1993). [If angular momentum transfer and condensation of gas take place at the same time (which they must), magnetohydrodynamics and plasma processes could play an important role in the chemical composition and mass distribution of the dust and planetesimals.]

The T-Tauri phase characterized by intense $H\alpha$ and other Balmer emission lines is short-lived, lasting usually no more than 10^7 years (Strom *et al.*, 1993). Zuckerman *et al.* (1995) found that there is a trend for strong CO depletion surrounding young stars with ages > 10 million years. This observational result suggested that the total mass of the molecular hydrogen gas in individual protoplanetary disc systems should be less than one jovian mass. In other words, giant gaseous planets like Jupiter and Saturn should have formed within this time interval prior to the stage of rapid gas dissipation. Even though Uranus and Neptune are not largely made up of H_2 and He, they still contain about 10% of hydrogen gas by mass. This has been used as an argument that these two outer planets must have formed at the tail-end of the gas dissipation so that a fraction of the gas component in the solar nebula could still be captured by the accreting protoplanets (Fernández and Ip, 1996). In turn, this chain of inferences places an important constraint on the accretion time scale of Uranus and Neptune.

We should note, however, that the large-scale depletion of CO gas in protoplanetary disks does not necessarily have to be the sole result of sweeping by the T-Tauri wind; condensation and adsorption of the CO molecules onto dust grains below a surface temperature of 20 K could just be equally effective (Aikawa *et al.*, 1996; 1997). However, the accretion time scale of Uranus and Neptune can not be too much longer than a few 10^7 years if the 3:2 resonance of Pluto with Neptune is to be explained by the orbital migration effect during planetary formation (see Ida and Nagasawa, 1999). This point will be addressed in the next section.

In standard models of the solar nebula, the mass density distribution is determined by the conservation of angular momentum and viscous transport of the mass (Ruden and Lin, 1986; Cassen *et al.*, 1994). Figure 1a depicts how the injected mass from the collapsing interstellar cloud is redistributed radially inward and outward. The demarcation radius depends on the initial angular momentum ($J_{SN} \approx 1$ to 3×10^{52} g cm^2 s^{-1}) and mass ($M_{SN} \approx 0.01 - 0.1$ $M_\odot$) of the solar nebula (Ruzmaikina and Ip, 1994). Besides the radial motion of the gas component, the condensed grains in Keplerian motion will follow an inward drift pattern because of viscous interaction with the slower-moving solar nebula gas (Whipple, 1972; Weidenschilling, 1980; 1997). If the original outer radius of the solar nebula is at about 100 AU, the radial drift of the icy grains condensed, say, between 40 and 100 AU could force most of the solid particles to be accumulated at an inner zone between 20 and 40 AU (see Fig. 1b). In this scenario,

a

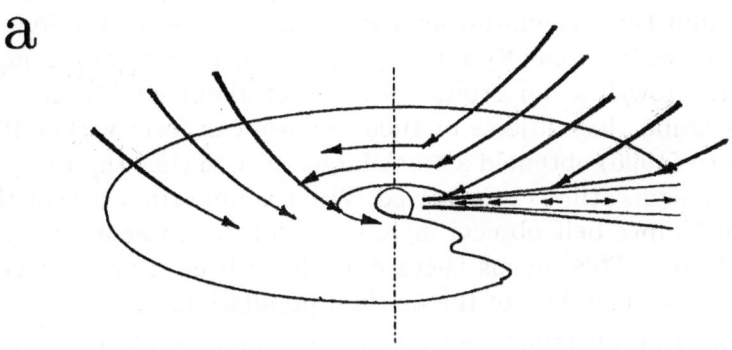

b

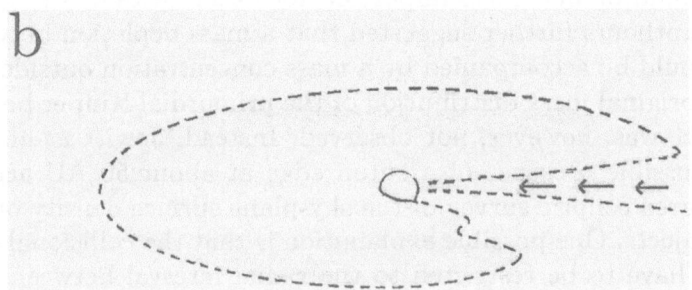

Figure 1. (a) Streamlines of gas entering the solar nebula from its parent cloud and joining the radial migration of material in the nebula. There is a particular radius within which nebular gas flows inward and beyond which it flows outward. This radius itself moves outward as the nebula evolves, overtaking outward flowing gas and reversing its direction of radial motion. From Cassen (1994). (b) A schematic view showing how the condensed grains and small planetesimals in the outer solar nebula will drift inward as a result of gas-dust aerodynamic interaction.

the solid mass density in the solar nebula would be highly enhanced (by a factor of 3) in the accretion region of Uranus and Neptune, thus partially solving the time scale issue raised by Safronov (1972).

With the condensed grains in place, icy planetesimals of larger sizes will form following collisional interaction at low relative speeds. How long it takes to build up a full size distribution with the largest members being of the size of Pluto depends sensitively on the total mass imbedded in the system and the coagulation models used. Because of the long orbital periods in the outer Solar System, a large mass ($\approx 30\ M_{\oplus}$) is needed to obtain Pluto growth at an orbital distance of about 40 AU. Kenyon and Luu (1998) found that objects of 1000 km size can form within 10^8 years whereas Stern (1996) obtained a much larger growth time (up to 10^9 years). This discrepancy is still to be resolved. What is important is that the total mass of the Kuiper belt objects including Pluto at present time is of the order of 0.1 $M_{\oplus}$. This means there must have been a very effective loss process for more than 99% of the original population.

Stern and Colwell (1997) investigated the erosive effect of the Kuiper belt via impact collision among the primordial planetesimals when Neptune started to gravitationally perturb the condensed matter into orbits of high eccentricities with e up to 0.3 between 30 and 50 AU. In their model calculations, they found that a Kuiper disk with original mass of 10 – 35 $M_{\oplus}$ in this radial interval can be reduced to as little as 0.1 – 0.3 $M_{\oplus}$ in 10^9 years if the average eccentricity $\langle e \rangle > 0.1$. Because Neptune's gravitational effect would be much less important for radial distances larger than 50 AU, these authours further suggested that a mass depletion between 30 and 50 AU should be accompanied by a mass concentration outside 50 AU reflecting the original mass distribution of the primordial Kuiper belt. This predicted effect was, however, not observed. Instead, Jewitt *et al.* (1998) inferred the possible presence of a cutoff edge at about 50 AU according to their large-area ecliptic survey of the sky-plane surface density of bright Kuiper belt objects. One possible explanation is that the collisional erosion effect did not have to be restricted to the radial interval between 30 and 50 AU as initially proposed by Stern and Cowell (1997). In Section 4, we will describe as an integral part of the multiple planetary accretion process (*i.e.*, not just limited to the stirring by Neptune), that the gravitational scattering effect of small planetesimals could be far more extensive and the outer Solar System out to a radial distance of a few hundred AU could be subject to intense high-velocity bombardment. Such an "external" impact scenario must be taken into consideration in the mass erosion of the Kuiper belt.

3. Orbital Migration

The presence of Pluto-sized objects will set the stage for further growth into full-sized protoplanets. It is worthwhile to mention here that a minimum value of 10^8 years for the Pluto growth time, as computed by Kenyon and Luu (1998), will by far exceed the gas dissipation time scale (T_d) of about 10^7 years discussed before (Zuckerman *et al.*, 1995). There is therefore no clear necessity to limit the formation time of Uranus and Neptune to be $< T_d$. Even so, a number of numerical simulations showed that the proto-Uranus and proto-Neptune could have formed rapidly given enough initial mass contained in a system of Pluto- to Mars-sized planetoids in their respective accretion zones.

In a series of works, Fernández and Ip (1981; 1983; 1984; 1987) considered the formation of Uranus and Neptune via collisional coagulation and accretion of a system of small planetesimals of 1 – 100 km sizes. The numerical code was based on the two-body encounter method developed by Öpik (1951) and Ip (1977). The basic idea is simply that coagulation will occur at collision with the angular momentum conserved and gravitational scattering or deflection will take place at close encounter between two objects. No fragmentation or pulverization of the colliding bodies is included. In spite of its simplicity, this approach was able to uncover several dynamical effects of potential importance to the evolution of the Solar System. First, it was found that the accretion process is self-limiting in the following sense. When a protoplanet has grown to a mass such that its capture cross section will be significantly enhanced by the gravitational effect, the mass growth rate would be exponential if the simulated system is in a closed box with fixed values of the relative velocities of the interacting bodies. However, in a three-dimensional orbital space, the growing protoplanet is also capable of gravitationally scattering the source population of planetesimals; the system will then begin to disperse because of the gradual increase in the relative speed. As a result, the so-called run-away growth will be truncated and followed by a long period of slow accretion.

In the first computational simulations (Fernández and Ip, 1981), the initial condition was characterized by the presence of two protoplanets in the Uranus and Neptune accretion zones, respectively, plus a swarm of small planetesimals of cometary size (1   10 km) accounting for the rest of the condensed materials. The starting masses of the two protoplanets are about 10% of their present values. In the model computation, no mutual coagulation of the small planetesimals is considered. Therefore, the size distribution of the interacting bodies is basically discontinuous with emphasis on the cometesimals of mass m_c on the low-mass end and the protoplanets on the high-mass end. Figure 2 shows the numerical results of two runs, one with $m_c = 10^{12}$ g and the other with $m_c = 10^{16}$ g. The truncation of the run-away growth effect can be seen to take place at the same time as the beginning of injection of the icy planetesimals into short-period orbits and hyperbolic orbits at about 4×10^7 years. This time hence earmarks the dispersal of the condensed materials in the planetary accretion zones. The resultant scattering of the small objects through the whole Solar System will lead to the extensive collisional bombardment of terrestrial planets,

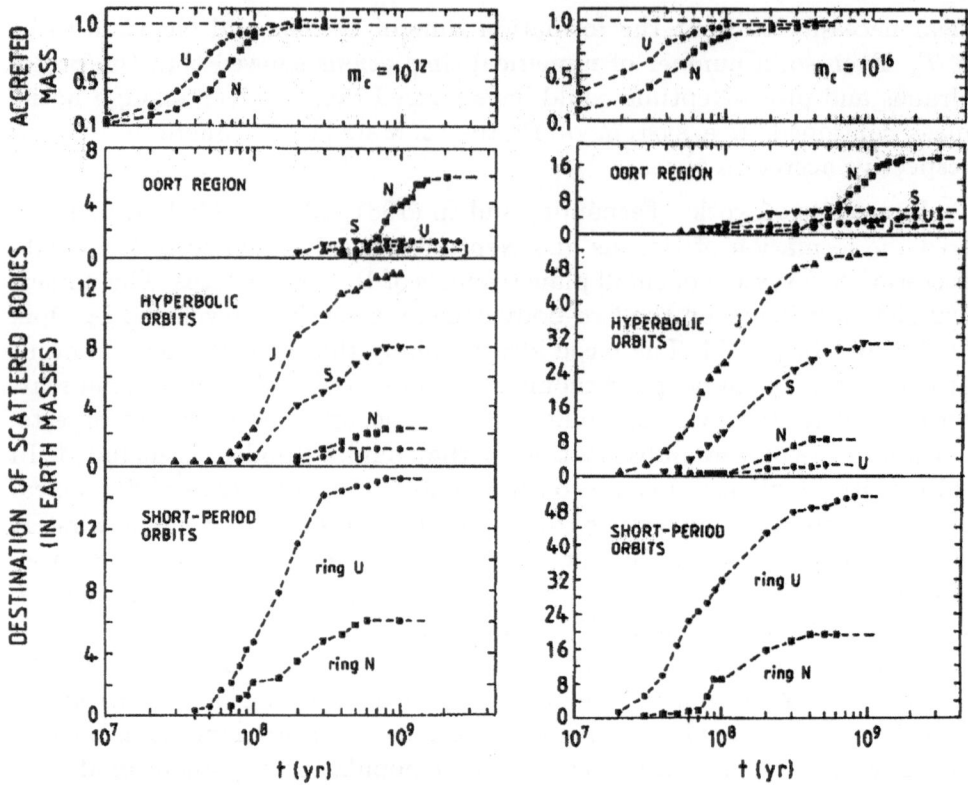

Figure 2. Evolutionary history of bodies in the accretion zones of Uranus and Neptune for two cometesimal masses. The accreted mass by Uranus and Neptune is expressed taking their present masses as unity. The bodies ejected to hyperbolic orbits or to the Oort region are plotted according to the giant planet that finally controls their evolution. Cometesimals scattered into short-period orbits are plotted according to the accretion zone where they were originally found. From Fernández and Ip (1981).

planetary satellites and other small bodies (*e.g.*, the KBOs) over the next few hundred million years. This is an interesting result because it links the formation of Uranus and Neptune to the heavy bombardment effect of the Moon in a self-consistent framework (Wetherill, 1977; Fernández and Ip, 1983; Chyba, 1991; Chyba *et al.*, 1994).

Another important consequence of the gravitational scattering of the planetesimal population is that this effect provides a very efficient mechanism for the transfer of angular momentum among the major planets. In anticipation of Jupiter's role as the main contributor of comets, Safronov (1972) theorized that Jupiter's orbit will shrink inward due to the loss of angular momentum carried by the comets flung into the Oort cloud reservoir. What was previously unexpected is that in a multi-planet system, the ran-

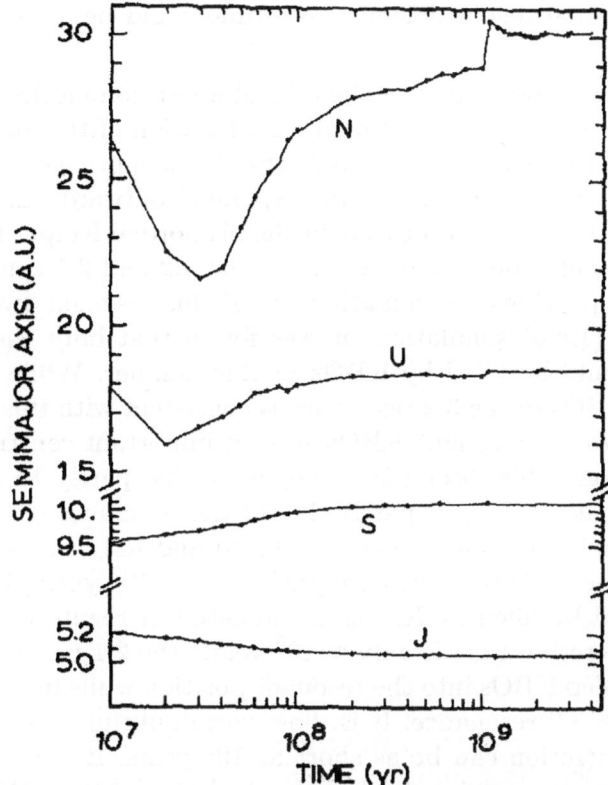

Figure 3. An example of the time variation of the semimajor axes of the four outer planets as a result of exchange of angular momentum with planetesimals. The initial semimajor axes are $a_J = 5.203$ AU, $a_S = 9.54$ AU, $a_U = 20$ AU and $a_N = 30$ AU. From Fernández and Ip (1984).

dom gravitational encounters of the stray bodies (as if in a ping-pong game) could permit the indirect transfer of angular momentum from Jupiter to the outer three planets (Saturn, Uranus and Neptune). As a result, the inward drift of Jupiter's orbit will be accompanied by the outward expansion of the orbits of the other three outer planets. Figure 3 shows an example of such orbital migration effect from the numerical simulations by Fernández and Ip (1984). In this idealized case, the orbits of both Neptune and Uranus move inward in the initial phase because of the gravitational scattering of small planetesimals in their own accretion zones to high-angular momentum orbits. This feature is consistent with Safronov's dynamical model. The subsequent turn-around and the outward orbital movement are caused by the scattering of the planetesimal population into Jupiter-crossing orbits. From then on, Uranus and Neptune will gain angular momentum from

Jupiter. The radial displacement of Neptune could be as much as 5 – 10 AU.

An interesting application of the orbital migration mechanism was proposed by Malhotra (1993, 1995) in the explanation of the locking of Pluto and a number of KBOs into stable 3:2 orbital resonance with Neptune. The basic idea is to have Neptune's orbit expanded outward during planetary accretion in such a way that objects in the primordial Kuiper belt would be swept into slots of orbital resonaces (*i.e.*, the 3:2 and 2:1 commensurabilities). Figure 4 provides a schematic view of this resonant sweeping effect. In Malhotra's (1995) simulations it was found that both the 3:2 and 2:1 resonances should be filled by KBOs in this manner. While the observed clustering of KBOs in the 3:2 resonance is consistent with this scenario, the small number of 2:1 resonant KBOs sets an important constraint on their orbital evolution. This discrepancy might be due partly to observational bias (Jewitt *et al.*, 1998), or partly due to the dynamics of orbital migration. For example, Ida and Nagasawa (1999) and Ida *et al.* (1999) showed that, if the time scale of orbital migration is $\gtrsim 10^7$ years, both resonant positions would be filled by KBOs as indicated in Figure 4. On the other hand, if the accretion time is only $\approx 10^6$ years, the 2:1 resonance would be too weak to sweep KBOs into the resonant position while full capture is still possible for the 3:2 resonance. It is, however, doubtful that the time scale of planetary accretion can be as short as 10^6 years. It is more likely that some other processes are involved in the depletion of the KBOs beyond 50 AU. One potential candidate is the heavy bombardment event caused by the large-scale gravitational scattering of the icy planetesimals in the final stage of planetary accretion. This will be discussed in the next section.

4. Megaimpact and Heavy Bombardment

Instead of using small planetesimals as the source population, another class of models made the fundamental assumption that the initial system was characterized by a number of planetoids of Mars- to Earth-size. To some extent, this is to simulate the final stage of planetary accretion in which collisions of the protoplanets with large planetoids appears to be quite frequent. Important examples are the large tilts ($I = 98°$) of the rotational axis of Uranus and other giant planets ($I = 26°$ for Saturn and 28° for Neptune) except Jupiter. According to Safronov (1972), the retrograde rotation of Uranus must have been caused by the impact with an object with a mass of about 7% (*i.e.*, one Earth mass) of the planetary mass. Ip (1989a) and Fernández and Ip (1996) used the modified Öpik approach to compute the accretionary growth of Uranus and Neptune via the collisional interaction of a system of large planetoids. It was shown in these model cal-

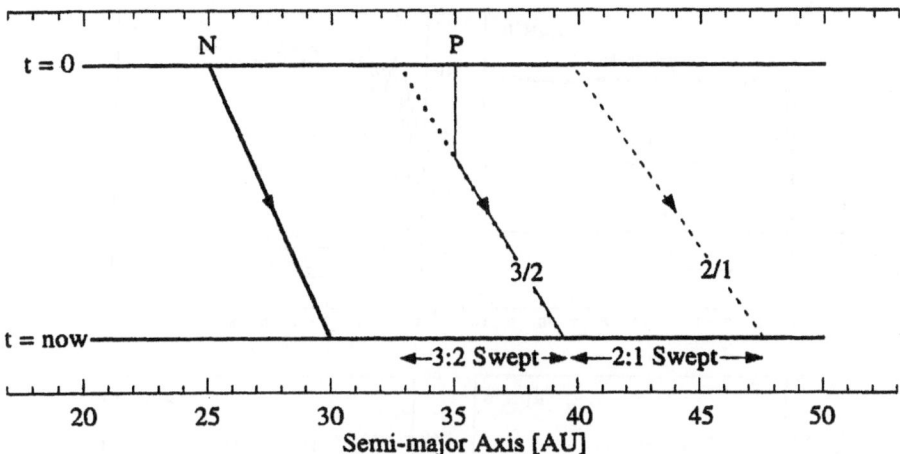

Figure 4. A schematic view showing the radial migration of Neptune (N) and the corresponding trapping of Pluto (P) and KBOs into 3:2 and 2:1 mean motion resonances. Figure originally from Malhotra (1995).

culations that the basic growth time scale and the orbital migration effect remain approximately the same as in the previous models adopting small planetesimals as the seed particles.

In these Monte Carlo simulations, the cross section for collision was artificially enhanced to reduce the gravitational scattering effect at the initial phase of planetary accretion. The Öpik scheme is also limited to two-body close encounters and distant perturbations are excluded from the numerical computation. To remedy these shortcomings, Brunini and Fernández (1999) used a direct integration method developed by Wisdom and Holman (1991) to follow the collisional and gravitational interaction of a system of N planetoids ($N = 250$, 500 and 1000) of equal mass in the Uranus and Neptune accretion zone. As before, Jupiter and Saturn are assumed to be fully grown with the orbital distances being the same as their present values. Brunini and Fernández found that, in agreement with Ip (1989a) and Fernández and Ip (1996), two to three large planets of the masses of Uranus and Neptune usually formed in the orbital region of these two outer planets. The accretion time scale is on the order of a few 10^7 years which is several times smaller than that derived by Fernández and Ip (1981). The reason why the time scale is shorter is because distant perturbations can now bring into crossing orbits that were initially non-intersecting (and thus non-interacting under the Öpik scheme). Moreover, the orbital migration phenomenon discussed earlier shows up in many of the numerical runs even though the orbital variations tend to have large-amplitude fluctuations (see

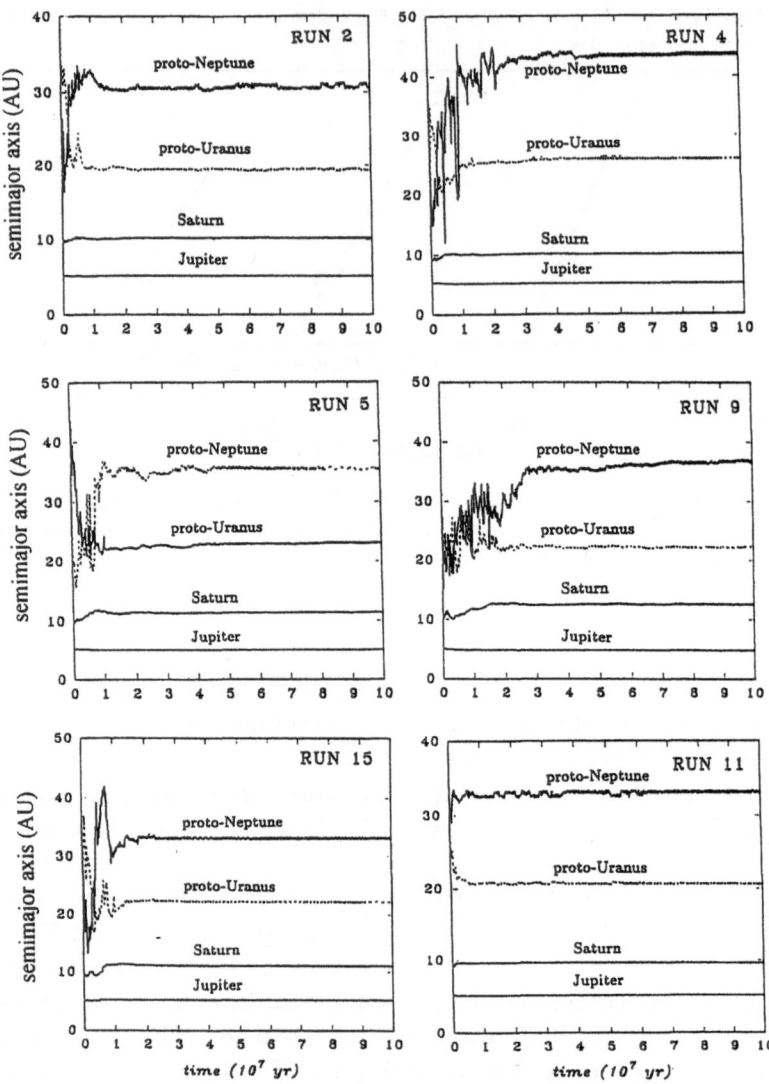

Figure 5. Evolution of the semimajor axis of the four Jovian planets for some of the runs where two massive proto-Uranus and proto-Neptune were formed. From Brunini and Fernández (1999).

Fig. 5). The latter effect is because of the stochastic interaction of the protoplanets with large planetoids up to an Earth mass.

In the actual situation, the accretion dynamics must be determined by the collisional impact and gravitational scattering of a wide spectrum of objects with sizes ranging from 1 km or less to a few thousand km. After the

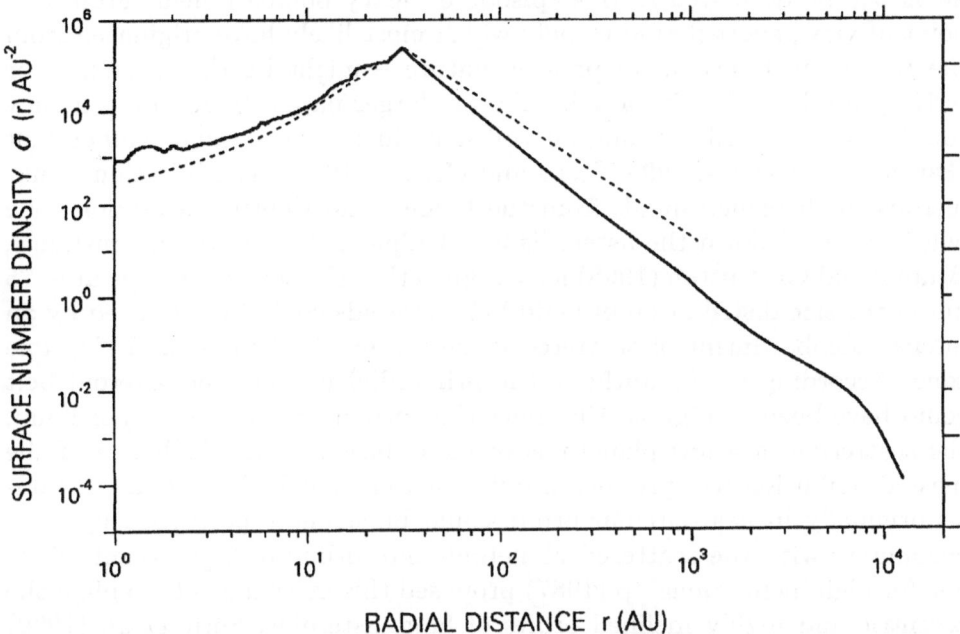

Figure 6. The cumulative surface number density (in arbitrary units) of particles scattered from the Uranus and Neptune accretion zone during the heavy bombardment event. The broken curve is from Levison and Duncan (1997) shown here for comparison. From Ip (1999).

formation of Uranus and Neptune, these accretionary remnants will remain in the Solar System with a dispersal time scale of 100 – 200 million years (Fernández and Ip, 1981; 1983). The injection of these residual materials of volatile composition into the inner Solar System will lead to the extensive bombardment of the terrestrial planets and the Moon causing the so-called late heavy bombardment event (Neukum and Ivanov, 1994; Chyba *et al.*, 1994). A significant fraction of the residual materials will be flung outward across the Kuiper belt far beyond the orbit of Neptune (Fernández and Ip, 1981; 1983; Ip, 1989a; Levison and Duncan, 1997); see Figure 6. This implies that the Kuiper belt objects must also have suffered from extensive impact erosion and fragmentation in this process (Fernández and Gallardo, 1996; Ip, 1999). If the total mass of cometary-size objects ejected by the proto-Uranus and proto-Neptune is $> 0.1 M_\oplus$, there is a high probability that KBOs of 100-km radius will be catastrophically fragmented by high-velocity impacts resulting in a gravitationally bound rubble-pile internal structure. At the same time, the original population of km-sized planetesimals will be destroyed in this episode of heavy bombardment. From this point of view, short-period comets which most likely have originated from the Kuiper belt instead of pristine nature – might be the secondary or tertiary product of collisional breakup of larger parent bodies even if they could have survived the mutual collisions in the primordial Kuiper belt (Farinella and Davis, 1996; Stern and Colwell, 1997). The injection of numerous small planetesimals from the Uranus and Neptune accretion zone will help grind down the asteroids and Kuiper belt objects. For instance, Brunini and Gil-Hutton (1998) have argued that the change in slope at ~ 75 km in the size distribution of main belt asteroids could be produced by an intense bombardment of scattered comets from the Uranus and Neptune zone. According to the authors, the primordial mass of the asteroid belt could have been as high as 10^3 times the current one. On the other hand, the scattering of a few planetoids of large masses at the high end of the mass distribution will produce a different dynamical effect. That is, bodies originally in near-circular orbits could be strongly perturbed by close encounters with the scattered planetoids into orbits of high eccentricities and/or high inclinations. Ip (1987) proposed this mechanism to explain the eccentric and highly inclined orbits of large asteroids. Petit *et al.* (1999) examined the consequence of gravitational interaction of the Kuiper belt objects with two planetoids of one Earth mass. These authors found that the temporary storage of these scattered planetoids could be responsible for the excitation of the orbital eccentricities and inclinations of the KBOs to the observed values after their formation in near-circular orbits. The temporary storage of Earth-sized objects in the outer Kuiper belt could in principle also lead to a strong modification/sculpting of the corresponding mass distribution and orbital structure (see Ip and Fernández, 1991). We will return to this specific point in the next section.

5. The Interrelation between the Cometary Oort Cloud and Kuiper Belt

A natural consequence of the gravitational accretion and scattering process of Uranus and Neptune is that a significant amount of materials will be injected into orbits of high eccentricities and very large semi-major axes ($> 10^4$ AU). According to Fernández and Ip (1981), Neptune is the main contributor with injected mass up to $6 - 16$ $M_\oplus$ while Uranus, Saturn and Jupiter combined together would scatter $4 - 10$ $M_\oplus$ into quasi-parabolic orbits. (Due to its large mass, Jupiter is very effective in ejecting particles into hyperbolic orbits and is hence not a good supplier of comets.) Because of angular momentum transfer, these long-period orbits should originally have low inclinations with respect to the ecliptic plane. Repeated interac-

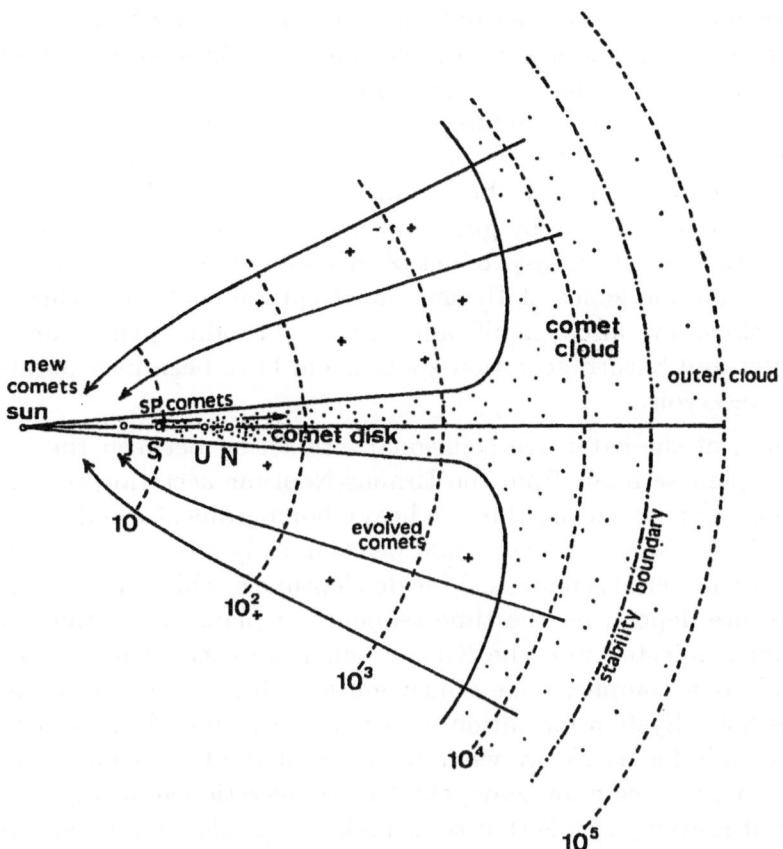

Figure 7. A schematic picture of the current spatial distribution of comets around the Solar System if they formed in the protoplanetary disk. A comet disk might still be present, stretching from the region of the outer planets to a few 10^3 AU. Farther away, the action of external perturbers should have by now randomized the orbital planes of comets. New comets probably come from the Oort cloud, whereas most short-period comets may come from the comet disk (*e.g.*, the Kuiper belt). The spherical volume of a few thousand AU in radius should contain evolved long-period comets (crosses). The dashed circles indicate distances to the Sun in AU. From Fernández and Ip (1991).

tions with passing stars and the galactic tidal effect over the age of the Solar System will eventually isotropize the inclination distribution in the outer cometary reservoir or the Oort cloud. The inner region ($< 10^3$ AU) where stellar perturbations are less frequent will remain relatively flat; see Figure 7. It is still a major question whether the Oort cloud of planetary scattering origin can bridge smoothly with the Kuiper belt formed in situ, or there should be a wide gap separating the Kuiper belt from the inner Oort cloud. The picture of the formation of the Oort cloud may have to be revised somewhat if the early Solar System was embedded in a galactic environment much denser than the current one. Fernández (1997) has argued that if this was the case, the Oort cloud might have ended up with a dense core of comets of no more than a few 10^3 AU radius. This tightly bound comet core could have been the source from which the external, loosely bound Oort cloud has been replenished through the age of the Solar System via perturbations by giant molecular clouds and very close stellar passages. An interesting by-product of this scenario is that not only bodies from the accretion zones of Uranus and Neptune could find their way to the Oort cloud, but also a significant number of residual planetesimals from the Jupiter and Saturn accretion zones could have been incorporated into the Oort reservoir.

Because of the extensive collisional interaction between the scattered planetary planetesimals from the Uranus-Neptune accretion zone and the Kuiper belt objects during the late heavy bombardment event, the Kuiper belt serving as the last barrier could be chemically mixed with volatile and non-volatile materials condensed inside Neptune's orbit. The efficiencies of mass capture depend on the time sequence of planet formation and condensed mass injected into the Kuiper belt from each planetary accretion zone. This is a complex issue which must be investigated when a global model of Solar System formation is in place in future. At this moment, it is not possible for us to say what fractions of the Oort cloud comets are from the Jupiter accretion zone, the Saturn accretion zone and so on. For example, it is even possible that some rocky asteroids could be injected into the Oort cloud when the primordial asteroidal belt was being "cleansed" (Weissman and Levison, 1997).

The exchange of condensed materials among different planetary accretion zones seems to be a wide-spread phenomenon. One manifestation, interesting enough, might be the ocean water of the Earth. Ip and Fernández (1988) and more recently, Brunini and Fernández (1999) estimated that about 10 to 15 $M_\oplus$ of planetesimals of volatile composition could be injected into the terrestrial region from the Uranus-Neptune accretion zone. The capture of a fraction of this material could provide the source of our oceans and even life (Oró and Lazcano, 1997; Delsemme, 1997). The recent finding that the deuterium-to-hydrogen (D/H) isotope ratios of three comets (1P/Halley, C/1996 B2 Hyakutake and C/1995 O1 Hale-Bopp) are very close to each other (D/H $\approx 2.5 \times 10^{-4}$) but a factor of two larger than the corresponding value of the terrestrial ocean water has led to the idea that the ocean water might have come from more than one source. In other words, besides the Uranus and Neptune zone, icy materials with smaller D/H ratios from the Jupiter and Saturn accretion zones could have contributed significantly (Delsemme, 1999). This is, however, one of many possible explanations and we probably need a larger sample of cometary D/H ratios to trace the dynamical histories and origins of different classes of comets.

It is interesting to note that the recent observations of Neptune by the Infrared Space Observatory (ISO) has raised another important issue. That is, the spectral measurements of CH_3D has shown that the D/H ratio of Neptune's atmosphere is only a factor of three higher than the interstellar value of 2.5×10^{-5}. In comparison, the D/H ratios of the three comets as mentioned above are all about a factor of ten higher than the interstellar value. Drouart *et al.* (1999) discussed the scenario in which turbulent mixing in the solar nebula can cause reduction of the D/H ratios in the condensed ice of Uranus and Neptune (via mixing with the icy materials injected from the Jovian zone) from the cometary value. It is, however, not clear how the factor of about three difference might be accounted for if the Oort cloud comets and Neptune have formed at the same radial distance. One possible contributing effect might have to do with the accretional and megaimpact process of Uranus and Neptune. During the collisional formation, these protoplanets must have built up extensive hot atmospheric envelopes allowing isotopic exchange between H_2 and HDO, CH_3D and NH_2D (see Figure 8). The net result is expected to transform the initial D/H value of the source planetesimals to lower value. On the other hand, small icy planetesimals which have escaped being processed by atmospheric interaction will retain their original D/H ratio as observed in comets. This is one area which we plan to pursue in future.

6. Discussion

Even though astronomical observations and theoretical models over the past twenty years have given us important insights into the origin and evolutionary history of the Solar System, there are still many missing links to be filled in. For example, we have no clue what happened to the original mass ($\approx 35\ M_\oplus$) in the Kuiper belt with only less than 1% remaining in the present location. The clearance of this amount of mass must have had an important effect on the general structure of the planetary orbits because of conservation of angular momentum. Should exchange of angular momentum be effective between the primordial Kuiper belt objects and the scattered planetesimals from the Neptune and Uranus accretion zones, the Kuiper belt objects would have executed radial orbital drift. By the same token, what was the back reaction effect on Neptune's orbit (and Uranus') in case the total mass of the Kuiper belt objects was initially non-negligible? Furthermore, what is the dynamical effect of Pluto-sized objects moving in eccentric orbits with aphelion distances reaching 50 – 60

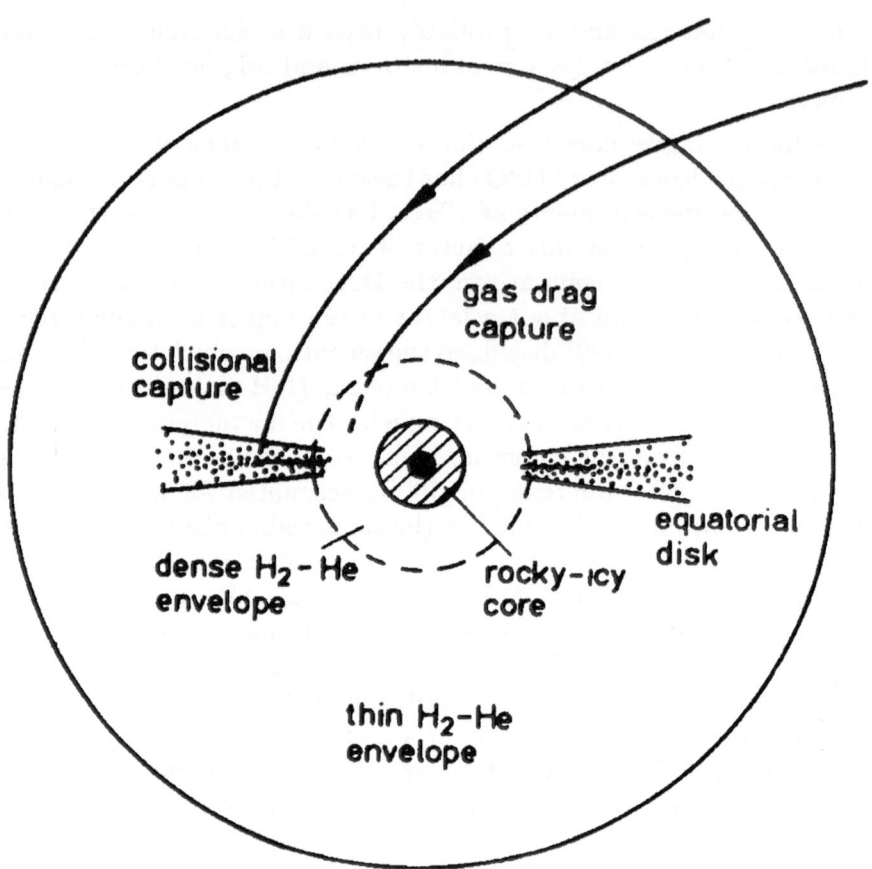

Figure 8. In the final stage of accretion proto-Uranus and proto-Neptune might have been surrounded by extended gaseous envelopes making capture by gas drag possible. Capture might have also occurred due to inelastic collisions with circumplanetary particles possibly distributed in a thin equatorial disk. The D/H ratio of the source material could be modified via isotopic exchange between H_2 and HDO, CH_3D and NH_2D.

AU? In the numerical results of model simulations, a few large planetoids could sometimes be placed in such orbits via gravitational scattering (Ip, 1989a). The long-term effects on the sculpting of the outer edge of the Kuiper belt and the ejection of short-period comets into the inner Solar System might constitute the echo of the deep past of our Solar System.

Last but not least, the regular orbital spacing of the major planets, namely, the mean-motion near-commensurability at the 3:1 resonance between Saturn and Uranus and the one at the 2:1 resonance between Uranus and Neptune could also be caused by the orbital migration effect described above (Fernández and Ip, 1996; Brunini and Fernández, 1999). In other words, the whole planetary system can be locked into orbital coupling as a result of the collisional accretion process. If this is a general property of planetary accretion, we would expect the existence of similar relations in other extra-solar systems.

References

Aikawa, Y., Miyama, S.M., Nakano, T., and Umebayashi, T. (1996) Evolution of molecular abundance in gaseous disks around young stars: Depletion of CO molecules, *Astrophys. J.* **467**, 684-697.

Aikawa, Y., Umebayashi, T., Nakano, T. and Miyama, S.M. (1997) Evolution of molecular abundance in protoplanetary disks, *Astrophys. J.* **486**, L51-L54.

Brunini, A., and Fernández, J.A. (1999) Numerical simulations of the accretion of Uranus and Neptune, *Planet. Space Sci.* **47**, 591-605.

Brunini, A. and Gil-Hutton, R. (1998) The cometary bombardment on the primitive asteroid belt, *Planet. Space Sci.* **46**, 997-1001.

Cassen, P. (1994) Utilitarian models of the solar nebula, *Icarus* **112**, 405-429.

Chyba, C.F. (1991) Terrestrial mantle siderophiles and the lunar impact record, *Icarus* **92**, 217-233.

Chyba, C.F., Owen, T.C. and Ip, W.-H. (1994) Impact delivery of volatiles and organic molecules to Earth, in *Hazards due to comets & asteroids*, (Ed. T. Gehrels), University of Arizona Press, pp. 9-58.

Delsemme, A.H. (1997) The origin of the atmosphere and of the oceans, in *Comets and the Origin and Evolution of life*, (Eds., P.J. Thomas, C.F. Chyba and C.P. McKay), Springer-Verlag, pp. 29-68.

Delsemme, A.H. (1999) The deuterium enrichment observed in recent comets is consistent with the cometary origin of seawater, *Planet. Space Sci.* **47**, 125-131.

Drouart, A., Dubrulle, B., Gautier, D. and Robert, F. (1999) Structure and transport in the Solar nebula from constraints on deuterium enrichment and giant planets formation, *Icarus* **140**, 129-155.

Duncan, M., Quinn, T. Tremaine, S. (1988) The origin of short-period comets, *Astrophys.J.* **328**, L69-L73.

Duncan, M. Levison, H., Budd, S. (1995) The dynamical structure of the Kuiper belt, *Astron. J.* **110**, 3073-3081.

Edgeworth, K.E. (1949) The origin and evolution of the solar sytem, *Mon. Not. Roy. Astron. Soc.* **109**, 600-609.

Edwards, S., Ray, T. and Mundt, R. (1993) Energetic mass outflows from young stars, in *Protostars and Planets III* (Eds. E.H. Levy and J.I. Lunine), University of Arizona Press, pp. 567-602.

Farinella, P. and Davis, D. (1996) Short-period comets: primordial bodies or collisional fragments?, *Science* **273**, 938-941.

Fernández, J.A. (1980) On the existence of a comet belt beyond Neptune, *Mon. Not. Roy. Astron. Soc.* **192**, 481-491.

Fernández, J.A. (1997) The formation of the Oort cloud and the primitive galactic environment. *Icarus* **129**, 106-119.

Fernández, J.A. and Gallardo, T. (1999) The origin of comets, in *Asteroids, Comets, Meteors 1999*, in press.

Fernández, J.A. and Ip, W.-H. (1981) Dynamical evolution of a cometary swarm in the outer planetary region, *Icarus* **47**, 470-479.

Fernández, J.A. and Ip, W.-H. (1983) On the time evolution of the cometary influx in the region of the terrestrial planets, *Icarus* **54**, 377-387.

Fernández, J.A. and Ip, W.-H. (1984) Some dynamical aspects of the accretion of Uranus and Neptune: The exchange of angular momentum with planetesimals, *Icarus* **58**,

109-120.

Fernández, J.A. and Ip, W.-H. (1987) Time-dependent injection of Oort cloud comets into Earth-crossing orbits, *Icarus* **71**, 46-56.

Fernández, J.A. and Ip, W.-H. (1991) Statistical and evolutionary aspects of cometary orbits, in *Comets in the Post-Halley Era*, (Eds. R.L. Newburn, Jr., M. Neugebauer, and J. Rahe), Kluwer Academic Publishers, pp. 487-533.

Fernández, J.A. and Ip, W.-H. (1996) Orbital expansion and resonant trapping during the late accretion stages of the outer planets, *Planet. Space Sci.* **44**, 431-443.

Fukui, Y., Iwata, T., Mizuno, A., Bally, J. and Lane, A.P. (1993) Molecular outflows, in *Protostars and Planets III* (Eds. E.H. Levy and J.I. Lunine), University of Arizona Press, pp. 603-639.

Gomes, R.S. (1997) Dynamical effects of planetary migration on the primordial asteroid belt, *Astron. J.* **114**, 396-401.

Ida, S. and Nagasawa, M. (1999) Review: Orbital distribution of the asteroid and Edgeworth-Kuiper belts and formation of the Solar System, *Planet. Space Science*, Submitted.

Ida, S., Tanaka, H., Bryden, G., and Lin, D. (1999) Migration of proto-Neptune and The orbital distribution of trans-Neptunian objects, preprint.

Ip, W.-H. (1977) On the early scattering processes of the outer planets, in *Comets-Asteroids-Meteorites: Interrelations, Evolution and Origin* (ed. A.H. Delsemme), University of Toledo Press, pp. 485-490.

Ip, W.-H. (1987) Gravitational stirring of the asteroid belt by Jupiter zone bodies, *Gerl. Beitr. Geophy.* **96**, 44-51.

Ip, W.-H. (1989a) Dynamical processes of macro-accretion of Uranus and Neptune: a first look, *Icarus* **80**, 167-178.

Ip, W.-H. (1989b) Dynamical injection of Mars-sized planetoids into the asteroidal belt from the terrestrial planetary accretion zone, *Icarus* **78**, 270-279.

Ip, W.-H. (1999) Heavy bombardment history of the Edgeworth-Kuiper objects, in *Cometary Nuclei in Space and Time*, (Ed. M.F. A'Hearn), ASP Conference Series, in press.

Ip, W.-H. and Fernández, J.A. (1988) Exchange of condensed matter among the outer and terrestrial protoplanets and the effect of surface impact and atmospheric accretion, *Icarus* **74**, 47-61.

Ip, W.-H. and Fernández, J.A. (1991) Steady-state injection of short period comets from the trans-Neptunian cometary belt, *Icarus*, **92**, 185-193.

Ip, W.-H. and Fernández, J.A. (1997) On dynamical scattering of Kuiper belt objects in 2:3 resonance with Neptune into short-period comets, *Astron. Astrophys.* **324**, 778-784.

Jewitt, D. (1999) Kuiper belt objects, *Ann. Rev. Earth, Planet Sci.* **27**, 278-312.

Jewitt, D. and Luu, J. (1993) Discovery of the candidate Kuiper belt object 1992 QB_1, *Nature* **362**, 730-732.

Jewitt, D. and Luu, J. (1995) The Solar System beyond Neptune, *Astron. J.* **109**, 1867-1876.

Jewitt, D.C. , Luu, J.X., and Trujillo, C. (1998) Large Kuiper belt objects: the Mauna Kea 8k CCD survey, *Astron. J.* **115**, 2125-2135.

Kenyon, S.J. and Luu, J.X. (1998) Accretion in the early Kuiper belt I. Coagulation and velocity evolution, *Astron. J.* **115**, 2136-2160.

Koenigl, A. and Ruden, S.P. (1993) Origin of outflows and winds, in *Protostars and Planets III* (Eds., E.H. Levy and J.I. Lunine), University of Arizona Press, pp. 641-687.

Kuiper, G.P. (1951) On the origin of the Solar System, in *Astrophysics*, Ed. J.A. Hynek, McGraw-Hill, New York, pp. 357-424.

Levison, H.F. and Duncan, M.J. (1997) From the Kuiper belt to Jupiter-family comets: The spatial distribution of ecliptic comets, *Icarus* **127**, 13-32.

Malhotra, R. (1993) The origin of Pluto's orbit, *Nature* **365**, 819-821.

Malhotra, R. (1995) The origin of Pluto's orbit: implications for the Solar System beyond Neptune, *Astron. J.* **110**, 420-428.

Neukum, G. and Ivanov, B.A. (1994) Crater size distributions and impact probabilities on Earth from lunar, terrestrial-planet, and asteroid cratering data, in *Hazards due to comets & asteroids*, (Ed. T. Gehrels), University of Arizona Press, pp. 359-416.

Öpik, E.J. (1951) Collision probabilities with the planets and distribution of interplanetary matter, *Proc. Roy. Irish Acad. Ser.*, A 54, 165-199.

Oró, J. and Lazcano, A. (1997) Comets and the origin and evolution of life, in *Comets and the Origin and Evolution of Life*, (Eds., P.J. Thomas, C.F. Chyba, and C.P. McKay), Springer Verlag, pp. 3-28.

Petit, J.-M., Morbidelli, A., and Valsecchi, G. (1999) Large scattered planetesimals and the excitation of the small body belts, *Icarus*, in press.

Ruden, S.P. and Lin, D.N.C. (1986) The global evolution of the primordial solar nebula, *Astrophys. J.* **308**, 883-901.

Ruzmaikina, T.V. and Ip, W.-H. (1994) Chondrule formation in the accretion shock, *Icarus* **112**, 430-447.

Safronov, V.S. (1972) *Evolution of the Protoplanetary Cloud and Formation of the Earth and the Planets*, Israel Program for Scientific Translation, Jerusalem.

Stern, S.A. (1996) On the collisional environment, accretion time scales, and architecture of the massive, primordial Kuiper belt, *Astron. J.* **112**, 1203-1211.

Stern, S.A. and Colwell, J.E. (1997) Collisional erosion in the primordial Edgeworth-Kuiper belt and the generation of the 30-50 AU Kuiper gap, *Astron. J.* **490**, 879-882.

Strom, S.E., Edwards, S. and Skrutskie, M.F. (1993) Evolutionary time scales for circumstellar disks associated with intermediate- and solar-type stars, in *Protostars and Planets III* (Eds., E.H. Levy and J.I. Lunine), University of Arizona Press, pp. 837-866.

Uchida, Y. and Shibata, K. (1985) Magnetodynamical acceleration of CO and optical polar flows from the region of star formation, *Publ. Astron. Soc. Japan* **37** , 515-535.

Wetherill, G.W. (1977) Evolution of the earth's planetesimal swarm subsequent to the formation of the earth and moon, *Proc. Lunar Sci. Conf.* **8**, 1-16.

Whipple, F.L. (1964) Evidence for a comet belt beyond Neptune, *Proc. Nat. Acad. Sci.* **51**, 711-718.

Whipple, F.L. (1972) On certain aerodynamic processes for asteroids and comets, in *From Plasma to Planet* (Ed., A. Elvius), Almqvist & Wiksell, Stockholm/ Wiley, New York, pp. 211-232latex.

Wisdom, J. and Holman, M. (1991) Symplectic maps for the N-body problem, *Astron. J.* **102**, 4, 1528-1638.

Weidenschilling, S.J. (1980) Dust to planetesimals: settling and coagulation in the solar nebula, *Icarus* **44**, 172-189.

Weidenschilling, S.J. (1997) The origin of comets in the solar nebula: a unified model, *Icarus* **127**, 290-306.

Weissman, P.R. and Levison, H.F. (1997) Origin and evolution of the unusual object 1996 PW: Asteroids from the Oort cloud?, *Astrophys. J.* **488**, L133-L136.

Zuckerman, B., Forveille, T. and Kastner, J.H. (1995) Inhibition of giant-planet formation by rapid gas depletion around young stars, *Nature* **373**, 494-496.

N-BODY SIMULATIONS OF MOON ACCRETION

S. IDA

Dept. of Earth and Planetary Science, Tokyo Institute of Technology, Ookayama, Meguro-ku, Tokyo 152-8551, Japan

E. KOKUBO

Dept. of Earth Science and Astronomy, University of Tokyo, Komaba, Meguro-ku, Tokyo 153-8902, Japan

AND

T. TAKEDA

Dept. Earth and Planetary Science, Tokyo Institute of Technology, Ookayama, Meguro-ku, Tokyo 152-8551, Japan

Abstract. We review the dynamical processes in accretion of the Moon from an impact-generated disk, based on the results of direct N-body simulations. The important processes are tidal inhibition of accretion within the Roche limit, development of density spiral arms, radial migration of disk mass associated with angular momentum transfer by the spiral arms, formation of particle aggregates by self-gravity, and interaction of formed moonlets with the disk. As a result of the disk evolution, a single large moon is formed at about 3–4$R_\oplus$ on a time scale of a month to a year, in most cases. The mass of the formed moon is regulated by the total mass and angular momentum of the initial disk.

1. Introduction

Planets are formed through accretion of many small planetesimals (*e.g.*, Safronov, 1969; Hayashi *et al.*, 1985; Lissauer and Stewart, 1993; Ohtsuki *et al.*, 1993). While planetesimals orbit the proto-sun, they gravitationally perturb each other and sometimes collide to accrete. Lissauer (1987) suggested through analytical argument that the planet mass when the runaway accretion of the planet terminates, called "isolation mass", is much smaller than the present mass in the terrestrial planet region. N-body sim-

M. Ya. Marov and H. Rickman (eds.), Collisional Processes in the Solar System, 203–222.

ulations by Kokubo and Ida (1998, 2000) showed a larger isolation mass
than Lissauer (1987) suggested; however, the isolation mass is still only
10% of the Earth mass ($M_\oplus$). As a result, a few tens of protoplanets with
$\sim 0.1 M_\oplus$ would be formed in the terrestrial planet region through runaway accretion (Kokubo and Ida, 1998, 2000). These protoplanets have
nearly circular and noninclined orbits due to dynamical friction and collisional damping (Wetherill and Stewart, 1989; Kokubo and Ida, 1998) and
are well separated, so that their orbits are stable (Kokubo and Ida, 1998,
2000). However, long term orbital perturbations may pump up the orbital
eccentricities and the protoplanets may start orbit crossing to collide with
each other (Chambers *et al.*, 1996; Chambers and Wetherill, 1998). Thus,
the Earth may have collided with Mars-sized protoplanets (mass $\sim 0.1 M_\oplus$)
in the final formation stage.

Hartmann and Davis (1975) and Cameron and Ward (1976) proposed
that Earth's Moon was formed by a giant impact of a Mars-sized body
with the proto-Earth. This model may account for the dynamical and
geochemical characteristics of the Moon: large angular momentum of the
Earth-Moon system, depletion of volatiles and iron, while the other lunar
formation models have difficulty explaining some of the above characteristics (*e.g.*, Stevenson, 1987). Hydrodynamic simulations of the giant impact
event (Kipp and Melosh, 1987; Benz *et al.*, 1986, 1987, 1989; Cameron and
Benz, 1991; Cameron, 1997) show that the outcome of the impact is formation of a circumterrestrial debris disk rather than direct formation of
the Moon and also predict that an impact by a Mars-sized body with angular momentum of nearly the current Earth-Moon system ejects enough
material to form a circumterrestrial debris disk as massive as the present
Moon. The debris disk consists mostly of the outer part (silicate mantle)
of the impactor in many cases. If a single body accretes from the debris
disk, an iron-free, present-sized moon may be obtained. Usually, the disk is
centrally-condensed and most of the disk mass is within the Roche limit.
The radius of the Roche limit is $\sim 3R_\oplus$ (where $R_\oplus$ is the Earth radius) (see
Sect. 2.1). The disk material might be vaporized because of high impact
energy (*e.g.*, Cameron and Benz, 1991). The disk might evolve viscously
during the vapor phase (Stevenson, 1987; Thompson and Stevenson, 1988).
As the disk cools, solid material is eventually condensed and a solid particle
disk is formed.

Here we review the dynamical processes of accretion of the Moon from
the particulate disk. Canup and Esposito (1996) simulated the lunar accretion with a statistical model that included tidal inhibition of accretion
within the Roche limit (see Sect. 2.1). They suggested that multiple small
moonlets are likely outcomes. However, global and collective effects, which
play important roles in the disk evolution, were not included in their model.

Accretion simulations utilizing direct N-body orbit integrations were performed by Ida, Canup and Stewart (1997, hereafter ICS97), and Kokubo, Ida, and Makino (2000, hereafter KIM00). The results are as follows. Inelastic collisions of the particles damp initially large orbital eccentricities and inclinations of the particles to moderate values in a few orbital periods. The self-gravitational instability tends to develop in such a "cold" disk (*e.g.*, Toomre, 1964), so that clumps are formed. However, the tidal forces inhibit formation of gravitationally-bound aggregates within the Roche limit. The temporal clumps are sheared by Keplerian differential rotation and result in density spiral arms. The spiral arms quickly transfer angular momentum and spread disk material. Since accretion is allowed only outside the Roche limit, such global mass transfer supplies the material that can be incorporated into lunar accretion. When the accreting moon becomes as massive as the present Moon, its gravity becomes strong enough to scatter most of the inner disk material onto the proto-Earth. As a result, a single large moon is formed at about 3–4$R_\oplus$. On a longer time scale, the formed moon gradually migrates outward by tidal interaction with the Earth (*e.g.*, Mignard, 1980), sweeping residual outer particles (Canup, Levison, and Stewart, 1999).

In Sect. 2, we briefly summarize the tidal inhibition of accretion near the Roche limit and self-gravitational instability, both of which are important processes in the evolution of a protolunar disk. In Sect. 3, we discuss lunar accretion processes in detail utilizing the results of the N-body simulations. Section 4 summarizes the results.

2. Basic Dynamical Processes in Disk Evolution

2.1. CLASSICAL ROCHE LIMIT

If the tidal effect of the Earth overwhelms self-gravity of a body orbiting the Earth, the body cannot be gravitationally bound and is sheared apart by the tidal force. Since the tidal effect increases with decreasing distance from the Earth, the gravitational binding is tidally inhibited in the vicinity of the Earth. The limit where the tidal force equals the self-gravitational force for an ideal-fluid body is called the Roche Limit.

The radius of the Roche Limit (a_R) is roughly given as follows. Suppose that a fluid sphere with mass M and radius R orbits circularly a central body with mass M_c and radius R_c. The orbital distance of the mass center of the sphere is a and the angular velocity is Ω $(= GM_c/a^3)$. The specific self-gravitational force on the surface of the fluid sphere is GM/R^2. The central body's gravitational force on the surface at the farthest point from the central body is $GM_c/(a+R)^2$, while the centrifugal force there is $(a+R)\Omega^2 = GM_c(a+R)/a^3$. The difference between the two forces (tidal force) at the farthest point is approximately

$$F_{\text{tidal}}(a + R) = \frac{GM_c(a + R)}{a^3} - \frac{GM_c}{(a + R)^2} \simeq \frac{3GM_c R}{a^3}. \tag{1}$$

If the self-gravitational force GM/R^2 exceeds $F_{\text{tidal}}(a + R)$, the fluid body is gravitationally bound. (A similar argument holds at $a - R$.) The binding condition is

$$1 > Q_R = 3 \left(\frac{M_c}{M} \right) \left(\frac{R}{a} \right)^3 = \frac{9}{4\pi} \left(\frac{\Omega^2}{G\rho} \right), \tag{2}$$

where ρ is the mean internal density of the orbiting sphere $(M = (4\pi/3)\rho R^3)$. The above condition is equivalent to

$$a > \left(\frac{3\rho_c}{\rho} \right)^{1/3} R_c, \tag{3}$$

$$R < R_H = \left(\frac{M}{3M_c} \right)^{1/3} a, \tag{4}$$

or

$$\rho > \frac{9}{4\pi} \left(\frac{M_c}{a^3} \right), \tag{5}$$

where ρ_c is the mean internal density of the central body $(M_c = (4\pi/3)\rho_c R_c^3)$. The radius R_H is the Hill radius (between the fluid sphere and infinitesimal fluid element at $a + R$) explained in Sect. 2.2. The quantities in the right hand side of Eqs. (3) and (5) represent the "Roche radius" and "Roche density", respectively, except for a numerical factor of the order of unity (see below). A more rigorous argument taking into account the deformation of the orbiting body under the tidal effect (*e.g.*, Chandrasekhar, 1969) gives the Roche radius as

$$a_R = 2.46 \left(\frac{\rho_c}{\rho} \right)^{1/3} R_c. \tag{6}$$

Note that if ρ is constant, a_R is independent of size R or mass M of the sphere. In the present Earth-Moon system, $\rho_c = \rho_\oplus = 5.5$ g/cm^3 and $\rho = 3.3$ g/cm^3, so that Eq. (6) reads $a_R = 2.9R_\oplus$ where $R_\oplus$ is the Earth radius.

2.2. SELF-GRAVITATIONAL INSTABILITY OF A DISK

Next we consider a self-gravitating fluid disk. The spatial density of the disk (ρ_{disk}) is

$$\rho_{\text{disk}} \sim \frac{\Sigma}{c_s/\Omega}, \tag{7}$$

where Σ is the surface density of the disk, c_s is the sound velocity, and c_s/Ω is the disk thickness. If we substitute ρ_{disk} for ρ in Eq. (2), the condition is

$$1 > Q_R \sim \frac{c_s \Omega}{G\Sigma}. \tag{8}$$

Linear stability analysis for a disk perturbed by axisymmetric perturbation (e.g., Toomre, 1964; Binney and Tremaine, 1987) shows that the condition for self-gravitational instability is

$$1 > Q = \frac{c_s \Omega}{\pi G\Sigma}, \tag{9}$$

where Q is Toomre's stability parameter. Although spherical contraction is considered in Eq. (8) while ring-mode contraction in Eq. (9), both conditions are very similar.

The linear theory gives a dispersion relation for the disk,

$$\omega^2 = \Omega^2 - 2\pi G\Sigma k + c_s^2 k^2, \tag{10}$$

where $k = 2\pi/\lambda$, λ is the wavelength of the axisymmetric perturbation, and ω is the growth rate of a perturbation (amplitude of the perturbation $\propto \exp(i\omega t)$). When $Q < 1$, $\omega^2 < 0$ for some k. Then the perturbation increases with time and clumps are formed. The tidal force prevents the clump formation for small k (large λ), while the pressure gradient does for large k (small λ). The largest unstable wavelength (the critical wavelength) is

$$\lambda_c = \frac{2\pi^2 G\Sigma}{\Omega^2}. \tag{11}$$

For a disk consisting of particles, the above relations hold, replacing c_s by the radial velocity dispersion (v_r) of the disk particles (e.g., Binney and Tremaine, 1987). With a_R, Q is rewritten as

$$Q \sim 0.1 \left(\frac{\rho_{\text{disk}}}{\rho}\right)^{-1} \left(\frac{a}{a_R}\right)^{-3}. \tag{12}$$

Since the spatial density ρ_{disk} is limited by the internal density ρ of the particles, Q cannot go down to ~ 1 in the vicinity of the central body.

N-body simulations of Saturn's ring (e.g., Salo, 1995; Daisaka and Ida, 1999) suggest that when a non-axisymmetric perturbation is included, clumping due to self-gravity occurs when $Q \sim 2$ rather than $Q \sim 1$ for axisymmetric perturbations. This may be because if bodies collide in the azimuthal rather than radial direction, the tidal force, which prevents the clump formation, is less effective.

2.3. TIDAL INHIBITION OF GRAVITATIONAL BINDING

We are interested in a problem similar to the classical Roche problem, namely, that two gravitating solid bodies (mass M_1 and M_2) collide with each other in the Earth's tidal potential. Constant internal density (ρ) of the bodies is assumed. The critical orbital radius at which the mutual gravitational force of the colliding bodies is equal to the tidal force is similar to a_R, but depends on mass ratio, M_1/M_2.

Ohtsuki (1993) addressed this problem in terms of the Jacobi energy in Hill's framework. The Hill coordinate system is defined so that the x axis points radially outward, the y axis is tangent to the circular orbit, and the z axis is normal to the orbital plane. The angular velocity of the coordinate system is the Keplerian orbital frequency, $\Omega = (GM_c/a^3)^{1/2}$, where a is the reference orbital radius, which can be taken to be the orbital radius of the mass-center of the two bodies. For more details, see $e.g.$, Nakazawa and Ida (1988). The linearized equations of motion of relative motion between the bodies in the Hill coordinates are

$$\begin{aligned}
\ddot{x} &= & 2\dot{y}\Omega &+ 3x\Omega^2 &- \frac{G(M_1+M_2)x}{r^3} \\
\ddot{y} &= & -2\dot{x}\Omega & &- \frac{G(M_1+M_2)y}{r^3} \\
\ddot{z} &= & &- z\Omega^2 &- \frac{G(M_1+M_2)z}{r^3},
\end{aligned} \quad (13)$$

where (x, y, z) give the relative position and $r = (x^2 + y^2 + z^2)^{1/2}$. The $(2\dot{y}\Omega)$ and $(-2\dot{x}\Omega)$ terms give the Coriolis forces, the terms proportional to $1/r^3$ are the mutual gravity of the orbiting bodies, and the $(3x\Omega^2)$ and $(-z\Omega^2)$ terms represent the tidal force (see Eq. (1)). Equations (13) have an energy integral called the Jacobi energy:

$$E_{\mathrm{J}} = \frac{1}{2}(\dot{x}^2 + \dot{y}^2 + \dot{z}^2) - \frac{3}{2}x^2\Omega^2 + \frac{1}{2}z^2\Omega^2 - \frac{G(M_1+M_2)}{r} + \frac{9}{2}R_{\mathrm{H}}^2\Omega^2, \quad (14)$$

where R_{H} is the mutual Hill radius defined by

$$R_{\mathrm{H}} = \left(\frac{M_1+M_2}{3M_c}\right)^{1/3} a. \quad (15)$$

The contours of E_{J} with $(\dot{x}, \dot{y}, \dot{z}) = (0, 0, 0)$ (the Hill potential) are shown in the Hill coordinates in Fig. 1. The last term in Eq. (14) is added so that E_{J} with zero velocity is zero at the Lagrangian points, $(x, y, z) = (\pm R_{\mathrm{H}}, 0, 0)$.

The Hill radius R_{H} is the characteristic radius of the potential well called the Hill sphere (the region surrounded by a thick line in Fig. 1). Since the Hill potential well is closed, if $E_{\mathrm{J}} < 0$, the two bodies in the Hill sphere are

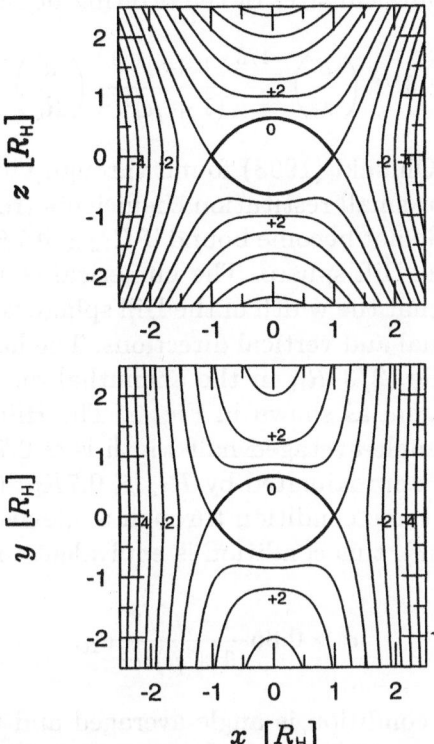

Figure 1. The contours of the Hill potential for the $z = 0$ plane (lower panel) and the $y = 0$ plane (upper panel); labels of the contours are the potential with time scaled by Ω^{-1} and length scaled by R_H. The area marked by a thick line is the Hill sphere. The Hill potential takes a negative value within the Hill sphere; however, the contours there are omitted.

gravitationally bound and those outside the Hill sphere cannot approach more closely than the Hill radius. Therefore the necessary and sufficient condition for the gravitational binding of the colliding bodies is

$$\text{I.} \quad E'_J < 0,$$
$$\text{and} \tag{16}$$
$$\text{II.} \quad (x, y, z) \text{ is within the Hill sphere,}$$

where E'_J is the Jacobi energy after the collision.

If we assume $M_2 \ll M_1 (= M)$ and $(x, y, z) = (R, 0, 0)$, condition (16) with $(\dot{x}, \dot{y}, \dot{z}) = (0, 0, 0)$ is the same as Eq. (3). With a_R, Eq. (3) reads as $a > 0.59\, a_R$. This condition is generalized to the cases with $y, z \neq 0$ and arbitrary M_1/M_2 as follows.

The sum of the physical sizes of the orbiting bodies (R_{12}) is

$$R_{12} = 3^{1/3} \left(\frac{\rho}{\rho_c}\right)^{-1/3} \frac{1 + \mu^{1/3}}{(1 + \mu)^{1/3}} \left(\frac{a}{R_c}\right)^{-1} R_H, \qquad (17)$$

where $\mu = M_1/M_2$. Ohtsuki (1993) found through numerical orbital inter-grations that even for small restitution coefficients (rebound velocity ~ 0), the colliding bodies do not become bound if $R_{12} \gtrsim 0.7 R_H$, because the phys-ical sizes overflow the Hill sphere. The numerical factor 0.7 rather than 1 comes from the fact that the width of the Hill sphere is smaller than the Hill radius in the azimuthal and vertical directions. The half-width is R_H in the radial direction, while $(2/3) R_H$ in the azimuthal direction and $\simeq 0.64 R_H$ in the vertical direction, as shown in Fig. 1. The Hill "sphere" is actually lemon-shaped. The angle-averaged half-width is $\simeq 0.7 R_H$. Thereby, condi-tion II in Eq. (16) is approximated by $R_{12} \lesssim 0.7 R_H$. (Canup and Esposito (1995) obtained the same condition through angle-averaging of the Jacobi energy.) With Eq. (17), this condition is equivalent to

$$a > 0.85 \frac{1 + \mu^{1/3}}{(1 + \mu)^{1/3}} a_R. \qquad (18)$$

Although the above condition is angle-averaged and the rebound velocity generally has a nonzero value, it reflects well the actual binding events and indicates well the dynamical properties of collisions in the tidal field, for example, that the binding between comparable-sized bodies is more restricted than that between different-sized bodies (see Fig. 2), as shown by KIM00's N-body simulations, which check the exact condition (16) at every collision. Canup and Esposito called the range of $0.85\, a_R \lesssim a \lesssim 1.4\, a_R$ where binding is mass ratio dependent the "Roche zone".

3. N-Body Simulation

In the evolution of a protolunar disk, global effects such as radial migra-tion of lunar material associated with rapid angular momentum transfer by the spiral arms, formation of particle aggregates (Moon seeds) by self-gravitational instability, and interaction of formed moonlets with the disk are important as well as the tidal inhibition of gravitational binding. N-body simulation automatically calculates such global gravitational effects and takes into account the tidal inhibition. On the other hand, its compu-tational cost is in general expensive.

The main result of the N-body simulations by ICS97 and KIM00 is that in most cases a single large moon is formed just outside the Roche limit on a nearly non-inclined circular orbit on a time scale of a month to a

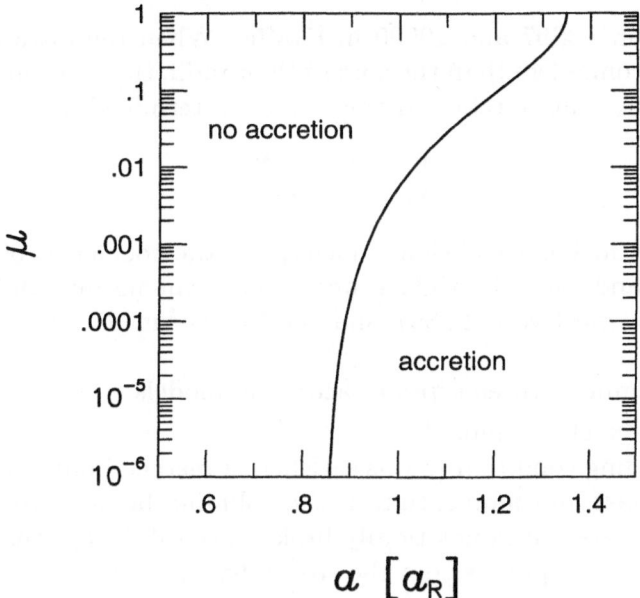

Figure 2. The critical orbital radius for gravitational binding as a function of mass ratio (Eq. (18)) for completely inelastic collisions ($\epsilon = 0$ or $v' = 0$)). Here a_R is the classical Roche limit given by Eq. (6).

year. Here we explain the basic physics which leads to the above conclusion, summarizing the results of these N-body simulations.

3.1. METHOD OF CALCULATION

In an N-body simulation, the orbits of particles are calculated by numerically integrating the equation of motion,

$$\frac{dv_i}{dt} = -GM_\oplus \frac{x_i}{|x_i|^3} - \sum_{j \neq i}^{N} Gm_j \frac{x_i - x_j}{|x_i - x_j|^3}, \quad (19)$$

where m_i, x_i, and v_i are the mass, the position, and the velocity of disk particle i, respectively, and G is the gravitational constant. The first and second terms of the r.h.s. of Eq. (19) represent the gravitational force by the Earth and mutual gravitational interaction of disk particles. ICS97 and KIM00 did not include orbital evolution due to tidal bulges raised on the Earth (*e.g.*, Mignard, 1980), since the time scale of the tidal orbital evolution is much longer than the accretional time scale (Canup and Esposito, 1996). For their numerical integrations, they used the predictor-corrector type Hermite scheme (Makino and Aarseth, 1992; Kokubo *et al.*, 1998).

Since the most expensive part of N-body simulation is the calculation of the mutual gravitational force, the number of integrated particles is limited to 1000–1500 in ICS97 and 10000 in KIM00. When the distance between two bodies becomes less than the sum of their radii, the two colliding bodies rebound with a relative rebound velocity v', determined by

$$
\begin{aligned}
v'_n &= -\epsilon_n v_n \\
v'_t &= \epsilon_t v_t,
\end{aligned}
\tag{20}
$$

where v is the incident collision velocity, ϵ is the coefficient of restitution ($0 \leq \epsilon \leq 1$), and the subscripts n and t represent normal and tangential components, respectively. ICS97 and KIM00 assumed $\epsilon_t = 1$, neglecting spins of bodies.

KIM00 examined three types of accretion models.

1. "perfect accretion" model:
   If the binding condition (16) is satisfied, a merged body is created conserving mass and momentum of the colliding bodies. Once a merged body is created, it is not tidally broken even if it happens to be scattered into the region within the Roche limit.
2. "rubble pile" model:
   No merged body is created. Outside the Roche limit, gravitationally bound aggregates are formed. If a bound aggregate is scattered to the region within the Roche limit, the aggregate is tidally broken up. On the other hand, if a bound aggregate is located well outside the Roche limit, the aggregate is no longer tidally broken.
3. "partial accretion" model:
   The reality may be between models 1 and 2. Melting due to collisional heating may result in a merged body with material strength rather than an only gravitationally bound aggregate, outside the Roche limit. In this case, the merged body would not easily be tidally broken, even when it is scattered to the region within the Roche limit. Hence, KIM00 considered a model between models 1 and 2 such that a merged body is created if $E'_J < 0$ and $R_{12} < 0.7 R_H$. The latter corresponds to the angle-averaged necessary condition for gravitational binding, discussed in section 2.3. ICS97 adopted this model.

KIM00 showed that a formed moon is almost similar in all the models, although in the rubble pile model, the moon gradually loses mass through stripping by tidal forces and collision with the other bodies and this results in a slightly smaller final moon mass than that in the other models. The disk evolution within the Roche limit is better understood in the rubble pile model, while statistics of a formed moon is better defined in the other models.

ICS97 and KIM00 start orbital integrations of the disk particles with various initial mass distributions, because the initial state is rather uncertain. They modeled the disk with power-law surface density distribution

$\Sigma \propto a^{-\beta}$ with cut-off $R_\oplus < a < a_{max}$ ($a_{max} \simeq 1.0$–$2.5a_R$) and power-law particle mass distribution $n \propto m^{-\alpha}$ ($0.5 < \alpha < 2.0$; $1 < \beta < 5$). The total mass of the initial disk is $M_{disk} \simeq 1.6$–$4M_L$ ($\simeq 0.02$–$0.05M_\oplus$), where M_L is the present lunar mass ($\simeq 0.0123M_\oplus$). The internal density of disk particles and proto-Earth are assumed to be $\rho = 3.3$ gcm^{-3} and $\rho_\oplus = 5.5$ gcm^{-3}.

3.2. DYNAMICAL EVOLUTION OF THE DISK

ICS97 and KIM00 found that in most cases, a single large moon is formed on a nearly non-inclined circular orbit just outside the Roche limit on a time scale of the order of 100–1000 T_K, where T_K is the Kepler period at the distance of the Roche limit ($a_R \simeq 2.9R_\oplus$) and $T_K \simeq 7$ hours. The result is essentially independent of the initial conditions of the disk and the adopted collision model, as long as a relatively massive ($M_{disk} \gtrsim M_L$) and compact ($a_{max} \lesssim a_R$) disk is considered.

Figure 3 shows the evolution of a protolunar disk of the run as simulated by KIM00. The initial disk parameters are $M_{disk} = 4M_L$ with $\alpha = \infty$ (equal mass), $\beta = 3$, $a_{max} = a_R$, and $\langle e^2 \rangle^{1/2} = \langle i^2 \rangle^{1/2} = 0.3$. The partial accretion model with $\epsilon_n = 0.1$ and $\epsilon_t = 1$ is adopted. Figure 3 shows snapshots in the r-z plane at $t = 0, 10, 30, 100, 1000\,T_K$. The mass of the material ejected from the gravitational field of the Earth is usually smaller than 5% of M_{disk}. At 1000 T_K, a single large moon forms around $a \simeq 1.3a_R$ with mass M_L on a nearly non-inclined circular orbit. 90% of the final mass has been already acquired by 100 T_K. In this rapid growth stage before $\sim 100\,T_K$, radial diffusion of disk material through angular momentum transfer supplies material for accretion outside the Roche limit: a significant fraction of the disk mass falls to the Earth while its counterpart is transported outward. The moon rapidly accretes in this stage. In the slow growth stage after $\sim 100\,T_K$, the moon gradually sweeps up or scatters the residual disk mass onto the Earth.

In the rapid growth stage, density spiral arms develop, which are responsible for the rapid angular momentum transfer. The fast accretion of the moon outside the Roche limit is collective gravitational instability rather than pairwise accretion. This is most easily seen by the rubble pile model. Figure 4 shows snapshots of the radial profile of the surface density in the x-y plane at $t = 0, 1, 5, 10, 20, 40\,T_K$. The initial conditions of the disk here are the same as shown above for the partial accretion model. The disk evolution in the rapid growth stage is as follows.

1. "Cooling" of the disk:
   The initial disk has large enough eccentricities and inclinations ("hot") to be gravitationally stable, that is, $Q > 1$, where Q is defined by Eq. (9). The mean collision time between the disk particles is of the order

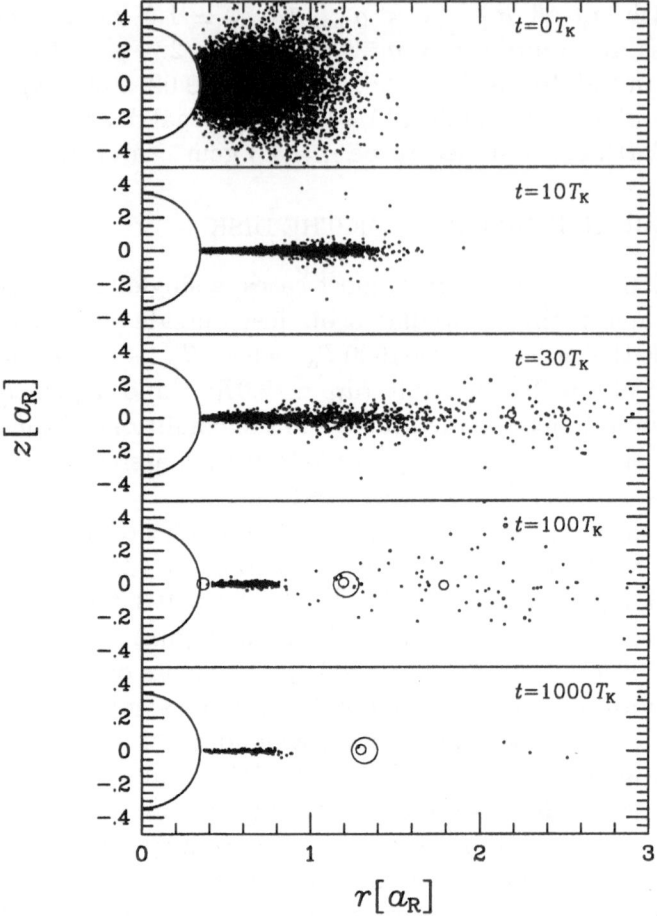

Figure 3.   Snapshots of the protolunar disk in the $r - z$ plane at $t = 0$, 10, 30, 100, $1000\,T_K$, in a partial accretion model. The semicircle centered at the coordinate origin stands for the Earth. Circles represent disk particles and their size shows the physical size of disk particles. After Kokubo, Canup, and Ida (2000).

of T_K in the disks we consider. Inelastic collisions damp eccentricities and inclinations of disk particles on a time scale similar to the mean collision time.

2. Self-gravitational instability:
   As the velocity dispersion decreases, Q decreases down to ~ 2 and the disk becomes gravitationally unstable except for very small a (Eq. (12)). The size of the clumps is roughly consistent with λ_c given by Eq. (11). For $a = 0.5a_R$ and $\Sigma = 0.03M_\oplus a_R^{-2}$, $\lambda_c \sim 0.1a_R$, which is consistent with the clumps observed in Fig. 4.

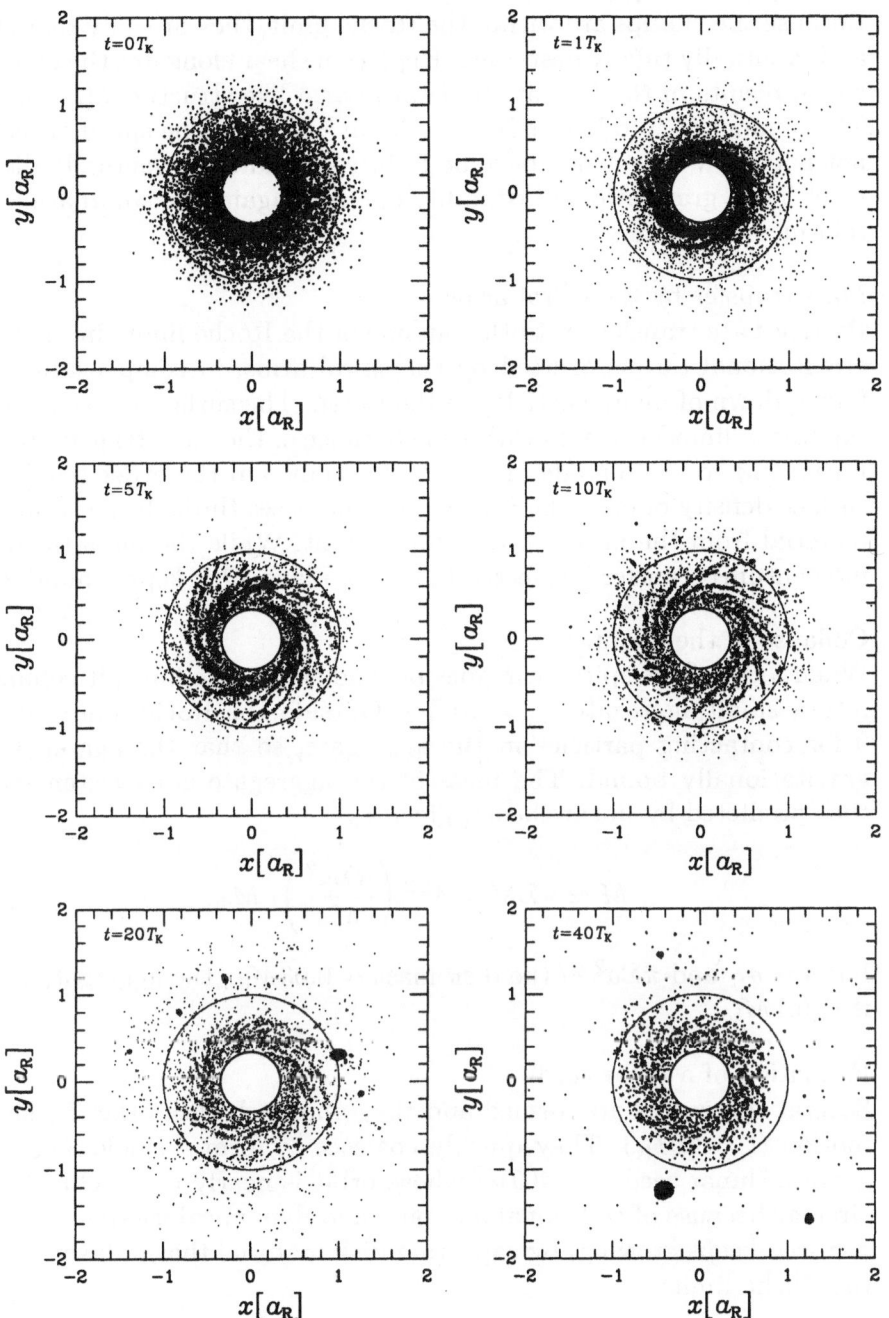

Figure 4. Snapshots of the disk evolution in the $x-y$ plane at $t = 0, 1, 5, 10, 20, 40\,T_K$, in a rubble pile model. After Kokubo, Canup, and Ida (2000).

3. Development of spiral arms:
   Because the clumps are within the Roche limit, they are only temporal
   and eventually tidally destroyed. Keplerian shear elongates the clumps,
   which results in the formation of spiral arm-like structures ($t = 5T_K$).
   The number of spiral arms is typically $m \sim 10$–20. The spiral arms are
   not pattern waves. They become tightly wound and eventually disap-
   pear. Then gravitational instability operates again and spiral arms are
   reformed.

4. Mass transfer by the spiral arms:
   Particles are transferred to the outside of the Roche limit through the
   gravitational torque exerted by the spiral arms, in compensation for
   falling down of many particles to the Earth. The surface density inside
   the Roche limit decreases with time because of the mass transfer to the
   Earth and to the outside of the Roche limit. On the other hand, the
   surface density outside the Roche limit increases through the supply of
   material from the inside of the Roche limit. While the mass (angular
   momentum) transfer is effectively taking place, Q is kept around 2.

5. Collapse of the clumps:
   When a tip of the spiral arm goes beyond the Roche limit, it collapses
   into a small aggregate ($t = 10T_K$). Outside the Roche limit, $E_J <$
   0 for contacting particles in the aggregate, so that the aggregate is
   gravitationally bound. The mass of the aggregate is consistent with
   that predicted by linear theory, given by

   $$M \simeq \pi \Sigma \lambda_c^2 = 4\pi^2 \left(\frac{\pi \Sigma a^2}{M_\oplus} \right)^3 M_\oplus, \tag{21}$$

   For $a = a_R$ and $\pi \Sigma a^2 \sim$ the disk mass $\sim 0.03 M_\oplus$, the aggregate mass
   is $\sim 0.1 M_L$.

6. Formation of a lunar seed:
   Several aggregates are formed and they approach each other by shear
   motion ($t = 20T_K$). They quickly coalesce to form a single large aggregate (lunar seed; $t = 40T_K$) whose orbit is nearly non-inclined and
   circular because of collisional damping and dynamical friction. The lunar seed keeps growing, by capturing disk material that diffuses out of
   the Roche limit.

KIM00 showed that the time scale of the rapid growth stage is of the
order of $100\,T_K$ ($\sim$ a month). This time scale is regulated by the angular momentum transfer. As the moon is primarily formed by the material
transferred beyond the Roche limit, the time scale of the moon formation is

almost equivalent to the time scale of the mass transfer due to the gravitational torque by spiral arms, because accretion of the moon is fast enough (Takeda *et al.*, 2000). The time scale of mass transfer is also equivalent to that of angular momentum transfer in the disk. The angular momentum transfer is dominated by the gravitational torque exerted by the spiral arms if the initial number of simulated bodies is larger than about a few thousands (Takeda *et al.*, 2000). The gravitational torque was evaluated by Lynden-Bell and Kalnajs (1972) under the WKB approximation, as

$$F_{\mathrm{g}} = \frac{\pi^2 G a m (\Delta\Sigma)^2}{k^2}, \tag{22}$$

where k is the radial wave number, $m = kR\tan\theta$ (θ is the pitch angle), and $\Delta\Sigma$ is the density contrast of the arms. Using Eq. (11) and results by the simulations, $\Delta\Sigma \sim \Sigma$ and $\tan\theta \sim 1$, the torque reads (KIM00; Takeda *et al.*, 2000)

$$F_{\mathrm{g}} \sim \frac{\pi^3 G a^5 \Sigma^3}{M_{\oplus}}, \tag{23}$$

The effective viscosity by the gravitational torque is (Lynden-Bell and Pringle, 1974)

$$\nu_{\mathrm{g}} = \frac{F_{\mathrm{g}}}{3\pi a^2 \Sigma \Omega}. \tag{24}$$

The time scale of the angular momentum transfer by the spiral arms is estimated as

$$T_{\mathrm{g}} \equiv \frac{(\Delta a)^2}{\nu_{\mathrm{g}}} \sim 10^2 \left(\frac{\Sigma}{0.01 M_{\oplus} a_{\mathrm{R}}^{-2}}\right)^{-2} \left(\frac{\Delta a}{0.5 a_{\mathrm{R}}}\right)^2 \left(\frac{a}{a_{\mathrm{R}}}\right)^{-9/2} T_{\mathrm{K}}, \tag{25}$$

where Δa is the length of diffusion due to angular momentum transfer. This time scale agrees well with the results of the N-body simulation by KIM00 and ICS97. Ward and Cameron (1978) obtained almost the same time scale by considering the energy dissipation in the clumps formed by gravitational instability.

The functional form of the time scale shows that the time scale of angular momentum transfer, in other words, the time scale of moon formation, depends not on the individual mass of disk particles but on the surface density of the disk. The mass of the first-born bound aggregates is determined by Eq. (21), which is a function of the surface density but not of the individual mass of disk particles. Thus the disk evolution and moon accretion are regulated by the disk surface density distribution. Actually, ICS97 and KIM00 showed similar results, although the initial individual mass of disk particles differed by a factor 10 between the two studies.

After $\sim 100 \, T_{\mathrm{K}}$, growth of the lunar seed becomes slow. The final stage of the disk evolution is as follows.

7. Clearing of the inner disk:
   As the Moon becomes massive, it gradually pushes the residual inner disk toward the Earth through disk-Moon gravitational interactions. The Hill radius of the Moon is $R_H \simeq 0.16(M/M_L)^{1/3}a$. The region of the disk within a radial distance of several R_H from the moon is strongly perturbed (*e.g.*, Ida, 1990), so that almost the entire region of the disk is affected by the gravity of a moon with $M \gtrsim M_L$. As a result, the inner disk is almost completely cleared. By recoil, the seed is slightly moved outward.

The disk evolution explained above would not depend on the initial conditions of the disk nor the accretion model. A single large moon just outside of the Roche limit is the typical outcome of the evolution of the protolunar disk, as long as the disk is relatively massive ($M_{disk} \gtrsim M_L$) and compact ($a_{max} \lesssim$ 1–2 a_R). Note, however, that if the initial disk is radially well extended across the Roche limit, in situ accretion is possible outside the Roche limit, which may result in a multiple moon system.

3.3. DYNAMICAL CHARACTERISTICS OF THE FORMED MOON

Utilizing the simulation result, the final mass of the Moon is semi-analytically estimated (ICS97). KIM00 surveyed the disk evolution with various initial disk mass distributions and various initial size distributions of the disk particles. Different normal coefficients of restitution, 0.1 and 0.01, and different accretion models ("perfect", "partial", and "rubble pile") were also tested. However, the result that a single large moon is formed at $a \simeq 1.3\,a_R$ is essentially the same in all the simulations. The result is also essentially the same as that of ICS97.

The mass of the Moon scaled by the initial disk mass at $t = 1000\,T_K$ is plotted against the initial specific angular momentum of the disk J_{disk}/M_{disk} in Fig. 5 (KIM00). Similar results were also obtained by ICS97. The initial specific angular momentum of the disks was surveyed over $0.62\sqrt{GM_\oplus a_R} \lesssim J_{disk}/M_{disk} \lesssim 1.0\sqrt{GM_\oplus a_R}$, where $\sqrt{GM_\oplus a_R}$ is the specific angular momentum of a body circularly orbiting at a_R. The efficiency of incorporation of disk material into a moon is 10–55%, which increases linearly with J_{disk}/M_{disk}. This relation shows that in a smaller J_{disk}/M_{disk} disk, in other words, in a more compact disk, a larger amount of mass must fall to the Earth to transfer mass across the Roche limit and thus the mass of the Moon is small. The mass fraction of the escapers also increases with J_{disk}/M_{disk}. The fraction is, however, usually less than 10% and thus has little effect on the lunar mass.

ICS97 explained the relation of the lunar mass M and the angular momentum of the protolunar disk J_{disk} obtained by the N-body simulation,

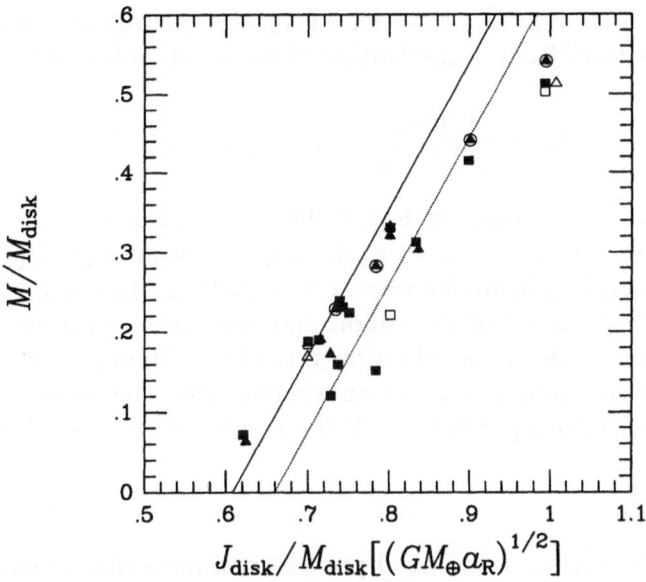

Figure 5.   The mass fraction of the material incorporated into the Moon to the initial disk mass M/M_{disk} is plotted as a function of the initial specific angular momentum of the disk $J_{\text{disk}}/M_{\text{disk}}$. The triangles correspond to the results of N-body simulations with $N = 10000$ with the initial disk mass $M_{\text{disk}} = 2M_{\text{L}}$ and the squares the results with $M_{\text{disk}} = 4M_{\text{L}}$. The filled triangles and squares mean that the coefficient of restitution $\epsilon_n = 0.1$ and the open ones $\epsilon_n = 0.01$. The circles indicate that the Moon has a second largest moonlet whose mass is larger than 20% of the lunar mass on a horseshoe orbit. In the cases in which the Moon has particles on horseshoe orbits, the sum of the mass of the Moon and the horseshoe orbiters are plotted. The theoretical estimate given by Eq. (27) is also shown for $M_{\text{esc}} = 0$ (solid line) and $M_{\text{esc}} = 0.05M_{\text{disk}}$ (dotted line). After Kokubo, Canup, and Ida (2000).

using the conservation of angular momentum. The conservation of the angular momentum of the protolunar disk gives

$$
\begin{aligned}
J_{\text{disk}} \simeq \quad & M\sqrt{GM_\oplus(1-e^2)a} \\
+ \; & (M_{\text{disk}} - M - M_{\text{esc}})\sqrt{GM_\oplus(1-e_{\text{fall}}^2)a_{\text{fall}}} \\
+ \; & M_{\text{esc}}\sqrt{GM_\oplus(1-e_{\text{esc}}^2)a_{\text{esc}}},
\end{aligned}
\qquad (26)
$$

where a, e, a_{fall}, e_{fall}, a_{esc}, and e_{esc} are the typical semimajor axis and eccentricity of the Moon, the Earth impactors, and the escapers, respectively. Here we assume that the inclinations of bodies are much smaller than unity (in radians) and the Moon accretion is completed. Note that $(1 - e_{\text{esc}}^2)a_{\text{esc}} = q_{\text{esc}}(1 + e_{\text{esc}})$ where q_{esc} stands for the perigee distance of the escapers (a similar relation holds for the Earth impactors). The mean values of each orbital element obtained by KIM00 are $\bar{a} = 1.3a_{\text{R}}$, $\bar{e} = 0.04$, $\bar{q}_{\text{fall}} = 0.3a_{\text{R}}$, $\bar{e}_{\text{fall}} = 0.2$, $\bar{q}_{\text{esc}} = 1.3a_{\text{R}}$, and $\bar{e}_{\text{esc}} = 1.1$. The relation

$\bar{q}_{esc} \simeq 1.3\bar{a}$ reflects the fact that mass is ejected mainly due to gravitational scattering by the Moon. Substituting these mean values into Eq. (26), we obtain

$$M \simeq \frac{1.9 J_{disk}}{\sqrt{GM_\oplus a_R}} - 1.1 M_{disk} - 1.9 M_{esc}. \tag{27}$$

This estimate is also shown in Fig. 5. The results of the N-body simulation agree well with the above analytic estimate except for high J_{disk}/M_{disk} cases where accretion is still under way at $t = 1000\,T_K$. In the high J_{disk}/M_{disk} cases, sum of the mass of the Moon and the mass of the particles bound to the Earth outside the semimajor axis of the Moon, which would likely be the finallunar mass, is more consistent with the analytical estimate. Therefore, Eq. (27) reproduces well the results of N-body simulations.

4. Summary

We have reviewed the dynamical processes of accretion of the Moon from an impact-generated disk, utilizing the results of direct N-body simulations performed by ICS97 and KIM00. We considered a relatively massive ($M_{disk} \gtrsim M_L$) and compact ($a_{max} \lesssim 2a_R$) disk.

Within the Roche limit with radius $\sim 3R_\oplus$, accretion is inhibited by tidal forces; however, density spiral arms develop. Relatively rapid (*i.e.,* $\sim 100 T_K$) radial migration of disk mass is induced by the torque due to the spiral arms. The disk material transferred outside the Roche limit gravitationally collapses to form particle aggregates. (To lift part of the disk material across the Roche limit, a significant amount of the other disk material falls onto the Earth.) The aggregates collide with each other and a single large moon is formed at about 3–4$R_\oplus$. The typical time scale of the spreading of the disk mass is $\sim 100\,T_K$ ($\sim$ a month). During the mass spreading time scale, most mass of the Moon accretes. After that, the residual inner disk is scattered onto the Earth by the Moon's gravity on a time scale $\sim 1000\,T_K$ ($\sim$ a year). This result hardly depends on the initial condition of the disk as long as a relatively massive and compact disk is considered. Using the result that the Moon is usually formed around $a \sim 1.3\,a_R$, the mass of the Moon is predicted simply by the conservation of angular momentum of the initial disk. The accretion efficiencies (the fraction of disk material incorporated into the Moon) range from 10% to 55%.

References

Benz, W., Slattery, W.L., and Cameron, A.G.W. (1986) The origin of the moon and the single impact hypothesis I, *Icarus* **66**, 515-535.
Benz, W., Slattery, W.L., and Cameron, A.G.W. (1987) The origin of the moon and the single impact hypothesis II, *Icarus* **71**, 30-45.

Benz, W., Cameron, A.G.W., and Melosh, H.J. (1989) The origin of the moon and the single impact hypothesis III, *Icarus* **81**, 113-131.

Binney, J., and Tremaine, S. (1987) *Galactic Dynamics*, Princeton University Press, Princeton.

Cameron, A.G.W. (1997) The origin of the moon and the single impact hypothesis V, *Icarus* **126**, 126-137.

Cameron, A.G.W., and Benz, W. (1991) The origin of the moon and the single impact hypothesis IV, *Icarus* **92**, 204-216.

Cameron, A.G.W., and Ward, W.R. (1976) The origin of the Moon, in *Lunar and Planetary Science VII*, Lunar and Planetary Institute, Houston, pp. 120-122.

Canup, R.M., and Esposito, L.W. (1995) Accretion in the Roche zone: Co-existence of rings and ringmoons, *Icarus* **113**, 331-352.

Canup, R.M., and Esposito, L.W. (1996) Accretion of the Moon from an impact-generated disk, *Icarus* **119**, 427-446.

Canup, R M., Levison, H.F., and Stewart, G.R. (1999) Evolution of a terrestrial multiple-moon system, *Astron. J.* **117**, 603-620.

Chambers, J.E., Wetherill, G.W., and Boss, A.P. (1996) The stability of multi-planet systems, *Icarus* **119**, 261-268.

Chambers, J.E., and Wetherill, G.W. (1998) Making the terrestrial planets: N-Body integrations of planetary embryos in three dimensions, *Icarus* **136**, 304-327.

Chandrasekhar, S. (1969) *Ellipsoidal figures of equilibrium*, Yale University Press.

Daisaka, H., and Ida, S. (1999) Spatial structure in dense planetary ring induced by self-gravitational instability, *Earth, Planet and Space*, **51**, 1195-1213.

Hartmann, W.K., and Davis, D.R. (1975) Satellite-sized planetesimals and lunar origin, *Icarus* **24**, 504-515.

Hayashi, C., Nakazawa, K., and Nakagawa, Y. (1985) Formation of the solar system, in *Protostars and Planets II*, eds. D.C. Black and M.S. Matthews, Univ. of Arizona Press, pp. 1100-1153.

Ida, S. (1990) Stirring and Dynamical Friction of Planetesimals in the Solar Gravitational Field, *Icarus* **88**, 129-145.

Ida, S., Canup R.M., and Stewart, G.R. (1997) Lunar accretion from an impact-generated disk, *Nature* **389**, 353-357.

Kipp, M.E., and Melosh, H. J. (1987) A numerical study of the giant impact origin of the Moon: The first half hour, in *Lunar and Planetary Science XVIII*, Lunar and Planetary Institute, Houston, pp. 491-492.

Kokubo, E., and Ida, S. (1998) Oligarchic growth of protoplanets, *Icarus* **131**, 171-178.

Kokubo, E., and Ida, S. (2000) Formation of Protoplanets from Planetesimals in the Solar Nebula, *Icarus* **143**, 15-27.

Kokubo, E., Canup, R.M., and Ida, S. (2000) Lunar accretion from an impact-generated disk, in *Origin of the Earth and Moon*, eds. R.M. Canup and K. Righter, Univ. of Arizona Press, in press.

Kokubo, E., Ida, S., and Makino, J. (2000) Evolution of a circumterrestrial disk and formation of a single moon, submitted.

Kokubo, E., Yoshinaga, K., and Makino, J. (1998) On a time-symmetric Hermite integrator for planetary N-body simulation, *Mon. Not. R. Astron. Soc.* **297**, 1067-1072.

Lissauer, J.J., and Stewart, G.R. (1993) Growth of planets from planetesimals, in *Protostars and Planets III*, eds. E.H. Levy and J.I. Lunine, Univ. of Arizona Press, pp. 1061-1088.

Lissauer, J.J. (1987) Time scales for planetary accretion and the structure of the protoplanetary disk, *Icarus* **69**, 249-265.

Lynden-Bell, D., and Kalnajs, A.J. (1972) On the generating mechanism of spiral structure, *Mon. Not. R. Astron. Soc.* **157**, 1-30.

Lynden-Bell, D., and Pringle, J.E. (1974) The evolution of viscous discs and the origin of the nebular variable, *Mon. Not. R. Astron. Soc.* **168**, 603-637.

Makino, J., and Aarseth, S.J. (1992) On a Hermite integrator with Ahmad-Cohen scheme for gravitational many-body problems, *Publ. Astron. Soc. Jpn.* **44**, 141-151.

Mignard, F. (1980) The evolution of the lunar orbit revisited, II, *Moon and Planets* **23**, 185-201.

Nakazawa, K., and Ida, S. (1988) Hill's approximation in the three-body problem, *Prog. Theor. Physics Supp.* **96**, 167-174.

Ohtsuki, K. (1993) Capture probability of colliding planetesimals: Dynamical constraints on the accretion of planets, satellites and ring particles, *Icarus* **106**, 228-246.

Ohtsuki, K., Ida, S., Nakagawa, Y., and Nakazawa, K. (1993) Planetary accretion in the solar gravitational field, in *Protostars and Planets III*, eds. E.H. Levy and J.I. Lunine, Univ. of Arizona Press, pp. 1089-1107.

Safronov, V. (1969) *Evolution of the protoplanetary cloud and formation of the Earth and planets*, Nauka Press, Moscow.

Salo, H. (1995) Simulations of dense planetary rings. III. Self-gravitating identical particles, *Icarus* **117**, 287-312.

Stevenson, D.J. (1987) Origin of the Moon, *Ann. Rev. Earth Plant. Sci.* **15**, 271-315.

Takeda, T., Tanaka, H., and Ida, S. (2000) in preparation

Thompson, C., and Stevenson, D.J. (1988) Gravitational instability in two-phase disks and the origin of the Moon, *Astrophys. J.* **333**, 452-481.

Toomre, A. (1964) On the gravitational stability of a disk of stars *Astrophys. J.* **139**, 1217-1238.

Ward, W.R., and Cameron, A.G.W. (1978) Disk evolution within the Roche limit, in *Lunar and Plan. Sci. IX*, Lunar and Planetary Institute, Houston, pp. 1205-1207.

Wetherill, G.W., and Stewart, G.R. (1989) Accumulation of a swarm of small planetesimals, *Icarus* **77**, 330-357.

VOLATILE INVENTORY AND EARLY EVOLUTION OF THE PLANETARY ATMOSPHERES

MIKHAIL YA. MAROV AND SERGEI I. IPATOV
Institute of Applied Mathematics
Miusskaya sq. 4, Moscow 125047, Russia

Abstract. Formation of atmospheres of the inner planets involved the concurrent processes of mantle degassing and collisions that culminated during the heavy bombardment. Volatile-rich icy planetesimals impacting on the planets as a late veneer strongly contributed to the volatile inventory. Icy remnants of the outer planet accretion significantly complemented the accumulation of the lithophile and atmophile elements forced out onto the surface of the inner planets from silicate basaltic magma enriched in volatiles. Orbital dynamics of small bodies, including near–Earth asteroids, comets, and bodies from the Edgeworth–Kuiper belt evolving to become inner planet crossers, is addressed to examine different plausible amounts of volatile accretion. The relative importance of comets and chondrites in the delivery of volatiles is constrained by the observed fractionation pattern of noble gas abundances in the atmospheres of inner planets. The following development of the early atmospheres depended on the amount of volatiles expelled from the interiors and deposited by impactors, while the position of the planet relative to the Sun and its mass affected its climatic evolution.

1. Introduction

Scenarios of formation and evolution of the planetary atmospheres are intimately related to the key problems of Solar System origin, involving the whole swarm of primordial bodies formed in the processes of mutual interaction. Unlike the outer gas envelopes of the giant planets which were preserved essentially non-modified, the atmospheres of the terrestrial planets appear to have experienced dramatic changes passing through the complicated phases of their formation. Therefore, the Earth and its neighboring planets Venus and Mars (barring Mercury, essentially deprived of atmo-

M. Ya. Marov and H. Rickman (eds.), Collisional Processes in the Solar System, 223–247.
© 2001 *Kluwer Academic Publishers. Printed in the Netherlands.*

sphere) are addressed to answer such fundamental questions as: how during the make up of these rocky planets did the bulk of their atmospheres and/or hydrospheres ultimately form; which physico-chemical mechanisms gave rise to the observed features and striking differences between them; and why did the planetary bodies located within a very limited space of our Solar System embark on different evolutionary paths?

2. Formation scenario

One may assume that in the formation of planetary atmospheres three fundamental mechanisms were at work with various degrees of effectiveness: fractionation of the original proto–planetary cloud, different degrees of degassing of composed matter, and heterogeneous accretion associated with late veneer deposits. The differentiation process, culminating in the early formation of the core and silicate-oxide mantle and accompanied by large-scale release of gases from the interior, ran in parallel to the final accretion of volatile rich material through numerous impacts. Differences in the follow up degassing mechanisms on Earth, Venus, and Mars were apparently regulated by their masses, which determined their thermal history as exhibited by many specific features of surface landforms, geology and internal structure, while the location relative to the Sun mostly influenced the climatic evolution.

Indeed, Earth and Venus differ sharply from the Moon, Mercury and Mars, which are smaller. The latter are characterized by a globally continuous non-segmented lithosphere which became stabilized in the early history of the Solar System, and their ancient surfaces have preserved numerous impact structures, thick lava flows and volcanism. In contrast, the distinguishing characteristic of Earth's geology is a segmented, transversely shifting lithosphere, whose individual blocks are involved in global spreading processes, divergence and subduction zones, and convergence zones (Head and Crumpler, 1987). The surface of Venus is relatively young and displays abundant volcanism and clearly defined evidence of tectonic activity, such as distinctive indications of regional stress–strain and extensive deformations along the horizontal as well as the vertical; it is tempting to link this with the presence of spreading in the crustal layer, continuing to modify the planet's relief (Basilevsky et al., 1997; Marov and Grinspoon, 1998). These features could be linked with the removal of heat from Venus' interior, likewise the Earth which loses most of its internal heat through this mechanism. It is not the case for single-plate bodies of smaller size like Mars, where the outflow of heat seems to have occurred mainly through conduction. Consequently, from the point of view of thermal history, both Earth and Venus may have experienced quite similar evolutionary processes since the accretion of both planets. At the same time, Venus, as well as Mars, are strictly different from Earth by their atmospheres, hostile climate, and the lack of liquid water on their surfaces.

Proceeding from the scenario of planetary accretion of cold material in the proto–planetary gas–dust nebula, it is supposed that the initial composition of this material, throughout the planetary formation zone, corresponded to the cosmic elemental abundances. While the most abundant elements, hydrogen and helium, were only retained by the massive cold planets, composed primarily of these gases at considerable distances from the segregated central condensation – the Sun, the terrestrial planets were formed by heavier and cosmically less abundant elements entering into the composition of an iron–silicate phase. Evidently, because the fractionation of elements in the course of their condensation is in an order inverse to their volatility, a certain change in the initial composition of the constituent material with heliocentric distance occured. In other words, the thermodynamic state in that region of the proto-planetary cloud where the planet formed, and corresponding mineralogical associations controlled by chemical equilibrium conditions, caused the selective accumulation of condensates as they emerge (see, e.g., Marov, 1986; Marov and Grinspoon, 1998). This process should result in a lack of low temperature volatiles in the region of terrestrial planets formation at temperatures around 1000 K. The complement of high temperature condensates in the vicinity of the Sun, the lightest atmophile elements and, in part, other volatiles were dissipated into outer space due to the small mass of these planets. Electromagnetic and corpuscular radiation may also have exerted a considerable influence. This process, usually associated with young stars passing through the T–Tauri evolutionary phase, would have swept away a portion of the original nebular material which did not accumulate, primarily volatiles not absorbed by the original particles.

Thus the primordial atmospheres of the inner planets produced through the early outgassing of the differentiation phase were apparently lost. The current, secondary atmospheres are formed by volatiles exhiled by low-melting point basaltic material, in response to the thermal energy of radioactive decay and density redistribution in the mantle provided by the potential energy of accretion, and owing to heavy impacts. Gases released upon impact of volatile-containing planetesimals during accretion, consolidating through collisions into protoplanets, could be an important source of atmophiles. Impactors incident, culminating at the period of heavy bombardment about 4.0 Gy ago, are thought to be responsible for both erosion of inner planet surfaces and supply of volatiles. The fundamental idea is that after formation of the main planetary mass, asteroid-size bodies consisting of the last low-temperature condensates (similar to the most primitive chondritic meteorites enriched in hydrated silicates and trapped gases), fell onto the inner planets during the final stage of accretion, devastating their surfaces and delivering a significant mass of volatiles. Such a mechanism of exogeneous material inventory, referred to as the model of heterogeneous accretion, has been considered in detail based on analysis of the concentrations of volatiles in terrestrial, planetary and meteoritic material (Turekian and Clark, 1975; Anders and Owen, 1977; Dreibus and Wänke, 1987, 1989).

Basically, the idea of heterogeneous accretion allows to remove the obvious contradiction posed by the cold condensation (or successive accumulation) model. It contrasts to the predicted increase of volatile content with increasing distance from the Sun and argues that there could not be proportionally more volatiles on Mars than on Venus and Earth because there was a considerable radial mixing of planetesimals located in the feeding zone of these planets. In particular, the portion of planetesimals that entered into Earth and Venus from various parts of the terrestrial feeding zone appears to be almost the same (Ipatov, 1993). Therefore, the overall picture can be supplemented with the idea that bodies formed at distances $a > 5$ AU had been the sources of volatiles, and in particular water, for the inner planets. This provides a basis for believing that the atmospheres of these planets were more likely formed from a relatively thin covering (veneer), which accumulated over a certain period of time and basically depended on the mass of the planet. In other words, differences in the positions of the terrestrial planets could not have played a significant role in the formation of volatile reserves. The degree of subsequent planetary degassing, including the surface layer of resultant meteoritic material, could have had a stronger effect.

Later on, however, comets – remnants of the later stages of outer planet formation, scattered by gravitational perturbations of the growing planets and implanted in the Oort cloud and Edgeworth–Kuiper belt (Ip and Fernández, 1988, and this volume) – were invoked as more effective carriers of icy matter mainly composed of volatiles and possibly comprising even organic materials. An important constraint on the relative importance of comets and chondrites in the delivery of volatiles is placed by the fractionation pattern of noble gas abundances in the atmospheres of inner planets. The basic argument was found in the laboratory experiments on trapping of gases in amorphous ice forming at low temperature (Bar–Nun et al., 1985; Owen and Bar–Nun, 1996). The relative abundance pattern of argon, krypton, and xenon in the atmospheres of inner planets (in particular, the relative abundance of Xe/Kr on Earth and Mars, which is about an order of magnitude lower compared to chondritic meteorites) could be explained by the mechanism of strongly temperature dependent gas trapping. In turn, the amount of water and other species in the planetary atmosphere should be scaled to the abundances of the noble gases trapped under temperatures ranging from less than 30 K to more than 100 K. For example, in the case of Earth, it appears that comets from the Uranus–Neptune region alone would deliver abundances of noble gases that would be several orders of magnitude too high to reproduce the observed volatile mixture; rather a mixture of comets formed at different places in the solar nebula is required (Chyba *et al.*, 1994). Outgassing from the interior was probably the most important supply mechanism of water and other gases in the formation of Earth's hydrosphere and atmosphere. Anyway, icy planetesimals from the outer solar nebula are identified as an external source of heavy noble gases and other volatiles, complementary to outgassing from the composed rocks as an internal source.

Much lower absolute abundances of the noble gases on Mars compared to Earth and Venus, which are compatible with the very thin contemporary Martian atmosphere, can be explained due to impact erosion associated with the time of heavy bombardment (Melosh and Vickery, 1989). Based on this idea, Owen and Bar–Nun (1996) suggested that collisional processes, including cometary volatiles inventory, were responsible for both the lower efficiency of volatiles delivery and stripping off a major part of the original atmosphere of this less massive planet. In other words, a much smaller amount of atmospheric gas was left behind by such collisions, compared to what the impactors would produce. Traces of cometary impactors are supported by the study of noble gas proportions in several SNC meteorites, which are accepted as coming from Mars and containing a mixture of Martian atmosphere with samples of internal (rocky) reservoir. The distribution of noble gas abundances in these meteorites, specifically the ratios of primordial isotopes of ^{36}Ar and ^{84}Kr to ^{132}Xe, brings stringent evidence that they were contributed by impacting icy planetesimals (Owen and Bar–Nun, this volume). In contrast to Earth and Mars, Venus is known to have an unusual noble gas pattern in its atmosphere involving, in particular, a high neon abundance and Ar/Kr ratio, both closer to the solar value. This can be explained by impact of comets from the Edgeworth–Kuiper belt, where they were formed under temperatures lower than 30 K and thus were able to trap noble gases in nearly solar proportions (Owen and Bar–Nun, 1996). The cometary source of neon is doubted, however, because temperatures below 25 K are required for it to be trapped in ice, and a relic of gas trapped in the rocks during the planetary formation is considered as an alternative source of its atmospheric inventory.

The concept of heterogeneous accretion, supported by the quantitative analysis of noble gas abundances and their fractionation, is critically dependent on orbital dynamics and physico-chemical properties of the various families of small bodies, specifically those evolving to become inner planet crossers. Evidence on their evolution and rate of potential collisions is of primary importance for evaluation of the volatile inventory and hence, contribution to the bulk of the planetary atmospheres. In the next Section we examine different plausible amounts of volatile accretion based on theoretical and experimental data available and set out to estimate whether the net mass of water stored on the planets could be accounted for by such an external source.

3. Early impactor delivery

It is generally agreed that the total mass of planetesimals in the feeding zone of the giant planets exceeded by a factor of several the total mass of solids that entered into the giant planets (Safronov, 1969). The mass M_{UN} of planetesimals in the feeding zone of Uranus and Neptune could exceed $100M_\oplus$, where $M_\oplus$ is the mass of the Earth (Ipatov, 1987, 1993). Following this scenario, most of these planetesimals could still move in this zone when Jupiter and Saturn had accreted the bulk of their masses. Zharkov and Kozenko (1990) suggested that Uranus and Neptune accreted a considerable portion of their masses in the feeding zone of Saturn, where they aquired also hydrogen envelopes. In a series of numerical experiments, Ipatov (1991, 1993) showed that the embryos of Uranus and Neptune could increase their semimajor axes from < 10 AU to their present values, moving permanently in orbits with small eccentricities, due to gravitational interactions with the planetesimals which migrated from beyond 10 AU to Jupiter. Later on, the idea of a narrow zone, where all giant planets have been formed, was supported and further developed by Thommes et al. (1999) (see also Malhotra, 1999).

The time interval, during which a substantial part of migrating planetesimals from the feeding zone of the giant planets evolved to cross Jupiter's orbit, is estimated to be $\Delta t_J \approx 0.2$ My; after that most of these planetesimals were ejected into hyperbolic orbits. Some Jupiter-crossing objects evolved in the sunward direction and could collide with the inner planets, the process culminating from 4.5 to 4 Gy ago. The total number of planetesimals from the feeding zone of Uranus and Neptune which collided with Earth can be figured out using the following formula: $m_{Ec} = M_{UN}p_J p_{JE}\Delta t_E/T_E$, where p_J is the fraction of objects from the feeding zone of Uranus and Neptune that reached the orbit of Jupiter; p_{JE} is the fraction of Jupiter-crossing objects that reached the orbit of Earth during their lifetimes; Δt_E is the mean time during which a Jupiter–crossing object crosses the orbit of Earth; and T_E is the characteristic time for such an object (crossing both the orbits of Jupiter and Earth) to impact the Earth. Figure 1 shows an example of migration of a Neptune-crossing object inside the Solar System

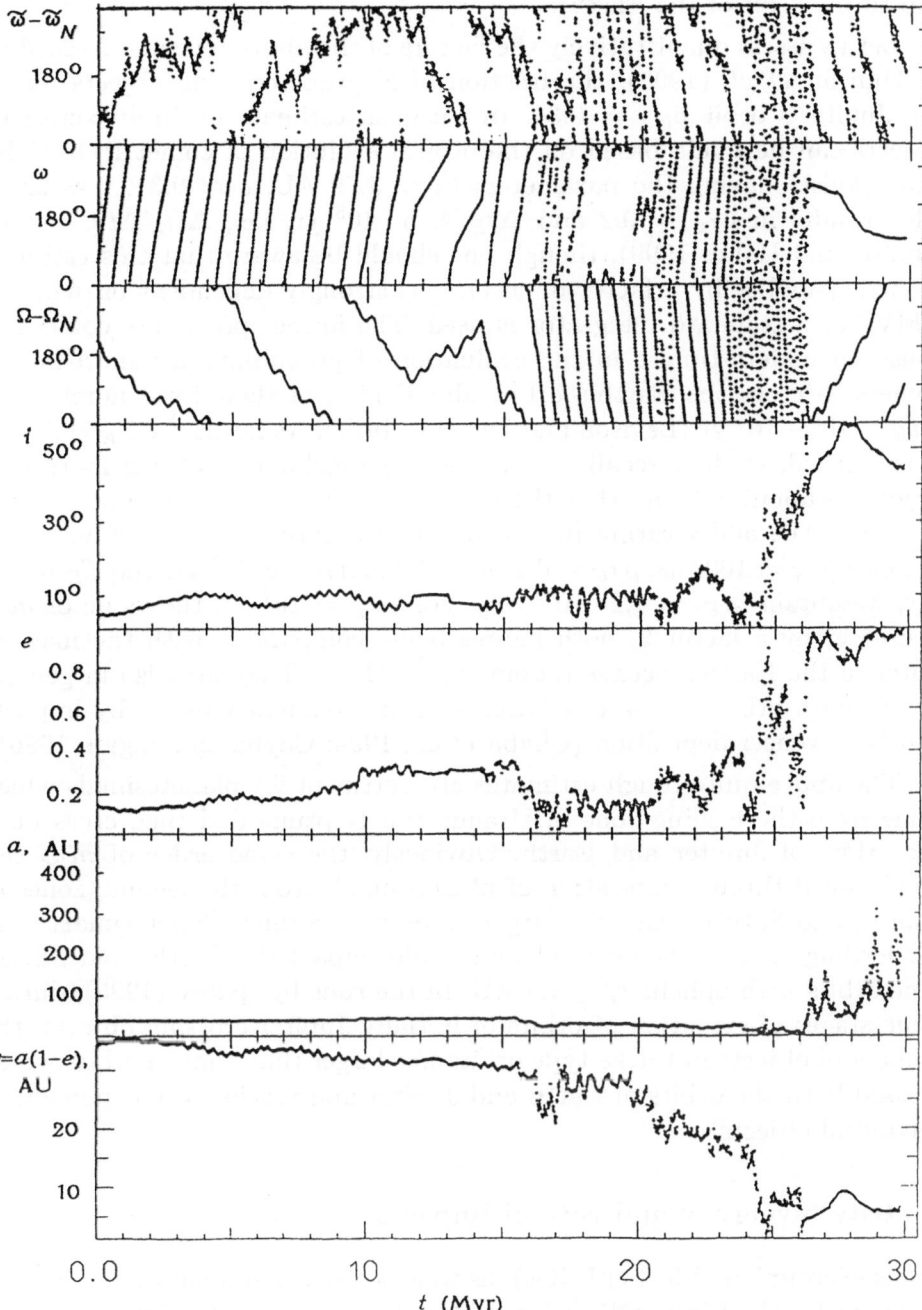

Figure 1. Time variations in the semimajor axis a, eccentricity e, perihelion distance $q = a(1-e)$, inclination i, the difference between the longitudes of the ascending node for the body and Neptune, $\Delta\Omega = \Omega - \Omega_N$, the argument of perihelion, ω, and the difference between the longitudes of perihelia for the body and Neptune, $\delta\tilde{\omega} = \tilde{\omega} - \tilde{\omega}_N$. Initial data: $a_0 = 40$ AU, $e_0 = 0.15$, $i_0 = 5°$, $\Omega_0 = \omega_0 = M_0 = 60°$.

(up to the orbit of the Earth).

Let us assess quantitatively the factors of the above formula. According to Duncan *et al.* (1995), the fraction of Neptune-crossing objects reaching Jupiter's orbit $p_{NJ} = 0.34$. In turn, an estimate of Jupiter-crossing objects can be made based on the orbital evolution of comet P/1996 R2 (Lagerkvist) having the parameters ($a \approx 3.79$ AU, $e \approx 0.31$, $i \approx 2.6°$). This results in $p_{JE} \approx 0.2$ and $\Delta t_E \approx 5 \cdot 10^3$ yr, i.e., $\Delta t_E / \Delta t_J \approx 1/40$ (Ipatov and Hahn, 1999), though one should be aware that this estimate for p_{JE} (smaller by a factor of several) is strongly dependent on whether RMVS or a standard integrator is used. The former procedure could also cause an underestimate in the evaluation of probability for short-period comets to cross the Earth's orbit: about 0.5% of their total number for $\Delta t_E = 10^3 - 10^5$ yr (Levison and Duncan, 1994). As far as the T_E estimate is concerned, we first recall that the average inclination of Earth-crossing objects is about 15° and that the larger a, the larger T_E. Then at $e = 0.7$, $a = 3.06$ AU, and i varing from 0 to 30°, we have $T_E = 4 \cdot 10^8$ yr. Now, taking $M_{UN} = 100 M_\oplus$, $p_{JE} = 0.2$, $p_J = 0.5$, $\Delta t_E = 5 \cdot 10^3$ yr, and $T_E = 400$ My, we obtain $m_{Ec} = 1.2 \cdot 10^{-4} M_\oplus$. For $\Delta t_E = 10^4$ yr, the value of m_{Ec} is greater by a factor 2, both figures being comparable with the mass of water in the Earth's oceans (about $2 \cdot 10^{-4} M_\oplus$). They are also in general accord with other estimates of accretion of a planetary ocean by impacts and later veneer deposition (Chyba *et al.*, 1994; Chyba and Sagan, 1996).

The above quite rough estimates are pertinent for planetesimals which came from the feeding zone of Uranus and Neptune and then cross both the orbits of Jupiter and Earth. Obviously, the same order of m_{Ec} can be obtained through migration of planetesimals from the feeding zones of Jupiter and Saturn. An even larger amount of former planetesimals from the feeding zone of the giant planets could impact the Earth from Encke-type orbits with aphelia $Q \le 4.5$ AU. In the runs by Ipatov (1995), during some stages of evolution of orbits of initially Jupiter-crossing objects, the number of objects in Encke-type orbits was larger than that in orbits which crossed both the orbits of Earth and Jupiter and reached a few percent of the initial objects.

4. Late inventory and rate of impacts

Trans–Neptunian objects (TNOs), as well as short- and long-period comets and near-Earth objects (NEOs), are identified as remnants of the planetary building blocks. They are distinguished because after the initial phase of planetary formation, these bodies served as a continuing source of volatiles throughout the time of Solar System evolution. The problem is how to evaluate accurately their potential contribution to the material delivery.

Evidently, although at present small bodies are still falling on planets, their number is much smaller than that during the planetary formation process. Let us address these plausible sources in more detail.

4.1. NEAR–EARTH OBJECTS

Among NEOs with perihelion distance $q \leq 1.3$ AU and aphelion distance $Q \geq 0.983$ AU there are at present more than 30 objects larger than 5 km in diameter, about 1500 larger than 1 km, and 135,000 larger than 100 m (Rabinowitz et al., 1994). The main sources of NEOs are considered to be the main asteroid belt and also the Edgeworth-Kuiper belt (EKB). Some bodies are injected in asteroid belt resonances (Gladman et al., 1997).

A support for the concept that the principal source of NEOs is the main asteroid belt was brought about by Farinella et al. (1993). Bodies coming from the main asteroid belt through chaotic dynamical routes result from perturbations associated with mean motion and secular resonances. At the same time the idea that about half the NEOs come from the EKB is supported in specifics of the present inclinations of their orbits (Wetherill, 1991). A major part of short-period comets also come from the EKB, some of them eventually evolving into asteroid-like bodies (Fernández, 1980; Marov, 1994).

One may admit that it is difficult to explain the number of objects of the Apollo and Amor groups and features of their orbits (for example, the mean inclinations which are larger than those in the main asteroid belt) if one considers only asteroidal sources (Wetherill, 1988, 1989; Weissman et al., 1989). Therefore, a significant part of NEOs may correspond to extinct comets. However, it is worth to note that NEOs that came from the main asteroid belt could also take such inclinations through the experienced resonances.

A cometary origin of NEOs, as well as meteorites, was earlier suggested by Öpik (1963). Wetherill (1988) estimated that extinct comets could provide 40% of NEOs if one short-period comet evolved to an Encke-type orbit every $5 \cdot 10^4$ yr. However, in order to evolve into an Earth-crossing (but not Jupiter-crossing) orbit, a comet must detach its aphelion from Jupiter's orbit. Using the sphere of action method (two problems of two bodies), Ipatov (1995) evaluated how N_o initially Jupiter-crossing bodies evolve under the gravitational influence of the planets. He found that in due course of the run about $0.04N_o$ bodies gained aphelia $Q < 4.5$ AU. Also there is a possibility of the gravitational "capture" of a comet due to an extremely close encounter with a terrestrial planet; a less important mechanism is associated with nongravitational forces due to gas sublimation from the cometary nucleus (Weissman, 1994).

The efficiency of volatile delivery ultimately depends on the variations in mass distribution of NEOs caused by their collisional fragmentation, specifically during their residence time in the main asteroid belt. The quantitative analysis is difficult to perform and we limit ourselves to a qualitative evaluation.

There are about 10^6 main-belt asteroids with diameter $d \geq 1$ km; the number N' of larger asteroids is proportional to $N^{-\alpha}$, with α between 2 and 2.5 (Binzel et al., 1991; Hughes and Harris, 1994). NEOs coming from the main asteroid belt eventually migrate into the resonant regions, from which they can reach the terrestrial planets, mainly via collisions with other asteroids (for small asteroids also due to the Yarkovsky effect). About half of the NEOs are Earth-crossers having orbits like Jupiter family comets, and some of them, likewise meteorites, are asteroidal fragments (Marov, 1986; Farinella et al., 1993). The mass ratio of an asteroid and an impactor for which catastrophic destruction occurs is about $s_0 = 10^4$ (Petit, 1993; Williams and Wetherill, 1994). In other words, an asteroid of diameter d can be destroyed by an impactor of diameter $d' \geq 0.046d$. The characteristic time T_{cN} to destroy an asteroid of diameter $d \geq 1$ km through collision with a body of diameter $d' \geq 100$ m ($s_0 = 10^3$) equals 5 Gy, and for $d = 1$ km and $d' = 46$ m ($s_0 = 10^4$), it is $T_{cN} \approx 1$ Gy (Ipatov, 1995). This can be reconciled with a similar collisional lifetime found by Bottke and Greenberg (2000, this volume) for ≈ 30 km asteroid disrupted by ≈ 2.5 km projectile.

Canavan (1993) obtained that $\alpha = 3$ for NEOs with $d < 40$ m and $\alpha = 2$ at 40 m $< d < 2$ km. If we assume that the size distribution of main-belt asteroids follows the same collisional criteria, then at $s_0 = 10^4$ for $d = 100$ m ($d' = 4.6$ m), $d = 10$ m ($d' = 46$ cm), and $d = 1$ m ($d' = 4.6$ cm), the values of T_{cN} will be smaller by a factor of $40/4.64 \approx 8.6$; 86; and 860, respectively, than those for $\alpha = 2$. In this case, the collisional lifetime of a 1 m body is only about 1 My; this means that numerous meteoroids are debris of larger asteroid-like bodies.

Let us note, however, that for Jupiter-crossing objects, the values of T_{cN} may be smaller than for asteroids of the same diameter in the main asteroid belt because, despite a larger semimajor axis, the values of s_0 for Jupiter-crossers being mostly icy bodies, may exceed significantly those for asteroids. Fernández (1990) estimated the probability of catastrophic collision for a long-period comet with a meteoroid to be about $2 \cdot 10^{-5}$ per perihelion passage. We assume the average time during which an object moves in a Jupiter-crossing orbit, to be about 0.2 My. Most of such objects with diameter greater than several meters would not be destroyed in the asteroid belt. If, however, such objects get orbits with aphelia inside the orbit of Jupiter, the dynamical lifetimes of some of them could exceed 10 My and objects with $d \leq 100$ m may be destroyed in the asteroid belt.

Most of the former comets with $d > 1$ km can survive, provided they are not destroyed by tidal forces at close encounters with planets and/or the Sun.

4.2. COMETS

Active and extinct periodic comets may account altogether for about 20% of the production of terrestrial impact craters larger than 20 km in diameter (Shoemaker et al., 1994). In a limiting case, about 40 active and 800 extinct Earth-crossing Jupiter-family comets having period $P < 20$ yr and nuclei ≥ 1 km, and about 140–270 active Earth-crossing Halley-family comets $(20 < P < 200$ yr) were estimated, some of the latter still assumed to come from the Oort cloud. While not specifically focusing here on this family, let us note that the study of dynamical evolution of Halley-type comet orbits for about ± 1 My revealed that some of them shift into Earth-crossing orbits for 1/3 to 2/3 of the considered time interval (Bailey and Emel'yanenko, 1996). The probabilities of collisions of short-period comets with Earth and Venus were estimated to be quite similar, whereas that with Mars is an order of magnitude smaller (Nakamura and Kurahashi, 1998). This means that the total mass of the matter delivered from the giant planets region, normalized to the mass of the planet, was nearly the same.

About 80% of all known comets are long–period comets, more than 80 members of the family being sungrazing comets. More than half of the close encounters of comets with Earth (up to 0.102 AU) correspond to long-period comets. Thus the number of collisions of active long-period comets with the inner planets may be of the same order of magnitude as that for active short-period comets, but short-period comets supply a larger number of extinct comets. Some long-period comets are as large as the Hale-Bopp comet which had a diameter of about 30 km, $q = 0.914$ AU and moved almost perpendicular to the plane of the Earth's orbit. With reference to the theoretical distribution of long–period comets, the mean impact probability per revolution is estimated as $(2 - 3) \cdot 10^{-9}$ (Marsden and Steel, 1994).

Beside comets and asteroids, numerous meteoroid–like bodies impact the terrestrial planets. About 98–99% of such bodies with masses less than 100 g in the vicinity of Earth are assumed to be of cometary origin, and in particular, the orbit of the Geminid meteor stream nearly coincides with the orbit of asteroid 3200 Phaethon (Whipple, 1983; Babadzhanov, 1987). Obviously, spectral variations in the collection of meteorites are considerably wider than those of asteroids (Britt et al., 1992) and, in particular, some spectral features in a body of cometary origin might be hidden in remote sensing observations. Note that water was found in the Monahans

(1998) H5 chondrite (Zolensky *et al.*, 1999) and it could exist in some other meteorites (Vilas and Zolensky, 1999). Morbidelli (1999) argues that during the formation of the main asteroid belt the bodies that migrated from this belt could deliver to the Earth even more water than comets.

4.3. TRANS–NEPTUNIAN OBJECTS

Although it is generally accepted that the main asteroid belt supplied a major part of NEOs, the bodies that migrated from the trans–Neptunian belt could play a significant role in the evolution of atmospheres of terrestrial planets, because these bodies consist mainly of water and volatiles. Since 1992 to August 1999, about two hundred TNOs with diameters $d \sim 100 - 400$ km were discovered and as much as 70,000 of such sized bodies were inferred to orbit the Sun in the 30–50 AU region, their total mass being estimated at about $0.06M_\oplus$ to $0.25M_\oplus$ (Jewitt *et al.*, 1996). They may be assumed to obey a differential size distribution in terms of radius r: $n(r)dr = c \cdot r^{-q}dr$, with $c = $ const and $q = 3$. The total number of bodies having diameter $d \geq 1$ km within the $30 \leq a \leq 50$ AU ranges from about 10^{10} (Jewitt *et al.*, 1996) or even 10^{11} (Jewitt, 1999) to $N = 5 \cdot 10^9$ (Jewitt and Fernandez, this volume)). The average values of eccentricity and inclination for 181 TNOs with $a \leq 50$ AU are equal to $e_{av} = 0.09$ and $i_{av} = 8.1°$, respectively. For five TNOs with $a > 50$ AU, $e_{av} = 0.58$ and $i_{av} = 23°$ were inferred. In addition, about 10^4 "scattered" objects such as 1996 TL$_{66}$ ($d \approx 500$ km) with a total mass of $0.5M_\oplus$ move in eccentric orbits between 40 and 200 AU (Luu *et al.*, 1997). Trujillo *et al.*(2000) consider that this mass equals $0.05M_\oplus$. Let us note that for the EKB bodies the energy of collision is smaller by a factor of 30 and the characteristic time in between collisions are greater than those for similar bodies in the main asteroid belt. Nevertheless, as it is easier to destruct an icy EKB body than a rocky body in the main asteroid belt, the collisional lifetimes of small bodies in the EKB may be of the same order as those in the main asteroid belt. Besides, the probability of destruction of an EKB body by scattered disk objects is usually larger than that by EKB bodies, because the mean energy of collision of a scattered disk object with an EKB body is greater (on the average, by a factor of 20) than for two colliding EKB bodies of the same masses. It is also interesting to mention that, according to Weissman (1995), the dynamically inactive region beyond 45 AU may extend out to 1000 AU or even more and contain up to several times 10^{13} objects with a total mass of several hundred Earth masses.

The number of TNOs migrating to the inner regions of the Solar System can be evaluated based on simple formulas and the results of numerical integration. Let $N_J = P_{NP}p_{JN}N$ be the number of former TNOs reaching

Jupiter's orbit for the given time span T, where N is the number of objects in the belt; P_N is the fraction of TNOs leaving the belt and migrating to Neptune's orbit during T; and p_{JN} is the fraction of Neptune-crossing objects which reach Jupiter's orbit for their lifetimes. Then the number of TNOs becoming Jupiter–crossers equals $N_{Jn} = N_J \Delta t_J / T$, where Δt_J is the average time during which the object crosses Jupiter's orbit. According to Duncan $et~al.$ (1995), the fraction P_N of TNOs that left the EKB during T=4 Gy under the influence of the giant planets is 0.1–0.2 and $p_{JN} = 0.34$. However, because in the generally dynamically stable region $36 \leq a \leq 39$ AU with small values of e and i the mutual gravitational influence of TNOs could play an important role in depopulating the region, the value $P_N = 0.2$ seems more justified (Ipatov 1998, 1999, 2000). Hence, taking T=4 Gy, $p_{JN} = 0.34$, $\Delta t_J = 0.2$ My, and $N = 10^{10}$ ($d > 1$ km), we have $N_J = 6.8 \cdot 10^8$ and $N_{Jn} = 3.4 \cdot 10^4$. Note that the number of Centaurs, coming from the EKB, turns out a bit less than $N_{Cn} = P_N T_C N / T$, where T_C is the mean Centaurs' lifetime. For $N = 7 \cdot 10^4$ ($d > 100$ km), $P_N = 0.2$, and $T_C = 10$ My, this yields $N_{Cn} = 35$. The smaller T_C, the smaller N_{Cn}. For Uranus–crossers $T_C \sim 10$ My, although for the Chiron-type objects $T_C \sim 1$ My.

We can now estimate the migration of TNOs to the Earth. Let the number of bodies approaching the Earth's orbit during the considered time span T be $N_E = N_J p_{JE} = P_N p_{JN} p_{JE} N$, and about $N_{Jn} p_{JE} \Delta t_E / \Delta t_J$ of the former TNOs cross the Earth's orbit now. Then the ratio of the number N_{NE} of Earth-crossing objects (ECOs) which came from the EKB, to the total number N_{ECO} of ECOs at the given time, is $P_{NE} = N_{NE}/N_{ECO} = P_N p_{JN} p_{JE} \Delta t_E N / (N_{ECO} T)$, where $N_{NE} = N_E \Delta t_E / T$. The number of collisions of such bodies with the Earth during T equals $N_{col} = N_E \Delta t_E / T_E$, where T_E is the characteristic time elapsed before such an ECO impacts the Earth. Note that the mean value of T_E for the Jupiter-crossing ECOs is several times larger than that for other ECOs, so the number of these particular projectiles is smaller as compared to their ECOs fraction. Only a small part of Jupiter-crossing ECOs has been observed, because they use to move in distant orbits; some of them are locked in 5:2, 7:3, 3:2 and other resonances with Jupiter (Ipatov and Hahn, 1999).

Let us assess the potential collisions with terrestrial planets, taking the Earth as an example. For $d \geq 1$ km, $N_{ECO} = 750$, T=4 Gy, $p_{JE} = 0.2$, and $\Delta t_E \approx 5000$ yr, we derive $N_E \approx 1.4 \cdot 10^8$ and $N_{NE} \approx 170$; $i.e.$, about 20% of the 1-km ECOs may be former TNOs, which are now in Jupiter-crossing orbits. For the above values and for $T_E = 400$ My, we obtain $N_{col} \approx 1750$. Although these figures should be regarded as very approximate estimates, they bring evidence that the number of TNOs impacting the Earth is not small. With increasing Δt_E, the values of P_{NE} and N_{col} become larger. If one denotes the fraction of Jupiter-crossers reaching Earth-crossing orbits with $Q \leq 4.5$ AU as k_E, and their mean residence time in such orbits as Δt_E^*, then the number of such objects proves to be $N_{NE}^* = N_E k_E \Delta t_E^*/T$. At $\Delta t_E^* k_E = 0.005$ My and $d \geq 1$ km, we have $N_{NE}^* \approx N_{NE}$. Therefore, the number of former Jupiter-crossers which migrate from distances 30–50 AU to inside Jupiter's orbit may be of the same order (or even more) than the number of objects crossing both the orbits of Jupiter and Earth. Hence the above values of N_{NE} and N_{col} may be considered as lower limits for the former TNOs. Their fraction among the NEOs may be even larger, if one admits $N = 10^{11}$ TNOs with $d \geq 1$ km (Jewitt, 1999) instead of $N = 10^{10}$ as taken in the above estimate. Bodies with $a > 50$ AU moving in highly eccentric orbits can also contribute to N_{col}.

4.4. CRATERING RECORD

In order to reconstruct the impactors' history the cratering record on the planetary bodies is of primary importance. From the investigation of lunar craters, Hartmann (1995) argues for a relatively uniform size distribution of interplanetary impactors of mixed origins back to 4 Gy ago and throughout the sampled Solar System. This generally complies with the evidence that the distribution of craters on Mars and Mercury is similar to the lunar distribution (between 0.1 and 3 Gy of age in the range 20 m $\leq D \leq 1$ km and 4 Gy for $D > 1$ km), while for Earth and Venus such a similarity is found for craters larger than 10 km and 20 km, respectively (Neukum and Ivanov, 1994). When assessing the crater record on the surface of inner planets and its origin, it is worth to take into account that the frequency of Earth impact by NEOs is estimated to be larger by a factor 2, 14, 24 and 30 as compared to Venus, Mars, Moon and Mercury, respectively (Bottke et al., 1994).

It has been shown that asteroidal impacts probably dominated the production of craters on Earth smaller than 30 km in diameter, whereas cometary impacts were responsible for the craters larger than 50 km (Shoemaker et al., 1990). This feature can be explained by the idea that TNOs can leave their residence EKB region essentially without collisions, in contrast to main belt asteroid bodies that experienced multiple collisions. The mean collisional lifetime T_{Ec} for Earth–crossing NEOs is estimated to be 240 My and the production rate for craters larger than 20 km in diameter to be $(4.9 \pm 2.9) \cdot 10^{-15}$ km^{-2}yr^{-1}, as follows from the astronomical evidence. This figure is consistent with the average cratering rate $(5.6 \pm 2.8 \cdot 10^{-15}$ km^{-2}yr$^{-1})$ estimated by Grieve and Shoemaker (1994) from the geologic record of craters with diameters $D \geq 20$ km for the last 120 My. In turn, Bottke et al. (1994) obtained a characteristic time $T_{Ec} = 134$ My and Ipatov (2000), using analytical formulas, calculated for the observed 417

Earth-crossing objects $T_{Ec} = 100$ My. Ipatov's estimate $T_{Ec} = 105$ My for 363 Apollo objects is in accord with $T_{Ec} = 120$ My calculated by Dvorak and Pilat-Lohinger (1999) based on the numerical integration of evolution of 54 Apollo objects. Obviously, the average time for Earth–crossers with $d > d_o$ to impact the Earth is T_{Ec}/N_o, where N_o is the number of impactors with such d. Basically, when dealing with the crater records it is assumed that the crater diameter D exceeds the diameter d of the projectile by a factor of 10–15. Altogether, these place an important constraint on the total number of impactors, which generally does not contradict the former evaluation.

4.5. TOTAL DELIVERY

We shall now attempt to assess the overall late delivery of volatiles. Based on the above consideration of migration and collisional probabilities, we focus on NEOs as a representative external source for such an estimate. Following the generally accepted concept (see Section 4.1) that about half of the NEOs are Earth-crossing objects, we further assume that for the earlier accepted differential size distribution $n(r)dr$ the number of NEOs with radii between r and $r + dr$ equals $N = c\, dr/r^3$, where the dimension of c is r^2. Integrating $n(r) \cdot (4/3)\, \pi\, \rho\, r^3 = (4/3)\, c\, \pi\, \rho\, dr$ from 0 to r_{max}, we obtain the total mass of ECOs $M_{eco} = (4\pi/3)\rho\, c\, r_{max}$, where ρ is the mean density of NEOs and r_{max} is the radius of the largest NEO (~ 20 km) or the largest ECO (~ 5 km). The value $r_{max} = 20$ km is used in the following. Integrating $n(r) = c\, dr/r^3$ from some radius r_* to r_{max}, we obtain that the number N_* of bodies with radius $r > r_*$ is $c\, k_r/(2r_*^2)$, where $k_r = 1 - (r_*/r_{max})^2$. Obviously, for $r_*/r_{max} = 0.1$, $k_r = 0.99 \approx 1$. For example, if $r_* = 0.5$ km, the mass $m_* = (4/3)\pi\rho r_*^3 = (\pi/6)k_\rho 10^{15} \approx 5.2 \cdot 10^{14} k_\rho$ g, where k_ρ is the ratio of the density ρ of a NEO to that of water (~ 2). Then the total mass of bodies with $r > r_*$ turns out to be $M_* = (4\pi/3)\rho\, c\, (r_{max} - r_*)$. On the other hand, if the mean mass of a body with $r > r_*$ equals M_*/N_*, it will exceed m_* by the factor of $(2r_{max}/r_*)r_{max}/(r_{max}+r_*) \approx 2r_{max}/r_*$. For example, for ~ 1500 NEOs with $r > 0.5$ km it yields $M_{eco} \approx 750(2r_{max}/r_*)m_* \approx 6 \cdot 10^4 m_* \approx \pi 10^{19} k_\rho$ g. The mean time elapsed for an ECO to collide with the Earth equals $T_{Ec} = 10^8$ yr. If we suppose that M_{eco} did not change significantly during $T = 4$ Gy, we obtain the total mass of bodies that collided with the Earth as much as $M_{Et} = (T/T_{Ec})M_{eco} = 40M_{eco}$. Assuming further that about $k_i = 20\%$ of them were icy bodies, the total mass of water delivered to the Earth during the last 4 Gy turns out about $16\pi 10^{19} \approx 5 \cdot 10^{20}$ g, i.e., about $3.6 \cdot 10^{-4}$ of the mass of the Earth's hydrosphere. However, most of the icy bodies (comets) probably impacted the Earth from Encke-type orbits, rather than from Jupiter-crossing orbits. Then, considering a mean dynamical lifetime T_d of typical ECOs equal to 10^7 yr, we obtain that during the last 4 Gy about $M_{teco} = (T/T_d)M_{eco} = 400M_{eco} \approx 2.5 \cdot 10^{22}$ g $\approx 4 \cdot 10^{-6}M_\oplus$ passed through the Earth's orbit with aphelia inside Jupiter's orbit.

Earlier we have estimated that about 1/5 of all ECOs can move in Jupiter-crossing orbits. Proceeding from the mass of bodies, which were both Jupiter- and Earth-crossers, this would yield an estimate of

$$0.2M_{eco}T/\Delta t_E = 0.2M_{eco} \cdot 4 \cdot 10^9/(5 \cdot 10^3) = 1.6 \cdot 10^5 M_{eco},$$

which at $\rho = 1$ g/cm^3 amounts to $5 \cdot 10^{24}$ g $\approx 0.0008M_\oplus$. Because, however, the values of T_{Ec} for Jupiter–crossers are larger at least by a factor of several than those for typical NEOs and $\Delta t_E \ll T_d$, the probability of collisions of such bodies with the Earth is very small and this source of impactors proves to be much smaller as compared to the typical NEOs. Overall, our evaluation of the total delivery basically testifies to a rather small contribution of the late inventory to the bulk of planetary volatiles, even if most of the ECOs are icy bodies.

5. Implication for atmospheric evolution

The above consideration brings support to the idea that early impactors delivery would play a key role in the formation and evolution of planetary atmospheres through the mechanism of heterogeneous accretion. It implies that a large amount of water and other volatiles has been delivered to the Earth and other inner planets from and beyond the zone of giant planet formation. Based on the estimate of the former Section, we can assess that the cumulative mass of icy bodies that impacted on the Earth more than 4 Gy ago could be of the order of the mass of the Earth's oceans ($1.37 \cdot 10^{24}$ g), whereas for the last 4 Gy the corresponding mass (for Earth-crossing bodies having orbits with aphelia inside Jupiter's orbit) appears to be three to four orders of magnitude less. This implies that, on average, scaling on the bulk of water, only a small fraction ($\sim 0.02\%$) of the overall inventory entered our planet after 4 Gy ago. For Venus and Mars the above mentioned scaling factors 1/2 and 1/14, respectively, should be introduced, as appropriate for the NEO collisions. Because the major part of the volatile inventory was delivered at the final stage of planetary formation, it also means that the later inventory could not influence dramatically the process of inner planetary atmosphere formation; for example, the mass of the Earth's atmosphere is $\sim 5 \cdot 10^{21}$ g, i.e., two orders of magnitude higher. An even higher threshold is found for Venus, whereas for Mars an easier atmospheric escape accompanying huge impacts should be taken into account.

Anyway, early impacts were concurrent with the formation of basaltic magma (in particular, its low melting-point, silicate fractions enriched in volatiles) and the accumulation of the lithophile and atmophile elements forced out onto the surface. Thus both the bulk of volatiles expelled from the interiors and deposited by impactors were responsible for the origin of early atmospheres, while the oxidation state of the upper mantle influenced its either reducing or oxidizing state and the following evolution (Marov, 1986; Marov and Grinspoon, 1998).

Obviously, given the current luminosity of the Sun, Earth under a relatively low effective temperature ($\approx$ 275 K) could have retained water, the bulk of which concentrated in the oceans. This temperature estimate is correct, however, only provided the lower solar luminosity before transition from a newly-formed star to the main sequence of the H-R diagram by 25 to 30% from the initial luminosity level (Newman and Rood, 1977) could be compensated by a strong greenhouse effect in the early Precambrian. Such an effect could be created either by ammonia (Sagan and Mullen, 1972), or more probably, a significantly greater amount of CO_2, compared to current levels, equivalent to a surface pressure on the order of 60 atm (Pollack and Black, 1979; Cess et al., 1980). The idea that the mean temperature of the terrestrial surface has been above the freezing point of water does not contradict modern geological and paleontological data, according to which primitive photogenic, autotrophic organisms arose on the Earth not less than $\sim$ 3.5 Gy ago and oceans even earlier. Indeed, as geochemical research in recent decades has shown, water basins could have existed not less than 3.8 Gy ago, soon after completion of the heavy bombardment. The most ancient stromatolites, stratified formations in limestone and dolomite masses, appeared to form as a result of the activity of blue-green algae colonies in such basins (Marov and Grinspoon, 1998).

In turn, carbon dioxide, under relatively low temperature conditions, accumulated in the terrestrial hydrosphere and in carbonates of sedimentary rock by bonding with metal oxides, which entered the composition of minerals in the oceanic crust and upper mantle and, to a significant extent through biogenic means, by the deposition of calciferous skeletons of marine organisms. The primary non-biogenic process appeared to occur in reactions of dissolved carbon dioxide with olivines (orthosilicates) and plagioclases (aluminosilicates), resulting in the formation of aqueous silicates containing hydroxyl groups, i.e., serpentine and kaolin (reactions of serpentinization and kaolinization). In this regard it is important to emphasize the role assigned to hydration reactions during the low temperature stage of proto-planetary condensation. Upon the interaction of olivine-pyroxene mineral groups with water vapor, hydrated silicates such as serpentines, talc, and tremolite are formed, which are most prevalent in carbonaceous chondrites. These silicates also serve as hidden reservoirs of water, subsequently deposited and driven out from the planet's interior. This makes it necessary to consider, at least in the case of the Earth, the origin of its atmosphere and hydrosphere as a genetically linked, single evolutionary process (Marov, 1986).

Similar estimates for Venus yield a surface temperature $\approx$ 325 K that is higher than the boiling point of water for an originally assumed not too dense atmosphere, given similar rates of degassing of mantle material and escape of the atmosphere into outer space for Venus and Earth. Thus it is reasonable to further assume that carbon dioxide, along with water vapor, gradually accumulated in the atmosphere. This in turn contributed to a further increase in the surface temperature due to the greenhouse effect and to the transport of even greater amounts of CO_2 and H_2O into the atmosphere due to an equilibrium state, characterized by relationships between mineral phases and volatiles at the surface. The most important of such a state is the carbonate-silicate interaction in the surface layer of the planetary crust, which is characteristic of a system with a positive feedback, where an initial perturbation is amplified quite rapidly, which in the case of Venus led to a runaway greenhouse effect. In such a case, with an increase in the Sun's luminosity and with a large atmospheric H_2O content, the Venus surface temperature for a certain period could even have exceeded the current value of 735 K.

Let us note, that the equilibrium between the partial pressure of carbon dioxide and carbonate concentrations in the crust, or between a planet's atmosphere and lithosphere, is characteristic of a heterogeneous natural system, including a large number of minerals and mineral associations in the crust and atmospheric gases in the atmosphere. This, in particular, affects the stability of carbonates and overall equilibrium of CO_2 atmosphere containing also sulfur- and halogen-bearing compounds. Three cycles, responsible for the global budget of sulfur components, were studied in detail on Venus (Lewis and Prinn, 1984; Marov and Grinspoon, 1998). One of these (the slow geological cycle) includes a sequence of transformations with the participation of pyrite (FeS_2), water, carbon dioxide, sulfur compounds, calcite ($CaCO_3$), anhydrite ($CaSO_4$, which is formed as a result of calcite reaction with sulfur dioxide) and FeO. Such an approach, including a complex calculation of the phase composition of the system, open with respect to H_2O, CO_2, CO, SO_2, HCl, HF and other volatile components of the Venus troposphere, gave an idea about the maximum possible changes in the initial mantle-derived rocks during chemical weathering. It allowed to arrive at a model of equilibrium mineral association of the soil, in which diffusion exchange processes of volatiles and petrogenic components are completed and a series of thermochemical reactions, with the participation of the above named components, along with the carbonate-silicate cycle, can in principle maintain the system in equilibrium.

Unlike the carbon dioxide, the situation with water on Venus is considerably more complicated to explain. Assuming "geochemical similarity" in the evolutionary processes of planetary interiors and the degassing of volatiles, the amount of water outgassed on Venus would have to correspond to the volume of the terrestrial hydrosphere. Moreover, on the surface of Venus water is not retained, since the temperature there is higher than even the critical value, 647 K, and this assertion remains valid for aqueous solutions (brines), for which the critical temperature is usually somewhat higher (675–700 K). As far as the atmosphere is concerned, given a mean relative water vapor content of 0.005%, the amount of water turns out to be $3.5 \cdot 10^{19}$ g. This exceeds the water content in the Earth's atmosphere $(1.3 \cdot 10^{19}$ g), but is almost five orders of magnitude less than the water reserves in the hydrosphere.

The view that Venus possessed quite a thick hydrosphere has been supported by the finding that the ratio of deuterium D to hydrogen H in Venus' atmosphere is $(1.6 \pm 0.2) \cdot 10^{-2}$, or two orders of magnitude greater than in the Earth's atmosphere (Donahue et al., 1982; de Bergh et al., 1991). Such a high deuterium concentration in Venus' atmosphere could be explained by the fractionation of these isotopes with nonthermal escape of hydrogen from the atmosphere, where it presumably accumulated through the vaporization of water removed from the interior (the "parent ocean") and subsequent dissociation of the water vapor by ultraviolet radiation. Meanwhile, Grinspoon and Lewis (1989) proposed a different interpretation for the measured D/H ratio. In their model the water content is essentially unchanged over the planet's entire geological history; consequently a nonthermal mechanism for hydrogen dissipation was maintained at about modern levels, while the most likely source of hydrogen is from the impact of cometary material – again one of the various mechanisms for heterogeneous accretion. However, the D/H ratio measured in comets Halley and Hyakutake is only about two times, and in comet Hale–Bopp between two and ten times that in the Earth's ocean. Interestingly, this latter difference, revealed by IR–spectroscopy for H_2O and HCN, reflects what is observed in interstellar clouds (see Owen and Bar–Nun, this volume). Despite yet very poor statistics for the D/H ratio in comets, a discrepancy with the observed deuterium concentration in Venus' atmosphere is obvious. This discrepancy could be removed, given specific assumptions on the degree of fractionation of hydrogen isotopes upon dissipation from the atmosphere. Grinspoon (1993) also proposed that, in the light of Magellan findings, volcanic outgassing may maintain the current water abundance in steady state with nonthermal escape. This model successfully produces the observed fractionation if the deuterium fraction of the source H on Venus is enhanced over the terrestrial value by a factor of 10. Alternatively, the fractionation observed could have resulted from escape of water outgassed during a global resurfacing episode within the last 1 Gy.

Evidently, different processes of atmospheric evolution occurred on Mars. Its equilibrium temperature (under the same assumed initial Bond albedo for all neighboring planets ≈ 0.07) is 223 K, *i.e.*, much lower than the freezing point of water. Therefore, water expelled from the interior could not be maintained on the surface of Mars in liquid state unless a rather dense atmosphere had formed and a substantial greenhouse effect had developed. This probably was the case in the early Martian history soon after completion of the heavy bombardment and until ~ 3.8 Gy ago, which is testified by numerous features associated with traces of liquid water, first of all river beds and mud-like patterns around impact craters. Unfortunately, there is no strict evidence on which greenhouse gases composed the ancient atmosphere of Mars and what caused its collapse to the contemporary thin gas envelope with a pressure less than one hundredth that of the Earth's value. As mentioned above, continuing impact erosion or even catastrophic asteroid impact may be invoked to explain the atmospheric sweep off, different from the earlier speculations about radiogenic isotope exhaustion, and hence, dramatic change in atmospheric–lithospheric equilibrium.

In contrast to Venus, the D/H ratio on Mars measured in the atmosphere and in SNC meteorites ($5.5\pm 2 \times$ terrestrial) is close to the cometary value, which testifies that the primary source of Martian water was comets (Owen and Bar–Nun, this volume). Although liquid water can not exist on contemporary Mars under such low pressure, there is a generally agreed concept, well supported by geological evidence, that a large water reservoir (up to 0.5 km depth of equivalent water distributed over the Martian surface, compared to the Earth's ocean of 3 km average depth) is preserved in its subsurface layer as a permafrost (Carr, 1998). As far as carbon dioxide is concerned, this is assumed to be mostly swept off and partially deposited in sedimentary rocks as carbonates; only a small fraction of CO_2 is left behind by these processes in the modern Martian atmosphere.

Whatever evolutionary model is used, the water abundance in the past and present and the possible mechanisms for its loss are the key questions in the evolution of Venus and Mars and their atmospheres. All terrestrial planets probably received approximately the same (per unit mass) amount of volatiles as a consequence of heterogeneous accretion during the final phase of accumulation and/or later. The opposite point of view, that Venus formed as a "dry" planet, seems less likely; rather, as earlier suggested, it received more volatiles from the outer regions of Solar System formation. The idea of a "parent ocean" on Venus, which remains very attractive, at the same time supports the decisive role of the heliocentric position of the inner planets in their climatic evolution. In any case, from this point of view it seems more natural to explain why Venus lost such a huge mass of water, comparable to the volume of the terrestrial hydrosphere, while Mars "preserved" a much more moderate amount of water in the form of ice in its cryosphere.

6. Summary

The atmospheres of inner planets formed as a part of the general process of Solar System origin and evolution, involving degassing of the mantle in due course of the differentiation of the interior, and continuing collisional processes culminating during the heavy bombardment about 4 Gy ago, which underlies the mechanism of heterogeneous accretion. Volatile-rich icy planetesimals (comets), considered as remnants of the later stages of outer planet formation, were mostly responsible for this late accretion. An important constraint on the relative importance of comets and chondrites in the delivery of volatiles is placed by the fractionation pattern of noble gas abundances in the atmospheres of inner planets, supported by laboratory experiments on the trapping of gases in amorphous ice forming at low temperature. Although the balance between internal degassing and input of volatile–rich materials from the outer regions of the Solar System remains unclear, heterogeneous accretion allows us to explain both the abundance and the observed ratios of different species in the current atmospheres of the planets.

The basic idea for numerous collisions and volatile inventory is rooted in orbital dynamics and physico-chemical properties of the various families of small bodies, specifically those evolving to become inner planet crossers. Different plausible amounts of volatile accretion were examined based on theoretical and experimental data available on small bodies evolution including near-Earth asteroids, comets, and trans-Neptunian bodies from the Edgeworth–Kuiper belt. The influx they provided yields an estimation of the probable amount of water and other volatile inventory stored on the inner planets from such a source of exogeneous origin, and therefore, their potential contribution to the bulk of the planetary atmospheres. In particular, this gives rise to a total mass of delivered water comparable to the Earth's oceans due to early impactors delivery and only a small fraction ($\sim 0.04\%$) of the overall inventory entering the inner planets during the last 4 Gy.

The process of an accretional veneer deposition is assumed to be concurrent with the formation of low melting-point silicate basaltic magma enriched in volatiles and the accumulation of the lithophile and atmophile elements forced out onto the surface. At present, the relative contribution of exogeneous and endogeneous sources is difficult to assess: apparently, both mechanisms, subsequent to core emergence, contributed to the formation of planetary gas envelopes. Anyway, the late volatile inventory could hardly influence significantly the atmospheric origin and evolution. Its following development depended on the amount of volatiles expelled from the interiors and deposited by impactors, and on the oxidation state of the upper mantle, influencing the either reducing or oxidizing state of the early atmosphere. In turn, the mass of the planet and its position relative to the Sun predetermined the successive processes of climatic evolution.

7. Acknowledgements

This work was supported by the Russian Foundation for Basic Research, projects no. 99-02-16008 and no. 96-15-96534. S. I. also acknowledges support given by the Federal scientific and technical program "Astronomy" (project number 1.9.4.1) and Belgian office for scientific, technical and cultural affairs.

References

Anders, E., and Owen, T. (1977) Mars and Earth: Origin and abundance of volatiles, *Science* **198**, 453–465.

Babadzhanov, P.B. (1987) *Meteory i ikh nablyudenie* (Meteors and observing them), Moscow: Nauka, in Russian.

Bailey, M.E. and Emel'yanenko, V.V. (1996) Dynamical evolution of Halley–type comets, *Mon. Not. R. Astron. Soc.* **278**, 1087–1110.

Bar–Nun, A., Herman, G., Laufer, D., and Rappaport, M.L., (1985) *Icarus* **63**, 317–332.

Basilevsky, A.T., Head, J.W., Shaber G.G., and Strom R.G., (1997) The resurfacing history of Venus. in *Venus II. Geology, Geophysics, Atmosphere and Solar Wind Enviroment*, eds. S.W. Bougher, D.M. Hunten, and R.J. Phillips, The University of Arizona Press, Tucson, Arizona, pp. 1047–1084.

Binzel, R.P., Barucci, M.A., and Fulchignoni, M. (1991) The origins of the asteroids, *Sci. Am.* **265**, 66–72.

Britt, D.T., Tholen, D.J., Bell, J.F. and Pieters, C.M. (1992) Comparison of asteroid and meteorite spectra classification by principal component analysis, *Icarus* **99**, 153–166.

Bottke, W.F., Jr., Nolan, M.C., Greenberg, R. and Kolvoord, R.A. (1994) Collisional lifetimes and impact statistics of near–Earth asteroids, in *Hazards due to comets and asteroids*, ed. T. Gehrels, The University of Arizona Press, Tucson & London, pp. 337–357.

Canavan, G.H. (1993) Value of space defenses, *Proc. near-Earth object intercept workshop*, eds. G.H. Canavan, J.C. Solem, and J.D.G. Rather, Los Alamos, pp. 261–274.

Carr, M. (1998), Mars: aquifers, oceans and the prospects for life, *Solar System Research* **32**, 453–463.

Cess, R.D., Ramanathan, V., and Owen, T. (1980) The Martian paleoclimate and enhanced atmospheric carbon dioxide, *Icarus* **41**, 159–165.

Chyba, C.F., Owen, T.C. and Ip, W.-H. (1994) Impact delivery of volatiles and organic molecules to Earth, in *Hazards due to comets and asteroids*, ed. T. Gehrels, The University of Arizona Press, Tucson & London, pp. 9–58.

Chyba, C.F., and Sagan, C. (1996) Comets as a source of prebiotic organic molecules for the early Earth, in *Comets and Origin of Life*, eds. P.J. Thomas, C.F. Chyba, and C.P. McKay, Springer-Verlag, pp. 147–173.

De Bergh, C., Bezard, B., Crisp. D., Maillard, J.-P., Owen, T., Pollack, J.B., and Grinspoon, D.H. (1991) Deuterium on Venus: Observations from Earth, *Science* **251**, 547–549.

Donahue, T.M., Hoffman, J.H., Hodges, R.R., and Watson, A.J. (1982) Venus was wet: A measurement of the ratio of deuterium to hydrogen, *Science* **216**, 630–635.

Dreibus, G, and Wänke, H. (1989) Supply and loss of volatile constituents during the accretion of terrestrial planets, in *Origin and Evolution of Planetary and Satellite Atmospheres*, eds. S.K. Atreya, J.B. Pollack, and M.S. Matthews, The University of Arizona Press, Tucson & London, pp. 268–288.

Dvorak, R., Pilat–Lohinger, E. (1999) On the dynamical evolution of the Atens and the Apollos, *Planet. Space Sci.* **47**, 665–677.

Duncan, M.J., Levison, H.F. and Budd, S.M. (1995) The dynamical structure of the Kuiper belt, *Astron. J.* **110**, 3073–3081.

Farinella, P., Gonczi, R., Froeschlé, Ch. and Froeschlé, C. (1993) The injection of asteroid fragments into resonances, *Icarus* **101**, 174–187.

Farinella, P., Froeschlé, C., Gonczi, R. (1994) Meteorite delivery and transport, in *Proc. of the IAU symposium No 160 "Asteroids, comets, meteors 1993"* (June 14–18, 1993, Belgirate, Italy), eds. A. Milani, M. DiMartino and C. Cellino, pp. 205–222.

Fernández, J.A. (1980) On the existence of a comet belt beyond Neptune, *Mon. Not. R. Astron. Soc.* **192**, 481–491.

Fernández, J.A. (1990) Collision of comets with meteoroids, in *Asteroids, Comets, Meteors III*, eds. C.-I. Lagerkvist, H. Rickman, B.A. Lindblad, and M. Lindgren, Uppsala University, Uppsala, pp. 309–312.

Gladman, B., Migliorini, F., Morbidelli, A., Zappalà, V., Michel, P., Cellino, A.,Froeschlé, Ch., Levison, H., Bailey, M., and Duncan, M. (1997) Dynamical lifetimes of objects injected in asteroid belt resonances, *Science* **277**, 197–201.

Grieve, R.A.F. and Shoemaker, E.M. (1994) The record of past impacts on Earth, in *Hazards due to comets and asteroids*, ed. T. Gehrels, The University of Arizona Press, Tucson & London, pp. 417–462.

Grinspoon, D.H. (1993) Implications of the high deuterium-to-hydrogen ratio for the sources of water in Venus' atmosphere, *Nature* **363**, 428–431.

Grinspoon, D.H., and Lewis, J. (1989) Cometary water on Venus: Implications of stochastic impacts, *Icarus* **74**, 21–35.

Hartmann, W.K. (1995) Planetary cratering 1. The question of multiple impactor populations: Lunar evidence, *Meteoritics* **30**, 451–467.

Head, J.W. and Crumpler, L.S. (1987) Evidence for divergent planet-boundary characteristics and crustal spreading on Venus, *Science* **238**, 1380–1385.

Hughes, D.W. and Harris, N.W. (1994) The distribution of asteroid sizes and its significance, *Planet. Space Sci.* **42**. 291–295.

Ip, W.-H., and Fernández, J.A. (1988) Exchange of condensed matter among the outer and terrestrial protoplanets and the effect on surface impact and atmospheric accretion, *Icarus* **74**, 47–61.

Ipatov, S.I. (1987) Accumulation and migration of the bodies from the zones of giant planets, *Earth, Moon, and Planets* **39**, 101–128.

Ipatov S.I. (1991) Evolution of initially highly eccentric orbits of the growing nuclei of the giant planets, *Sov. Astron. Letters* **17**, 113–119.

Ipatov, S.I. (1993) Migration of bodies in the accretion of planets, *Solar System Research* **27**, 65–79.

Ipatov, S.I. (1995) Migration of small bodies to the Earth. *Solar System Research* **29**, 261–286.

Ipatov, S.I. (1998) Migration of Kuiper–belt objects inside the solar system, *"Planetary systems – the long view"*, Proc. 9th Rencontres de Blois (June 22–28, 1997), eds. L.M. Celnikier and Tran Thanh Van, Editions Frontières, Gif sur Yvette, pp. 157–160.

Ipatov, S.I. (1999) Migration of trans–Neptunian objects to the Earth, *Celest. Mech. Dyn. Astron.* **73**, 107–116.

Ipatov, S.I. (2000) *Migration of celestial bodies in the solar system*, URSS Publishing Company, Moscow, in Russian, in press.

Ipatov, S.I. and Hahn, G.J. (1999) Evolution of the orbits of the P/1996 R2 and P/1996 N2 objects, *Solar System Research* **33**, 487–500.

Jewitt, D. (1999) Kuiper belt objects, *Annu. Rev. Earth. Planet. Sci.* **27**, 287–312.

Jewitt, D., Luu., J. and Chen, J. (1996) The Mauna Kea–Cerro–Tololo (MKCT) Kuiper belt and Centaur survey, *Astron. J.* **112**, 1225–1238.

Levison, H.F. and Duncan, M.J. (1994) The long–term dynamical behavior of short–period comets, *Icarus* **108**, 18–36.

Lewis, J.S., and Prinn, R.G. (1984) *Planets and Their Atmospheres: Origin and Evolution.* London: Academic Press.

Luu, J., Marsden, B.G., Jewitt, D., Trujillo C.A., Hergenrother, C.W., Chen, J. and Offutt, W.B. (1997) A new dynamical class of objects in the outer Solar System, *Nature* **387**, 573–575.

Malhotra, R. (1999) Chaotic planet formation, *Nature* **402**, 599–600.

Marov, M.Ya. (1986) *Planets of the Solar System*, 2nd edition. Nauka, Moscow (in Russian).

Marov, M.Ya. (1994) Physical Properties and Models of Comets, *Solar System Research* **28**, 302–368.

Marov, M.Ya., and Grinspoon, D. (1998) *The Planet Venus.* Yale University Press.

Marsden, B.G. and Steel, D.I. (1994) Warning times and impact probabilities for long–period comets, in *Hazards due to comets and asteroids*, ed. T. Gehrels, The University of Arizona Press, Tucson & London, pp. 221–239.

Melosh, J., and Vickery, A.M. (1989) Impact erosion of the primordial atmosphere of Mars, *Nature* **338**, 487–489.

Morbidelli, A. (1999) Private communication.

Nakamura, T. and Kurahashi, H. (1998) Collisional probability of periodic comets with the terrestrial planets: an invalid case of analytic formulation, *Astron. J.* **115**, 848–854.

Newman, M.J., and Rood, R.T. (1977) Implications of solar evolution for the Earth's early atmosphere, *Science* **198**, 1035–1037.

Neukum, G. and Ivanov, B.A. (1994) The record of past impacts on Earth, in *Hazards due to comets and asteroids*, ed. T. Gehrels, The University of Arizona Press, Tucson & London, pp. 359–416.

Öpik, E.J. (1963) Small bodies in the Solar System 1: Survival of cometary nuclei and the asteroids, *Adv. Astron. Astrophys.* **2**, 219–262.

Owen, T., and Bar–Nun, A. (1996) Comets, meteorites and atmospheres, *Earth, Moon, and Planets* **72**, 425–432.

Petit, J.-M. (1993) Modelling the outcomes of high–velocity impacts between small solar system bodies, *Celest. Mech. Dyn. Astron.* **57**, 1–28.

Pollack, J.B., and Black, D.C. (1979) Implications of the gas compositional measurements of Pioneer Venus for the origin of planetary atmospheres, *Science* **205**, 56–59.

Rabinowitz, D., Bowell, E., Shoemaker, E., and Muinonen, K. (1994) The population of Earth–crossing asteroids, in *Hazards due to comets and asteroids*, ed. T. Gehrels, The University of Arizona Press, Tucson & London, pp. 285–312.

Safronov, V.S. (1969) *Evolution of the protoplanetary cloud and formation of the Earth and planets*, Moscow: Nauka. Transl. Israel Program for Scientific Translation, 1972, NASA TTF-677.

Sagan, C., and Mullen, G. (1972) Earth and Mars: Evolution of atmospheres and surface temperature, *Science* **177**, 52–56.

Shoemaker, E.M., Wolfe, R.F. and Shoemaker, C.S. (1990) Asteroid and comet flux in the neighborhood of Earth, in *Global Catastrophes in Earth History*, eds. V.L. Sharpton and P.D. Ward, Geological Soc. of America, Special Paper 247 (Boulder: geological Soc. of America), pp. 155–170.

Shoemaker, E.M., Weissman, P.R. and Shoemaker, C.S. (1994) The flux of periodic comets near Earth, in *Hazards due to comets and asteroids*, ed. T. Gehrels, The University of Arizona Press, Tucson & London, pp. 313–335.

Thommes, E.W., Duncan, M.J., Levison, H.F. (1999) The formation of Uranus and Neptune in the Jupiter–Saturn region of the solar system, *Nature* **402**, 635–638.

Trujillo, C.A., Jewitt, D.C., and Luu, J.X. (2000) Population of the scattered Kuiper belt, *Astrophys. J.* **529**, L103-L106.

Turekian, K.K., and Clark, S.P., Jr. (1975) The non-homogeneous accumulation model for terrestrial planet formation and the consequence for the atmosphere of Venus, *J. Atmos. Sci.* **32**, 1257–1261.

Vilas, F., Zolensky, M.E. (1999) Water, water everywhere: The aqueous alteration history recorded in meteorites, *Meteorite!* **5**, N 1, 8–10.

Weissman, P.R. (1994) The comet and asteroid impact hazard in perspective, in *Hazards due to comets and asteroids*, ed. T. Gehrels, The University of Arizona Press, Tucson & London, pp. 1191–1212.

Weissman, P.R. (1995) The Kuiper belt, *Annu. Rev. Astron. Astrophys.* **33**, 327–357.

Weissman, P.R., A'Hearn, M.F., McFadden, L.A. and Rickman, H. (1989) Evolution of comets into asteroids, in *Asteroids II*, eds. R.P. Binzel, T. Gehrels and M.S. Matthews, Tucson: Univ. Arizona Press, pp. 880–919.

Wetherill, G.W. (1988) Where do the Apollo objects come from?, *Icarus* **76**, 1–18.

Wetherill, G.W. (1989) Cratering of the terrestrial planets by Apollo objects, *Meteoritics* **24**, 15–22.

Wetherill, G.W. (1991) End products of cometary evolution: cometary origin of Earth–crossing bodies of asteroidal appearance, *Proc. of the IAU Colloquium 121 "Comets in the post–Halley era"*, eds. R.L. Newburn, J. Rahe and M. Neugebauer, Amsterdam: Kluwer Acad. Publ., pp. 537–556.

Whipple (1983) 1983 TB and the Geminid meteors, *IAU Circ*, No. 3881 **107**.

Williams, D.P. and Wetherill, G.W. (1994) Size distribution of collisionally evolved asteroidal populations: Analytical solution for self–similar collision cascades, *Icarus* **107**, 117–128.

Zharkov, V.N. and Kozenko, A.V (1990) The role of Jupiter in the formation of the giant planets, *Sov. Astron. Lett.* **16**, no 2, (pp. 169–173 in Russian edition).

Zolensky, M.E., Bodnar, R.J., Gibson, E.K. Jr., Nyquist, L.E., Reese, Y., Shih, Chi-Yu, Wiesmann, H. (1999) Asteroidal water within fluid inclusion–bearing halite in an H5 chondrite, Monahans (1998), *Science* **285**, 1377–1379.

FROM THE INTERSTELLAR MEDIUM TO PLANETARY ATMOSPHERES VIA COMETS

TOBIAS C. OWEN

University of Hawaii, Institute for Astronomy, 2680 Woodlawn Drive, Honolulu, Hawaii 96822, USA

AND

AKIVA BAR-NUN

Department of Geophysics and Planetary Sciences, Tel-Aviv University, Ramat-Aviv, Israel

Abstract. Laboratory experiments on the trapping of gases by ice forming at low temperatures implicate comets as major carriers of the heavy noble gases to the inner planets. Recent work on deuterium in Comet Hale-Bopp provides good evidence that comets contain some unmodified interstellar material. However, if the sample of three comets analyzed so far is typical, the Earth's oceans cannot have been produced by comets alone. The highly fractionated neon in the Earth's atmosphere also indicates the importance of non-icy carriers of volatiles, as do the noble gas abundances in meteorites from Mars.

1. Introduction

Ever since the pioneering suggestion by Oró (1961) that comets could have been an important source of organic material on the early Earth, numerous investigations have suggested that comets could have brought in a variety of volatile elements and compounds (*e.g.*, Sill & Wilkening, 1978; Delsemme, 1991). Recent work on this question has focused on models of the dissipation of icy planetesimals from the Uranus-Neptune region as these planets finished forming (Ip & Fernández, 1988), or attempts to calculate the extent of the terrestrial cometary influx through analyses of the cratering record of the moon (Chyba, 1990). All of these studies suffer from the absence of evidence for a uniquely identifiable contribution in the Earth's volatile inven-

M. Ya. Marov and H. Rickman (eds.), Collisional Processes in the Solar System, 249–264.
© 2001 *Kluwer Academic Publishers. Printed in the Netherlands.*

tory. Until recently, there was little incentive for discovering such evidence since models invoking meteoritic sources for terrestrial volatiles appeared perfectly adequate, requiring no additional contributions (*e.g.*, Turekan & Clark, 1975; Anders & Owen, 1977).

The difficulty in identifying the source of the Earth's volatiles is compounded by the 4.5 billion year history of the planet, during which chemical reactions with the crust, escape of gases from the upper atmosphere, and the origin and evolution of life have completely changed the composition of the atmosphere and hydrosphere. The central problem of life's origin is intimately dependent on the early composition of the atmosphere, making our inability to define that composition all the more frustrating.

It is this dilemma that the heavy noble gases have the potential to resolve, since they are chemically inert, do not easily escape from the atmosphere, and are not involved in the activities of living organisms. The firmly entrenched idea that the atmospheric noble gas abundances are the result of delivery by meteorites stems from the recognition of a so-called "planetary pattern" in the noble gases found in these objects many years ago. This idea gained widespread support, despite the fact that it has never been possible to explain why the abundances of krypton and xenon in the meteorites are about the same, while xenon is much lower than krypton in the earth's atmosphere (Figure 1). Attempts to find the "missing" xenon buried in shells, clathrates, or ice have failed (Wacker & Anders, 1984; Bernatowicz, Kennedy & Podosek, 1985). The xenon isotopes in atmospheric and meteoritic xenon are also distinctly different (Pepin, 1989). It was these discrepancies coupled with the arguments for early cometary bombardment that encouraged us to pursue the idea that icy, rather than rocky planetesimals may have delivered these gases.

2. A Cometary Model for Delivery of Volatiles

Accordingly, we set out to determine whether or not the abundance patterns of the heavy noble gases in the atmospheres of the inner planets could be accounted for by comets. The first step was to apply the same laboratory techniques for trapping gases in amorphous ice deposited at low temperatures that had been used with N_2, CO, H_2, etc. (Bar-Nun *et al.*, 1985; Bar-Nun, Kleinfeld, & Kochavi, 1988; Laufer, Kochavi, & Bar-Nun, 1987) to a mixture of heavy noble gases: argon, krypton, and xenon. These experiments were designed to imitate the formation of comets in the outer solar nebula. The idea is that interstellar ice grains probably sublimated as they fell toward the mid-plane of the nebula and recondensed on cold refractory cores (Lunine, Engel, Rizk, & Horanyi, 1991). In this re-condensation process, they could trap ambient gas according to the local temperature.

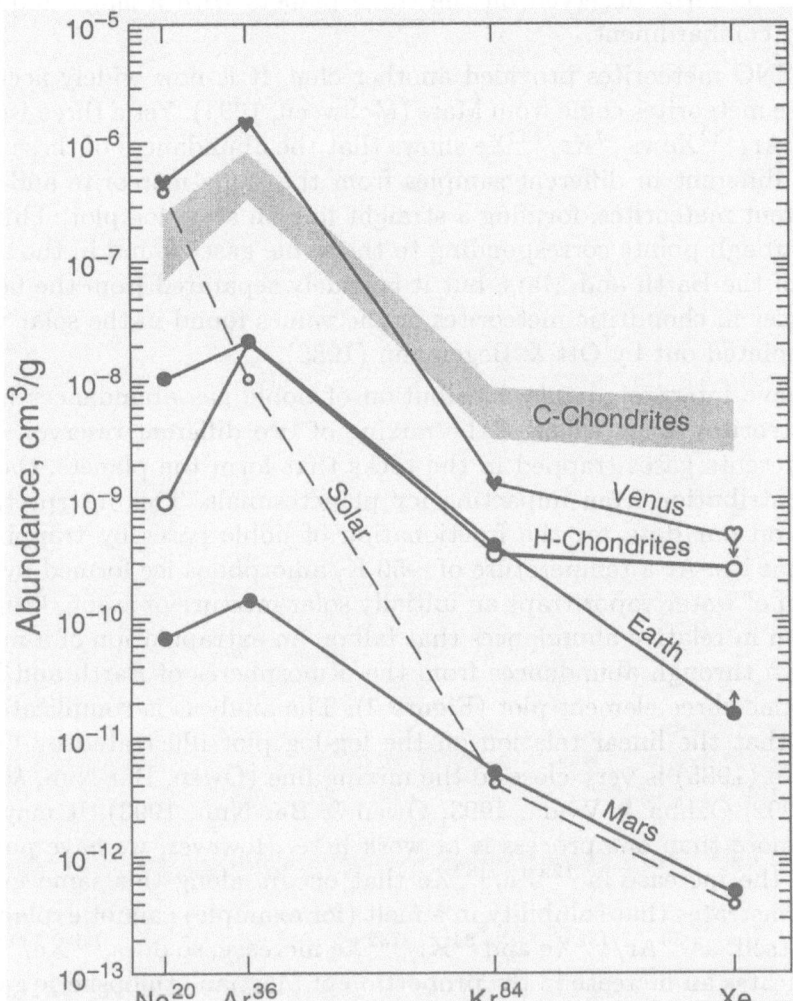

Figure 1. Chondritic meteorites contain about as much xenon as krypton. The meteoritic noble gas abundances therefore do not match the abundance patterns found in inner planet atmospheres, despite the apparent agreement for Ne, Ar, and Kr. (Solar values are normalized to ^{84}Kr on Mars.)

The laboratory results showed that temperature-dependent fractionation of the gas mixtures occurs, with increasing depletion of Ar for $T > 30$ K. This suggested that trapping in ice might indeed be responsible for the patterns of noble gas abundances found in the atmospheres of Mars, Earth, and Venus (Owen, Bar-Nun, & Kleinfeld, 1991). This initial analysis was not conclusive, however, since it did not seem possible to account for the variety of patterns observed on the three planets with a single model of cometary combardment.

The SNC meteorites provided another clue. It is now widely accepted that these meteorites come from Mars (McSween, 1994). Yet a three-isotope plot of $^{36}Ar/^{132}Xe$ vs $^{84}Kr/^{132}Xe$ shows that the abundances of these gases are very different in different samples from the same meteorite and from the different meteorites, forming a straight line on a log-log plot. This line passes through points corresponding to the noble gases found in the atmospheres of the Earth and Mars, but it is widely separated from the field of abundances in chondritic meteorites or the values found in the solar wind, as first pointed out by Ott & Begemann (1985).

We have interpreted this distribution of noble gas abundances in the SNC meteorites as an effect of the mixing of two different reservoirs: one that represents gases trapped in the rocks that form the planets, the second a contribution from impacting icy planetesimals. This interpretation is based on our data for the fractionation of noble gases by trapping in amorphous ice. At a temperature of ~50 K, amorphous ice formed by condensation of water vapor traps an initially solar mixture of argon, krypton, and xenon in relative abundances that fall on an extrapolation of a mixing line drawn through abundances from the atmospheres of Earth and Mars on the same three element plot (Figure 2). The analysis is complicated by the fact that the linear relation on the log-log plot illustrated by Ott & Begemann (1985) is very close to the mixing line (Owen, Bar-Nun, & Kleinfeld, 1992; Ozima & Wada, 1993; Owen & Bar-Nun, 1993). It may well be that more than one process is at work here. However, we have pointed out that the increase in $^{129}Xe/^{132}Xe$ that occurs along this same mixing line demonstrates that solubility in a melt (for example) cannot explain the data by itself: as $^{36}Ar/^{132}Xe$ and $^{84}Kr/^{132}Xe$ increase, so does $^{129}Xe/^{132}Xe$. This indicates an increase in the proportion of Martian atmopsheric gas (as opposed to gas from the interior of the planet) that is present in the samples plotted in Figure 2.

We have further tested the applicability of the laboratory data to natural phenomena by examining the abundances of CO and N_2 in comets (Owen & Bar-Nun, 1995a). We pointed out that the apparently mysterious depletion of nitrogen in comets probably results from the inability of ice to trap N_2 efficiently when the ice forms at $T > 35$ K. Our assumption is that most of the nitrogen that was present in the outer solar nebula was in the form of N_2, just as it is in the interstellar medium (van Dishoeck et al., 1993). Hence for comets to acquire a solar ratio of N/O, the condensing ice that formed the comets would have to trap N_2, which behaves much like Ar in the laboratory experiments (Bar-Nun, Kleinfeld, & and Kochavi, 1988). We also showed that the relative abundances of CO^+/N_2^+ derived

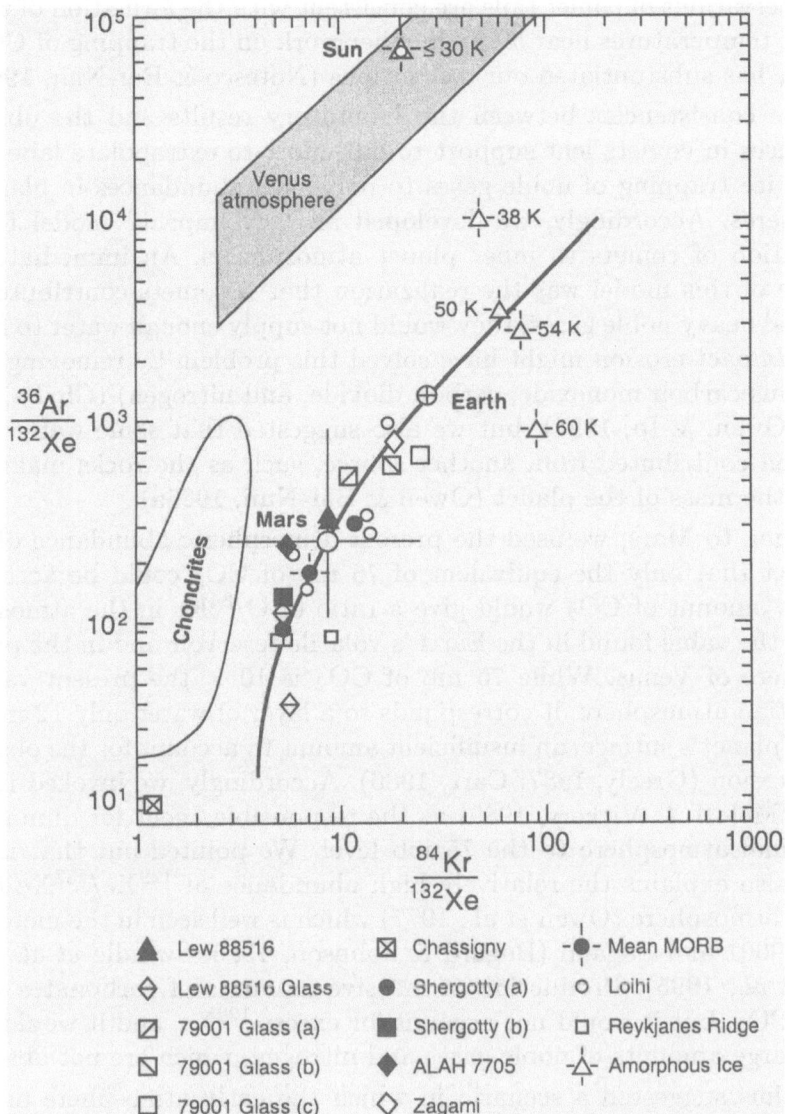

Figure 2.   In a three element plot, noble gas abundances in the atmospheres of Mars and Venus can be used to define a mixing line between internal and external volatile reservoirs. The internal reservoir lies below Mars on this plot, and consists of the rocks that formed the planet. We suggest icy planetesimals as the external reservoir, lying above the Earth at the opposite end of the line. The external reservoir is represented here by the noble gases trapped in amorphous ice in laboratory experiments (the open triangles). Noble gas abundances in Shergottites (meteorites from Mars) fall along this line, as do gases from mantle-derived rocks on Earth (Mean MORB, Loihi, Reykjanes Ridge). The gases on Venus could have been delivered by comets from the Kuiper Belt that formed at temperatures of 30–35 K. The abundance of Xe on Venus is not yet known, hence the stippled trapezoid.

from observations of comet tails are consistent with the formation of the icy nuclei at temperatures near 50 K. Further work on the trapping of CO and N_2 in ice has substantiated our conclusions (Notesco & Bar-Nun, 1996).

These consistencies between the laboratory results and the observed abundances in comets lent support to our effort to extrapolate laboratory work on ice trapping of noble gases to noble gas abundances in planetary atmospheres. Accordingly, we developed an "icy impact" model for the contribution of comets to inner planet atmospheres. An immediate consequence of this model was the realization that if comets contributed the terrestrial heavy noble gases, they would not supply enough water to fill the oceans. Impact erosion might have solved this problem by removing noble gases (plus carbon monoxide, carbon dioxide, and nitrogen) (Chyba, 1990; Chyba, Owen, & Ip, 1994), but we also suggested that some water might have been contributed from another source, such as the rocks making up most of the mass of the planet (Owen & Bar-Nun, 1995a).

Turning to Mars, we used the present atmospheric abundance of ^{84}Kr to predict that only the equivalent of 75 mb of CO_2 could be accounted for. This amount of CO_2 would give a ratio of $C/^{84}$Kr in the atmosphere equal to the value found in the Earth's volatile reservoir and in the present atmosphere of Venus. While 75 mb of CO_2 is 10 × the present value in the Martian atmosphere, it corresponds to a layer of water only 12 m thick over the planet's surface, an insufficient amount to account for the observed fluvial erosion (Greely, 1987; Carr, 1986). Accordingly, we invoked impact erosion (Melosh & Vickery, 1989) as the responsible agent for diminishing the original atmosphere to the 75 mb level. We pointed out that impact erosion also explains the relatively high abundance of ^{129}Xe/^{132}Xe in the Martian atmosphere (Owen et al., 1977) which is well seen in the meteorites EETA 79001 and Zagami (Bogard & Johnson, 1983; Swindle et al., 1986; Marti et al., 1995). Production of massive amounts of carbonates would remove CO_2, but it would not account for excess ^{129}Xe, and it would leave behind large amounts of noble gases and nitrogen, which are not observed.

We thus suggested a scenario in which the early atmosphere of Mars could have passed through several episodes of growth and decay, depending on the planet's bombardment history (Owen, 1992; Owen & Bar-Nun, 1995a). We suggested that it is possible to see the effects of cometary bombardment in the gases trapped in the SNC meteorites, although other interpretations are certainly possible (Owen & Bar-Nun, 1995b; Swindle, 1995).

3. Deuterium in Comets: A Tie to the Interstellar Medium

Clearly one of the assumptions in this model is that comets have retained a good memory of interstellar chemical conditions. One way to test this assumption is to investigate the abundances of stable isotopes of common elements, as isotope ratios should not change during the comet-formation process we have described.

The recent appearance of two bright comets has allowed us to add to the data collected by in situ measurements of Comet Halley using ground-based techniques. Observations at radio wavelengths have been especially helpful.

The value of D/H is particularly sensitive to physical processes, since the mass ratio of these two isotopes is the largest in the periodic table. In Halley's comet, two independent investigations led to the result that D/H $= 3.2 \pm 0.3 \times 10^{-4}$ in the comet's H_2O, twice the value found in standard mean ocean water (SMOW) on Earth (Balsiger, Altwegg, & Geiss, 1995; Eberhardt et al., 1995). These studies were carried out in situ , with mass spectrometers on the Giotto spacecraft. Observations of Comet Hyakutake at radio and infrared wavelengths from Earth allowed the detection of lines of HDO and H_2O that led to a value of D/H $= 3.3 \pm 1.5 \times 10^{-4}$ (Bockelée-Morvan et al., 1998).

In the case of Comet Hale-Bopp, it was possible to use these same remote sensing techniques to determine D/H in both H_2O (Meier, 1998a) and HCN (Meier, 1998b). The results were quite different: (D/H) H_2O = $3.2 \pm 1.2 \times 10^{-4}$, (D/H) HCN $= 2.3 \pm 0.4 \times 10^{-3}$. This is exactly the kind of difference that shows up in interstellar molecular clouds (Millar, Bennett & Herbst, 1989), providing a strong case for the preservation of interstellar chemistry in comet ices. If comets had formed from solar nebula gas that was warmed and homogenized by radial mixing, the values of D/H in both molecules would be much lower and identical, as the HCN and H_2O would exchange D with the huge reservoir of H_2 in the solar nebula, in which D/H $= 2 - 4 \times 10^{-5}$.

4. The Origin of Water on Earth and Mars

We might also expect to see signs of cometary bombardment in the water that is present on the inner planets. Note that all three of the comets that have been studied for this exhibit a value of D/H in ice that is twice the value found in ocean water on Earth. All of these comets came from the Oort Cloud, a spherical shell of cometary nuclei whose average radius is 50,000 times the Earth's distance from the sun. Three is certainly a tiny number compared to the 10^{11} comets that may be present in the Oort Cloud (Weissman, 1991), so one might speculate that other comets may have different values of D/H. In particular, we have not yet determined deuterium abundances in any comets from the Kuiper Belt, a disk of comet nuclei extending outward from Pluto. The Kuiper Belt is currently thought to be the likely source of short period comets, which are unfortunately too faint to yield values of D/H to present ground-based techniques. Obviously, we should continue to measure D/H in all comets that are bright enough, and we should take advantage of proposed spacecraft visits to short period comets to extend this analysis to that family as well.

Meanwhile, we shall assume that the three comets are indeed representative and see whether we can find traces of cometary water on the inner planets. Mars is the best case, because the atmosphere is so thin. The total amount of water that exchanges seasonally through the Martian atmosphere from pole to pole is just 2.9×10^{15} gm, equivalent to a single comet nucleus (density $\rho = 0.5$ g/cm^3) with a radius of only 1 km. Hence the impact of a relatively small comet could have a significant effect on the surface water supply. Indications of such an effect can be found in studies of D/H in hydrous phases of the SNC meteorites. Watson, Hutcheon, & Stolper (1994) reported a range of values of D/H in kaersutite, biotite, and apatite in Chassigny, Shergotty, and Zagami, with the highest values, 4 to 5.5 times the terrestrial ocean standard, occurring in Zagami apatite. Infrared spectroscopy from Earth has determined D/H in Martian atmospheric water vapor to be $5.5 \pm 2 \times$ terrestrial (Owen et al., 1988; Krasnopolosky et al., 1997). The overlap between the SNC and Mars atmosphere values indicates that some mixing between atmospheric water vapor and the crustal rocks must occur. It is therefore arresting to note that the lowest values of D/H measured in the SNC minerals do not cluster around 1 × terrestrial, but rather around 2 × terrestrial. The significance of this lies in the fact that D/H in the water in our three comets is also 2 × terrestrial. It thus appears possible that most of the near-surface water on Mars was contributed primarily by cometary impact, rather than by magma from the planet or by meteoritic bombardment.

This interpretation is consistent with the geochemical analysis by Carr and Wänke (1992), who concluded that Mars is much drier than the Earth, with roughly 35 ppm water in mantle rocks as opposed to 150 ppm for Earth. These authors suggested that one possible explanation for this difference is the lack of plate tectonics on Mars, which would prevent a volatile-rich veneer from mixing with mantle rocks. This is exactly what the D/H values in the SNC minerals appear to signify. In whole rock samples, D/H in the Shergottites is systematically higher than in Nakhla and Chassigny (Leshin, Epstein, & Stolper, 1996). This could be a result of the larger fraction of atmospheric H_2O incorporated in the Shergottites by shock, a process Nakhlites evidently avoided (Bogard, Hörz, Johnson, 1986; Ott, 1988; Drake, Swindle, Owen, and Mussellwhite, 1994).

Furthermore, higher values of $\Delta^{17}O$ are found in these samples of Nakhla and Chassigny compared with the Shergottites, reinforcing the idea that the

Shergottites sampled a different source of water. Finally, the oxygen isotope ratios in water from whole rock samples of the SNCs are also systematically different from the ratios found in silicates in these rocks (Karlsson et al., 1992) again suggesting a hydrosphere that is not strongly coupled to the lithosphere.

On Earth, the oxygen isotopes in sea water match those in the silicates, indicating thorough mixing for at least the last 3.5 BY (Robert, Rejon-Michel, & Javoy, 1992). The Earth has also lost relatively little hydrogen into space after the postulated hydrodynamic escape, so the value of D/H we measure in sea water today must be close to the original value (Yung & Dissly, 1992). If Halley and Hyakutake and Hale-Bopp are truly representative of all comets, then we can't make the oceans out of melted comets alone. This is a very different situation from Mars. It suggests that water from the inner reservoir, the rocks making up the bulk of the Earth, must have mixed with incoming cometary water to produce our planet's oceans.

Explanations for the relatively high value of O/C (12 ± 6) in the terrestrial volatile inventory also suggest such mixing. The solar value of O/C = 2.4, which was also the value found in Halley's comet (Geiss, 1988; Krankowsky, 1991). Impact erosion on the Earth could remove CO and CO_2 while having little effect on water in the oceans or polar caps thereby raising the value of O/C (Chyba, 1990; Owen, & Bar-Nun, 1995a). This process would not affect the value of D/H, however. The average value of D/H in chondritic meteorites is close to that in sea water, so mixing meteoritic and cometary water would not lead to the right result. We need a contribution from a reservoir with D/H $< 1.6 \times 10^{-4}$. Lecluse & Robert (1994) have shown that water vapor in the solar nebula at 1 AU from the Sun would have developed a value of D/H $\approx 0.8 \times 10^{-4}$ to 1.0×10^{-4}, depending on the lifetime of the nebula (2×10^5 to 2×10^6 years). An ocean made of roughly 35% cometary water and 65% water from the local solar nebula (trapped in planetary rocks) would satisfy the D/H constraint and would also be consistent with the observed value of O/C $= 12 \pm 6$.

To accept this idea, we should be able to demonstrate that water vapor from the solar nebula was adsorbed on grains that became the rocks that formed the planets. Our best hope for finding some of that original inner-nebula water appears to be on Mars, where mixing between the surface and the mantle has been so poor. The test is thus to look for water incorporated in SNC meteorites that appear to have trapped mantle gases, to see if D/H $< 1.6 \times 10^{-4}$.

The best case for a such a test among the rocks we already have is Chassigny, which exhibits no enrichment of ^{129}Xe and thus appears not to have trapped any atmospheric gas. However, there is no evidence of water with low D/H in this rock (Leshin et al., 1996). It may be that contamination by terrestrial water has masked the Martian mantle component in Chassigny. This is a good project for a sample returned from Mars by spacecraft, where such contamination can be avoided.

5. The Importance of Neon

We have concentrated our analysis on water and the heavy noble gases: argon, krypton, and xenon. Any model for the origin of the atmosphere must also account for neon. This gas has about the same cosmic abundance as nitrogen relative to hydrogen (Anders & Grevesse, 1989), viz., 1.2×10^{-4} vs. 1.1×10^{-4}. Hence we expect any atmosphere that consists of a captured remnant of the solar nebula to exhibit a ratio of $Ne/N_2 \approx 2$. On Earth, Mars, and Venus, $Ne \ll N_2$. The neon is not only deficient in these atmospheres, the isotopes have been severely fractionated. The solar ratio of $^{20}Ne/^{22}Ne = 13.7$ (Anders & Grevesse, 1989), on Earth $^{20}Ne/^{22}Ne = 9.8$, on Venus 11.8 ± 0.7 (Istomin, Grechnev, & Kochnev, 1982), and Mars 10.1 ± 0.7 (Becker and Pepin, 1984). Concentrating on the Earth, we must ask how neon can be so severely fractionated, whereas nitrogen (mass 14 amu) is either not fractionated at all or at least much less than neon. (The atmospheric value of $^{14}N/^{15}N= 272$ may be compared with the cometary value of 323 ± 46 measured in Comet Hale Bopp (Jewitt et al., 1998).

The answer may again lie with the comets, albeit in a paradoxical manner. The laboratory work shows that neon is not trapped in ice that forms at temperatures above 20 K (Bar-Nun et al., 1988). As the overwhelming majority of the icy planetesimals that formed in the Solar System condensed at temperatures higher than this, we have assumed that comets carry no neon (Owen & Bar-Nun, 1993; Owen & Bar-Nun, 1995a). This assumption is supported by the apparent absence of neon in the atmospheres of Titan, Triton, and Pluto, where the upper limits on Ne/N2 are typically about 0.01 (Hunten et al., 1984). Triton and Pluto represent giant icy planetesimals that can be thought of as the largest members of the Kuiper Belt. Hence the absence of detectable neon in their atmospheres may be taken as a good indication that ice condensing in the outer solar nebula did not trap this gas, and thus we do not expect to find it in comets. This prediction is consistent with new observations of Comet Hale-Bopp carried out with the Extreme Ultraviolet Explorer satellite by Krasnopolsky et al. (1997). These authors established an upper limit of Ne/O < 1/200× solar.

If the comets don't carry neon, how did this gas reach the inner planets? Once again the meteorites don't help. Even if they brought in all the xenon, the neon they could deliver would be <10% of what we observe. Instead it seems likely that neon was brought in by the rocks. In fact, we have evidence that this was the case, because we can still find neon whose isotope abundances approach the solar ratio in rocks derived from the mantle (Craig & Lupton, 1975; Honda *et al.*, 1990). Unlike the other noble gases, neon cannot be subducted into the Earth's interior (Hiyagon, 1994). Thus it is not possible to dilute the original trapped gas with highly fractionated atmospheric neon. The record of original emplacement is preserved.

If neon, which diffuses so easily through solids, was retained by the Earth's rocks from the time of the planet's accretion, we can reasonably assume that some water was also kept in the interior, to emerge after the catastrophic formation of the moon, mixing with incoming water from comets to form the oceans we find today. This perspective supports the idea that we may yet find evidence of this original water in mantle-derived rocks on Mars.

Returning to the Earth, it appears that the atmospheric neon bears a record of an early fractionating process that sharply reduced the ratio of $^{20}Ne/^{22}Ne$ from the solar value. This process must have affected all of the other species in the atmosphere at the time. It may have been the massive, hydrodynamic escape of hydrogen produced by the reduction of water by contemporary crustal iron (Zahnle, Kasting, & Pollack, 1991; Dreibus & Wänke, 1989). The fact that we do not see evidence of such a large fractionation in atmospheric nitrogen in the atmosphere today suggests that the nitrogen reached the Earth after this process had ended. Cometary delivery of nitrogen (and other volatiles) but not neon offers an easy means of achieving this condition (Owen & Bar-Nun, 1995a). In this case, neon is a kind of atmospheric fossil, a remnant of conditions that existed on the Earth before the volatiles that produced the bulk of the present atmosphere were in place.

6. Conclusions

How unique is the Earth? This is a perennial question in efforts to estimate the possibilities for abundant life in the universe. We have argued here that the source of our planet's atmosphere can be found in a combination of volatiles trapped in the rocks that made the planet and a late-accreting veneer of volatile-rich material delivered by icy planetesimals. The volatiles composing the atmosphere include the carbon, nitrogen, and water essential to life. The close similarities between the elemental and isotopic abundances found in comets with those in the interstellar medium imply that icy planetesimals that form in any planetary system originating from an interstellar cloud will carry these same biogenic materials. Hence this model for the origin of our planet's atmosphere suggests that there is nothing unique about the inventory of volatiles that was delivered to the early Earth. Current studies of the Martian atmosphere, aided by study of the SNC meteorites, reinforce this idea by indicating a similar inventory on that planet.

Nevertheless, we are still in the stage of finding "similarities" and "indications." We do not yet have a rigorous proof of the validity of the icy impact model. We have stressed the constraints provided by the abundances and isotope ratios of the noble gases and determinations of cometary D/H. It is clear that meteoritic delivery of volatiles cannot satisfy the constraints set by our present knowledge of noble gas abundances and isotope ratios. However, the cometary alternative that we have emphasized will remain conjectural until noble gases are actually measured in a comet. Although laboratory studies strongly suggest that comets can deliver the correct elemental abundances, trapping of gas in ice does not affect isotopic ratios. We are therefore forced to assume that comets carry xenon whose isotopes resemble the distribution found in the terrestrial and Martian atmospheres rather than that found in the solar wind. It is not at all obvious why this should be the case. If the cometary xenon in fact resembles solar wind xenon, it will be necessary to invoke a fractionating process that acted to produce identical results for xenon on Mars and Earth followed by subsequent selective replacement of other volatiles (Pepin, 1991, 1994).

How can we move forward from this unsatisfactory situation? There are a number of possible sources of new data on the horizon:

1. MARS. The Japanese Nozomi Mission launched in 1998 carries instruments that will teach us much more about possible non-thermal escape processes on Mars. This knowledge will allow a more confident reconstruction of the early mass and composition of the Martian atmosphere. These parameters can then be used (again!) to test our understanding of the origin of our own atmosphere.

If present plans mature, the step forward achieved from Nozomi will soon be overshadowed by information obtained from Mars Sample Return Missions, scheduled to begin in 2010. These missions will bring back samples of Martian rocks and atmosphere for analysis on Earth, enabling far more accurate measurements of isotopic ratios than we can expect from missions to the planet. These measurements will include not only the noble gases, but also isotopes of carbon, nitrogen, and oxygen, the last in both H_2O and CO_2. With the kind of precision obtainable in laboratories on Earth, great progress should be achieved in unraveling the history of the Martian atmosphere from these isotope measurements, including estimates of the size and location of contemporary reservoirs of H2O and CO_2 (McElroy et al., 1977; Jakosky, 1991; Owen, 1992; Jakosky et al., 1994). Another goal of this research would be a search for low values of D/H in water from mantle-derived rocks, which should also contain neon with $^{20}Ne/^{22}Ne$ approaching the solar value of 13.7.

2. COMETS. Both NASA and ESA are planning missions that will

rendezvous with comets and deploy landers to explore their nuclei. These missions will have the capability to detect and measure the abundances of the heavy noble gases and their isotopes. This will be the most definitive test of the icy impact model. It is especially important to have this information from several comets, as we already know that the composition of comets can vary, both from the laboratory work on the trapping of gas in ice (Figure 2) and from observations of variations in carbon compounds in comets (Fink, 1992; A'Hearn *et al.*, 1995).

The ESA mission to Comet Wirtanen is called Rosetta, it should arrive at its destination in 2012. The first NASA mission to a comet is DS-1, scheduled to encounter P/Borrelly in September 2001. There is even hope for a new atmospheric probe to Venus in this time frame, which would tell us more about the apparently anomalous noble gas abundances on this planet. We await all these new results with great interest.

Acknowledgments

This paper is based on a paper of the same title published in *Faraday Discussions, 109, 453–462*, and reproduced in an updated and revised form with permission of the *Faraday Discussions*. This research was supported in part by the NASA Exobiology Program.

References

A'Hearn, M. F., Millis, R. L., Schleicher, D. G., Osip, D.J., & Birch, P. V. 1995, The ensemble properties of comets: Results from narrowband photometry of 85 comets, 1976–1992, *Icarus* 118, 223–270.

Anders, E., & Grevesse, N. 1989, Abundances of the elements: Meteoritic and solar, *Geochim. Cosmochim. Acta* 53, 197–214.

Anders, E., & Owen, T. 1977, Mars and Earth: Origin and abundances of volatiles, *Science* 198, 453–465.

Balsiger, H., Altwegg, K., & Geiss, J. 1995, D/H and $^{18}O/^{16}O$ ratio in the hydronium ion and in neutral water from in situ ion measurements in Comet P/Halley, *J. Geophys. Res.* 100, 5827–5834.

Bar-Nun, A., Herman, B., Laufer, D., & Rappoport, M. L. 1985, Trapping and release of gases by water ice and implications for icy bodies, *Icarus* 63, 317–332.

Bar-Nun, A., Kleinfeld, I., & Kochavi, E. 1988, Trapping of gas mixtures by amorphous water ice, *Phys. Rev. B.* 38, 7749–7754.

Becker, R. H., and Pepin, R. O. 1984, The case for a martian origin of the Shergottites: nitrogen and noble gases in EETA 79001, *Earth Planet. Sci. Lett.* 69, 225–242.

Bernatowicz, T. J., Kennedy, B. M., & Podosek, F. A. 1985, Xe in glacial ice and the atmospheric inventory of noble gases, *Geochim. Cosmochim. Acta* 49, 2561–2564.

Bockelée-Morvan, D., et al. 1998, Deuterated water in comet C/1996 B2 (Hyakutake) and its implications for the origin of comets, *Icarus* 133, 147–162.

Bogard, D. D., Hörz, F., Johnson, P. H. 1986, Shock-implanted noble gases: An experimental study with implications for the origin of Martian gases in Shergottite meteorites. LPSC XVII, *J. Geophys. Res.* 91, E99–E114.

Bogard, D. D., & Johnson, P. 1983, Martian gases in an Antarctic meteorite?, *Science*

221, 651–654.

Carr, M. H. 1986, Mars: A water-rich planet?, *Icarus* **56**, 187–216.

Carr, M. H., and Wänke, H. 1992, Earth and Mars: Water inventories as clues to accretional histories, *Icarus* **98**, 61–71.

Chyba, A. C. 1990, Impact delivery and erosion of planetary oceans in the inner solar system, *Nature* **343**, 129–133.

Chyba, C., Owen, T., & Ip, W.-H. 1994, Impact delivery of volatiles and organic molecules to Earth, in *Hazards Due to Comets & Asteroids* ed. T. Gehrels (Tucson: Univ. Arizona Press), pp. 9–58.

Craig, H., & Lupton, J. E. 1976, Primordial neon, helium, and hydrogen in oceanic basalts, *Earth Planet. Sci. Lett.* **31**, 369–385.

Delsemme, A. 1991, Nature and history of the organic compounds in comets: An astrophysical view, in *Comets in the Post-Halley Era*, ed. R. L. Newburn, Jr., M. Neugebauer, & J. Rahe (Dordrecht: Kluwer), pp. 337–428.

Drake, M. J., Swindle, T. D., Owen, T. & Musselwhite, D. L. 1994, Fractionated martian atmosphere in the nakhlites?, *Meteoritics* **29**, 854–859.

Dreibus, G., & Wänke, H. 1989, Supply and loss of volatile constituents during the accretion of terrestrial planets, in *Origin and Evolution of Planetary and Satellite Atmospheres*, ed. S. K. Atreya, J. B. Pollack, & M. S. Matthews (Tucson: Univ. Arizona Press), pp. 268–288.

Eberhardt, P., Reber, M., Krankowsky, D., & Hodges, R. R. 1995, The D/H and $^{18}O/^{16}O$ ratios in water from comet P/Halley, *Astron. Astrophys.* **302**, 301–316.

Fink, U. 1992, Comet Yanaka (1998r): A new class of carbon poor comet, *Science* **257**, 1926–1929.

Geiss, J. 1988, Composition in Halley's comet: Clues to origin and history of cometary matter, *Rev. Mod. Astron.* **1**, 1–27.

Greely, R. 1987, Release of juvenile water on Mars: Estimated amounts and timing associated with volcanism, *Science* **136**, 688–690.

Hiyagon, H. 1994, Retention of solar helium and neon in IDPs in deep sea sediment, *Science* **263**, 1257–1259.

Honda, M., McDougall, I., Patterson, D., Doulgeris, A., & Claugue, D. A. 1991, Possible solar noble-gas component in Hawaiian basalts, *Nature* **349**, 149–151.

Ip, W. H., & Fernandez, J. A. 1988, Exchange of condensed matter among the outer and terrestrial protoplanets and the effect on surface impact and atmospheric accretion, *Icarus* **74**, 47–61.

Istomin, V. G., Grechnev, K. V., & Kochnev, V. A. 1982, Preliminary results of mass-spectrometric measurements on board the Venera 13 and Venera 14 probe, *Pisma Astron. Zh.* **8**, 391–398.

Jakosky, B. M. 1991, Mars volatile evolution: Evidence from stable isotopes, *Icarus* **94**, 14–31.

Jakosky, B. M., Pepin, R. M., Johnson, R. E., & Fox, J. L. 1994, Mars atmospheric loss and isotopic fractionation by solar-wind-induced sputtering and photochemical escape, *Icarus* **111**, 271–288.

Jewitt, D., Matthews, H. E., Owen, T., & Meier, R. 1997, Measurements of $^{12}C/^{13}C$, $^{14}N/^{15}N$, and $^{32}S/^{34}S$ Ratios in Comet Hale-Bopp (C/1995 O1), *Science* **278**, 90–93.

Karlsson, H. R., Clayton, R. N., Gibson, E. K., Jr., & Mayeda, T. K. 1992, Water in SNC meteorites: Evidence for a martian hydrosphere, *Science* **255**, 1409–1411.

Krankoswky, D. 1991, The composition of comets, in *Comets in the Post-Halley Era*, ed. R. L. Newburn, Jr., M. Neugebauer, & J. Rahe (Dordrecht: Kluwer), pp. 855–879.

Krasnopolsky, V. A., Bjoraker, G. L., Mumma, M. J., & Jennings, D. E. 1997a, High resolution spectroscopy of Mars at 3.7 and 8 μm : A sensitive search for H_2O_2, H_2CO, HCl and CH_4, and detection of HDO, *J. Geophys. Res.* **102**, 6524–6534.

Krasnopolsky, V. A., Mumma, M. J., Abbott, M., Flynn, B. C., Meech, K. J., Yeomans, D. K., Feldman, P. D., & Cosmovici, C. B. 1997b, Detection of soft x-rays and a sensitive search for noble gases in comet Hale-Bopp (C/1995 01), *Science* **277**, 1488.

Laufer, D., Kochavi, E., & Bar-Nun, A. 1987, Structure and dynamics of amorphous water ice, *Phys. Rev. B* **36**, 9219–9227.

Lécluse, C., & Robert, F. 1994, Hydrogen isotope exchange reaction rates: Origin of water in the solar system, *Geochim. Cosmochin. Acta* **58**, 2927–2940.

Leshin, L. A., Epstein, S., & Stolper, E. M. 1996, Hydrogen isotope geochemistry of SNC meteorites, *Geochim. Cosmochim. Acta* **60**, 2635–2650.

Lunine, J. L., Engel, S., Rizk, B., & Horanyi, M. 1991, Sublimation and reformation of icy grains in the primitive solar nebula, *Icarus* **94**, 333–343.

Marti, K., Kim, J. S., Thakur, A. N., McCoy, T. J., & Keil, K. 1995, Signatures of the martian atmosphere in glass of the Zagami meteorite, *Science* **267**, 1981–1984.

McElroy, M. B., Kong, T. Y., & Yung, Y. L. 1997, Photochemistry and evolution of Mars' atmosphere: A Viking perspective, *J. Geophys. Res.* **82**, 4379–4388.

McSween, H. Y., Jr. 1994, What have we learned about Mars from SNC meteorites?, *Meteoritics* **29**, 757–779.

Meier, R., Owen, T., Matthews, H. E., Jewitt, D., Bockelée-Morvan, D., Biver, N., Crovisier, J., & Gautier, D. 1998a, A Determination of the HDO/H_2O Ratio in Comet C/1995 O1 (Hale-Bopp), *Science* **279**, 842–844.

Meier, R., Owen, T., Jewitt, D., Matthews, H. E., Senay, M., Biver, N., Bockelée-Morvan, D., Crovisier, J., & Gautier, D. 1998b, Deuterium in Comet C/1995 O1 (Hale-Bopp): Detection of DCN, *Science* **279**, 1707–1710.

Melosh, H. J., & Vickery, A. M. 1989, Impact erosion of the primordial atmosphere of Mars *Nature* **338**, 487–489.

Millar, T. J., Bennett, A., & Herbst, E. 1989, Deuterium fractionation in dense interstellar clouds, *Astrophys. J.* **340**, 906–920.

Notesco, G., & Bar-Nun, A. 1996, Enrichment of CO over N_2 by their trapping in amorphous ice and implications to Comet Halley, *Icarus* **122**, 118–121.

Oro, J. 1961, Comets and the formation of biochemical compounds on the primitive Earth, *Nature* **190**, 389–390.

Ott, U. 1988, Noble gases in SNC meteorites: Shergotty, Nakhla, Chassigny, *Geochim. Cosmochim. Acta* **52**, 1937–1948.

Ott, U., & Begemann, F. 1985, Are all the "martian" meteorites from Mars?, *Nature* **317**, 509–512.

Owen, T. 1992, The composition and early history of the atmosphere of Mars, in *Mars*, ed. H. H. Kieffer et al. (Tucson: Univ. Arizona Press), pp. 818–834.

Owen, T. & Bar-Nun, A. 1993, Noble gases in atmospheres, *Nature* **361**, 693–694.

Owen, T., & Bar Nun, A. 1995a, Comets, impacts and atmospheres, *Icarus* **116**, 215–226.

Owen, T., & Bar-Nun, A. 1995b, Comets, impacts and atmospheres II, Isotopes and noble gases, in *AIP Conf. Proc. 341, Volatiles in the Earth and Solar System*, ed. K. Farley (New York: AIP), pp. 123–138.

Owen, T., Bar-Nun, A., & Kleinfeld, I. 1991, Noble gases in terrestrial planets: Evidence for cometary impacts, in *Comets in the Post-Halley Era*, ed. R. L. Newburn, Jr., M. Neugebauer, & J. Rahe (Dordrecht: Kluwer), pp. 429–438.

Owen, T., Bar-Nun, A., & Kleinfeld, I. 1992, Possible cometary origin of heavy noble gases in the atmospheres of Venus, Earth and Mars, *Nature* **358**, 43–46.

Owen, T., Biemann, K., Rushneck, D. R., Biller, J. E., Howarth, D. W., & LaFleur, A. L. 1977, The composition of the atmosphere at the surface of Mars, *J. Geophys. Res.* **82**, 4635–4639.

Owen, T., Maillard, J. P., de Bergh, C., & Lutz, B. L. 1988, Deuterium on Mars: The abundance of HDO and the value of D/H, *Science* **240**, 1767–1770.

Ozima, M., & Wada, N. 1993, Noble gases in atmospheres, *Nature* **361**, 693.

Pepin, R. O. 1989, Atmospheric compositions: Key similarites and differences, in *Origin and Evolution of Planetary and Satellite Atmospheres*, ed. S. K. Atreya, J. B. Pollack, and M. S. Matthews (Tucson: Univ. of Arizona Press), pp. 293–305.

Pepin, R. O. 1991, On the origin and early evolution of terrestrial planet atmospheres and meteoritic volatiles, *Icarus* **92**, 2–79.

Pepin, R. O. 1994, Evolution of theMartian atmosphere, *Icarus* **111**, 289–304.

Robert, R., Rejon-Michel, A., & Javoy, M. 1992, Oxygen isotopic homogeneity of the Earth: New evidence, *Earth Planet. Sci. Lett.* **108**, 1–9.

Sill, G. T., & Wilkening, L. 1978, Ice clathrate as a possible source of the atmospheres of the terrestrial planets, *Icarus* **33**, 13–27.

Swindle, T. D. 1986, Xenon and other noble gases in shergottites, *Geochim. Cosmochim. Acta* **50**, 1001–1015.

Swindle, T. D. 1995, How many Martian noble gas reservoirs have we sampled?, in *AIP Conf. Proc. 341, Volatiles in the Earth and Solar System*, ed. K. Farley (New York: AIP), pp. 175–185.

Turekan, K. K., & Clark, S. P., Jr., 1975, The non-homogeneous accumulation model for terrestrial planet formation and the consequences for the atmosphere of Venus, *J. Atmos. Sci.* **32**, 1257–1261.

van Dishoeck, E. F., Blake, G. A., Draine, B. T., & Lunine, J. I. 1993, The chemical evolution of protostellar and protoplanetary matter, in *Protostars and Planets III*, ed. E. H. Levy & J. I. Lunine (Tucson: Univ. Arizona Press), pp. 163–244.

Wacker, J. F., & Anders, E. 1984, Trapping of xenon in ice and implications for the origin of the Earth's noble gases, *Geochim. Cosmochim. Acta* **48**, 2372–2380.

Watson, L. L., Hutcheon, I. D., & Stopler, E. M. 1994, Water on Mars: Clues from deuterium/hydrogen and water contents of hydrous phases in SNC meteorites, *Science* **265**, 86–90.

Weissman, P. R. 1991, Dynamic History of the Oort Cloud, in *Comets in the Post-Halley Era*, ed. R. L. Newburn, Jr., M. Neugebauer, & J. Rahe (Dordrecht: Kluwer), pp. 463–486.

Yung, Y., & Dissly, R. W. 1992, Deuterium in the Solar System, in *Amer. Chem. Soc. Symp. Ser. 502, Isotope Effects in Gas-Phase Chemistry*, ed. J. A. Kaye (Washington, DC: American Chemical Society), pp. 369–389.

Zahnle, K., Kasting, J. E., & Pollack, J. B. 1991, Mass fractionation of noble gases in diffusion-limited hydrodynamic hydrogen escape, *Icarus* **84**, 502–527.

DUST INFLUX TO TITAN FROM HYPERION

ALEXANDER V. KRIVOV

Astronomical Institute, St. Petersburg University, 198904 St. Petersburg, Russia

AND

MAREK BANASZKIEWICZ

Space Research Centre, Bartycka 18A, PL-00-716 Warsaw, Poland

Abstract. The outer part of the saturnian system, comprising four saturnian satellites – Phoebe, Iapetus, Hyperion, and Titan – is believed to excel in intricate mechanisms of production, evolution, and transport of dust between the involved moons. This paper is focused on the delivery of the dust material to Titan, with the neighboring Hyperion being the most effective dust supplier. Hypervelocity impacts of dust particles coming from Phoebe, as well as bombardments by interplanetary micrometeoroids should eject the surface material of Hyperion to the planetocentric space. We discuss the complex dynamics of the Hyperion ejecta, resulting from the interplay between the resonant gravity of Titan, solar radiation pressure, and plasma drag force. It is shown that unlike Hyperion, the motion of which is stabilized by a strong 4:3 mean motion resonance with Titan, a significant part of the Hyperion debris either is free of resonance initially or is liberated from the resonance during the subsequent dynamical evolution. These particles get in unstable orbits and experience multiple close approaches to Titan. Most of the grains larger than several μm in size finally collide with Titan. We show that the dust influx to Titan from Hyperion may exceed markedly the direct influx of interplanetary grains. It is argued that the influx of water-containing particles from Hyperion may play an important role in the chemistry of Titan's atmosphere, making a significant contribution to the budget of the oxygen-bearing compounds.

M. Ya. Marov and H. Rickman (eds.), Collisional Processes in the Solar System, 265–276.

1. Introduction

Titan and Hyperion may be listed among the most interesting objects in the saturnian system. Titan, having the size of a small planet, is one of the largest satellites in the Solar system and the only one to own a dense nitrogen atmosphere. Small icy Hyperion, a neighbor of Titan in the saturnian system, is known as the largest irregularly-shaped planetary satellite and should also be noted for a number of other intriguing features, such as its peculiar rotational state. The upcoming extensive exploration of the saturnian system by the Cassini spacecraft has renewed interest in these moons, especially Titan, which is the target of the dedicated Huygens probe, part of the Cassini mission.

Despite the striking dissimilarity between the two satellites, they are thought to be intimately related, primarily due to a strong mean-motion resonance, in which both moons orbit Saturn. Farinella *et al.* (1983, 1990) have expressed the idea that the present Hyperion is most likely the core of its much larger, round-shaped precursor, disrupted in a huge break-up event. Using analytical and numerical arguments, they have shown that the resonant gravity of Titan could have prevented the collisional fragments of proto-Hyperion from being reaccreted. This provides a natural explanation of the irregular shape of the satellite, in contrast to the other moons of Saturn, which were disrupted many times, but could revive from debris rings in the form of "rubble piles". Moreover, the relation between the two moons is probably not unilateral. Not only the uncommon features of Hyperion may be accredited to its interactions with Titan, the latter has also been influenced, and is affected now, by the neighboring Hyperion. The influx of the proto-Hyperion fragments is thought to be in close relationship with the origin and evolution of Titan's dense atmosphere, as well as with the origin of large impact basins on Titan's surface (Farinella *et al.*, 1997).

Recently, Banaszkiewicz and Krivov (1997) and Krivov and Banaszkiewicz (2000) suggested that, in addition to its own unusual properties, Hyperion may play the role of an important dust supplier to Titan. The idea relies on the fact that Hyperion, as all atmosphereless bodies in the Solar system, experiences continuous bombardment by hypervelocity projectiles populating interplanetary space, primarily interplanetary dust particles (IDPs). Furthermore, there should exist additional classes of projectiles, which are specific for Hyperion, due to its particular position among the outer saturnian moons (Fig. 1). An extended region from Titan to the outermost saturnian moon Phoebe shows nontrivial mechanisms of production, evolution, and transport of dust material between the involved satellites. Soter (1974) first suggested that the dust grains ejected from Phoebe, when spiraling toward the primary because of the Poynting-Robertson effect, would be swept out by the inner moons, primarily Iapetus, possibly explaining the observed albedo asymmetry of this moon (*e.g.*, Wilson and Sagan, 1996). About 20% of the Phoebe material should drift past Iapetus and may reach Hyperion, the next satellite of Saturn (Burns *et al*, 1996). Since the orbits of Phoebe particles are retrograde while that of Hyperion is prograde, the impact velocities would be as large as $\sim 10\,\mathrm{km\,s^{-1}}$, which results in a considerable ejecta production from Hyperion's surface.

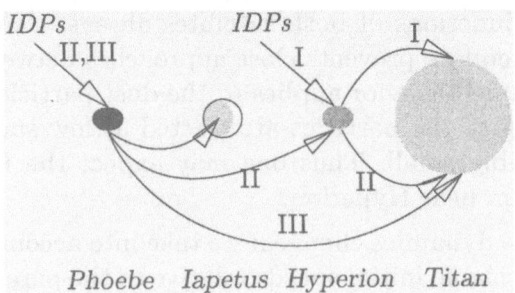

Figure 1. "Dust road map" in the outer saturnian system: I – dust ejected by IDPs from Hyperion and arriving at Titan; II – dust ejected by IDPs from Phoebe, impacting Hyperion and ejecting secondary particles which also arrive at Titan; III – dust ejected by IDPs from Phoebe and reaching Titan directly.

Hyperion, a 250 km-sized satellite, is not massive enough to reaccrete all of the ejecta from its surface. A sizeable fraction of the icy debris, ejected by both Phoebe projectiles and interplanetary impactors, start traveling in the circumplanetary space. Our preliminary analysis of the dust evolutionary paths (Banaszkiewicz and Krivov, 1997; Krivov and Banaszkiewicz, 2000) has revealed a complexity of Hyperion icy ejecta dynamics, driven by gravitational, radiative, and plasma drag forces, and has led us to a conclusion that much of the Hyperion dust finally collides with Titan, contaminating its dense nitrogen atmosphere with water ice. We have also examined the dust distributions in the Hyperion–Titan system and estimated the expected dust influx to Titan to show that Hyperion may dominate the dust influx to Titan. Based on our preceding studies, we present in this paper a general picture of expected dust production from Hyperion, dynamics of the Hyperion ejecta, spatial distributions of dust, and possible influence of the Hyperion dust on the properties of Titan's atmosphere.

2. Qualitative Picture of the Hyperion Ejecta Dynamics

The dynamical evolution of the particles can be qualitatively outlined as follows. A dust grain lost by Hyperion is subject to several forces, of which the following are the most important: Saturn's and Titan's gravity, solar radiation pressure, and plasma drag (PD) force. For a while, neglect the latter two forces and consider the restricted 3-body problem "dust grain–Titan–Saturn". Even in this approximation, the dynamics are not simple. Indeed, Hyperion is known to be locked in a strong 4:3 mean motion resonance with Titan. The resonance is characterized by the critical argument: $\phi = 4\lambda - 3\lambda_T - \varpi$, where λ and λ_T are the longitudes of Hyperion and Titan and ϖ is the longitude of pericenter of Hyperion's orbit. The critical argument librates around 180° with an amplitude $\approx 36°$. The resonant locking ensures that conjunctions of both satellites always occur near Hyperion's apocenter and therefore prevents close approaches between the two moons. Obviously, the same behavior applies to the dust particles originating from Hyperion, as long as the particles are ejected at low speeds and the other forces are negligibly small. Thus one may expect the Hyperion ejecta to form a dust swarm near Hyperion.

How will these dynamics change if we take into account that some of the grains have appreciable initial speeds relative to the parent satellite and are subject to the perturbing forces – the radiation pressure and the PD force? Most notably, are there mechanisms that have the ability to destroy the resonant locking? These mechanisms would be of major interest, because the particles disengaged from the resonance would get an opportunity to experience stochastic encounters with Titan, which would have immediate and principal consequences for the dynamics. It turns out that taking into consideration the non-zero ejection velocities, yet without inclusion of non-gravitational forces, may change the dynamics drastically. If a particle is ejected with an appreciable speed, of the order of several tens $\mathrm{m\,s^{-1}}$ relative to the parent moon, the resonant locking does not occur (Farinella *et al.*, 1983, 1990). Another possibility is that the particles may get into orbits which are more distant from the center of "resonant stability island" than the orbit of Hyperion is ("shallow resonance"). For such orbits, the libration amplitude is much higher that Hyperion's 36°. An amplitude of 100° (or a libration interval [80°, 280°]) was reported by previous workers (see Farinella *et al.*, 1983) as a critical one: with this amplitude, the grain's orbital pericenter at the conjunction with Titan is too close to the satellite to allow stability. How much of the ejecta is fast enough to be either unlocked from the resonance at all or at least to move in a shallow resonance? Estimates, based on plausible velocity distributions of the impact ejecta at the Hyperion surface, show that a mass fraction of the ejecta faster than the mean escape velocity for Hyperion ($u_{esc} \approx 140\,\mathrm{m\,s^{-1}}$) is $\sim 10^{-3}$ (Krivov and Banaszkiewicz, 2000). In turn, nearly half of these (escaping) grains should have speeds of $\gtrsim 100\,\mathrm{m\,s^{-1}}$ near the boundary of Hyperion's action sphere, which is enough for the grains to avoid resonance locking.

Now, we consider the rest of the particles, ejected at relatively low

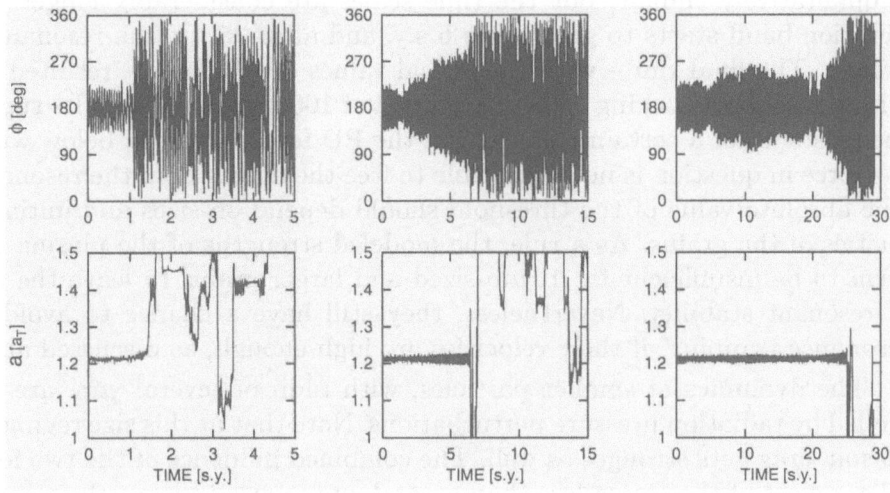

Figure 2. Breaking resonance by the plasma drag force. Shown is a trajectory of a $10\,\mu$m test particle launched from the edge of the action sphere of Hyperion with zero initial velocity. The particle is exposed to plasma drag forces of various strengths: 30 (left), 10 (middle), and 3 (right) times its actual (modeled) value. Top: critical argument (in degrees), bottom: semimajor axis (in units of the semimajor axis of Titan's orbit).

speeds and starting their planetocentric motion in the resonant lock. Can the resonance be broken later, by non-gravitational forces? A candidate mechanism to destroy the stable resonant motion is a PD force, with an estimated strength of 0.1 to 0.3% of the radiation pressure force (see Sect. 3). To find out whether such a small dissipative force is able to release the particles from the resonance, we have performed a series of numerical experiments. A $10\,\mu$m-sized particle was launched from the edge of the action sphere of Hyperion with zero initial velocity. First we followed this trajectory with the normal strength of the PD force predicted by our model. In the subsequent runs, we did the same, but with the PD force artificially magnified by the factors of 3, 10, and 30. The results are shown in Fig. 2. The PD force magnified by a factor of 30 breaks down the resonance in 1 s.y. (saturnian year), which is seen as a sharp transition of the critical argument from libratory to circulatory regime. This is accompanied by an expected slow monotonic increase in the semimajor axis during the first 1 s.y. of the motion. The PD force which is ten-fold the actual one breaks down the resonance as well, but does it more gently. After 2 s.y., the libration band starts to grow and after 5 s.y. broadens to $[80°, 280°]$ which, as discussed above, leads to instability and indeed, after 7 s.y. the critical argument gets into the circulatory mode – the resonance is broken. For the PD force strength exceeding the actual value by the factor of 3, the picture is qualitatively similar, but the time scales are considerably longer. The libration band starts to grow after 6 s.y. and after 25 s.y. the resonance is broken. The final run – with the actual values of PD force – resulted in a strong resonance locking over the interval of 100 s.y.! These results suggest the existence of a certain threshold in the PD force strength, below which, the force in question is no longer able to free the debris from the resonance. The absolute value of the threshold should depend on sizes and initial velocities of the grains. As a rule, the modeled strengths of the plasma drag seem to be insufficient for 10 μm-sized and larger grains to leave the zone of resonant stability. Nevertheless, they still have a chance to avoid the resonance trapping, if their velocities are high enough, as discussed above.

The dynamics of smaller particles, with radii of several μm, are controlled by radiation pressure perturbations. Note that at this size regime the plasma drag gets stronger as well. The combined influence of the two forces allows most of the ejecta less than several μm in size to leave the resonant zone in a small fraction of a s.y., regardless of their initial velocities.

3. Numerical Results and Statistics

From the qualitative picture outlined above, we turn now to a more detailed study of the Hyperion ejecta dynamics. As the main tool, we use numerical integrations of the full Newtonian equations of motion with the following perturbing forces: Titan's gravity, solar radiation pressure, and PD force. Calculation of the first two forces is straightforward. For the latter, we use a two-dimensional axially-symmetric plasma model compiled by Krivov and Banaszkiewicz (2000), predicting the PD strength to range from 0.1% to 0.3% of the radiation pressure for most locations near the Hyperion and Titan orbits.

In our particular problem, a major technical difficulty of the numerical simulations is a significant loss of accuracy during close encounters of grains with Titan (Farinella et al., 1997; Banaszkiewicz and Krivov, 1997). We therefore choose a classic technique of the 3-body problem and transform the equations of motion to a coordinate system centered on Titan and rotating around Saturn with the angular velocity of the satellite's orbital motion.

With this technique, we integrated the trajectories of grains of three sizes – 100, 10, and 2 μm – over the maximum time interval of 100 s.y. (100 trajectories for each size). The starting velocities (at the boundary of Hyperion's sphere of influence) were calculated randomly in accordance with a plausible ejecta velocity distribution at the surface. Describing the results, we will proceed from the largest grains to the smallest ones, giving the statistics of various scenarios and analyzing them from the dynamical

in a fraction of a saturnian year after ejection.

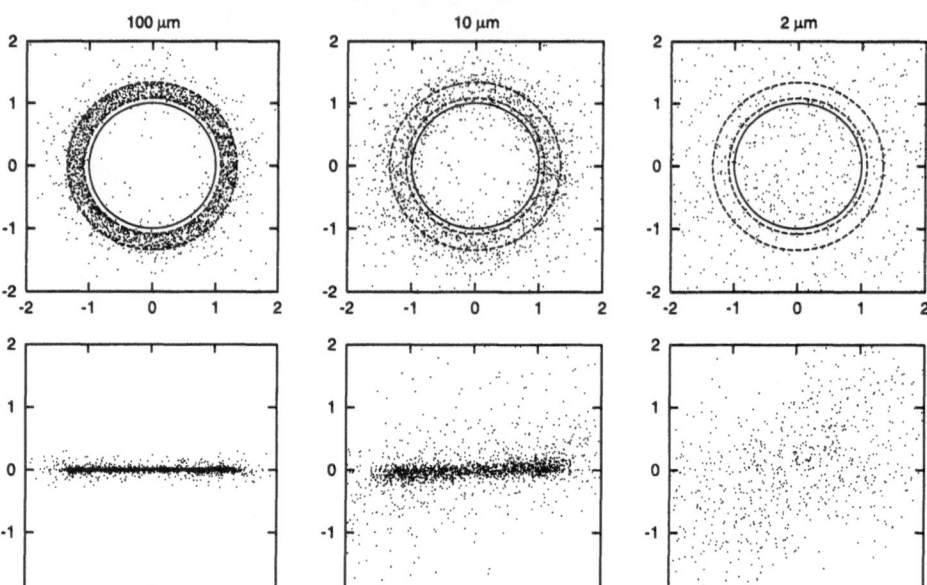

Figure 3.   Spatial distribution of dust in the Saturn-centered equatorial system (x-axis toward vernal equinox, z-axis toward Saturn's north pole). From left to right: 100, 10, and 2 μm. Top and bottom: XY and YZ-projections. Coordinates are in Titan's semimajor axis units. The solid circle in the upper panels is the orbit of Titan, dashed circles correspond to the apocentric and pericentric distances of Hyperion and therefore limit the zone of Hyperion's motion.

The dynamics of the Hyperion ejecta determine the geometry and dimensions of the steady-state dust cloud in the Hyperion-Titan system. Figure 3 depicts snapshots of the dust clouds, formed by different-sized grains, in the Saturn-centered reference frame. Most of the 100 μm grains find themselves in a narrow and flat ring between the pericentric and apocentric distances of Hyperion. The 10 μm particles are located in a torus which is more extended both radially and vertically. The 2 μm grains form a diffuse dust cloud of a spheroidal shape that extends from the innermost part of the saturnian system to far beyond Hyperion's orbit. An interesting feature seen in the plots is a tilt of the dust clouds off the equatorial plane of Saturn. The plane of symmetry of a cloud always intersects the equatorial plane along the direction toward the saturnian vernal equinox point. The tilt angle increases with decreasing grain size and reaches nearly 30° for 2 μm-sized grains.

These features are exactly the same as those predicted for the presumed dust torus around the orbit of Deimos, the outer moon of Mars (Krivov *et al*, 1996; Hamilton and Krivov, 1996; Krivov and Hamilton,

1997). In the cited papers, the planetocentric motion of a particle, perturbed by solar radiation pressure, was investigated analytically and numerically. This suggests that the radiation pressure is a significant, if not the dominant, dynamical factor that determines the geometry of dust clouds in the Hyperion-Titan system as well. This conclusion is confirmed by other analyses of the orbital evolution that we made – short-term variations in eccentricity and long-term high-amplitude variations in inclination. However, the strong gravity of Titan, which acts as a continuous disturber and occasionally modifies the orbits during close encounters, causes a tangible scatter of orbits in the space of orbital elements, making the dynamics more complicated.

4. Discussion

Is it possible to detect the dust swarms (like those shown in Fig. 3) observationally? Simple estimates (Krivov and Banaszkiewicz, 2000) suggest that the number density of the Hyperion grains in the region between the orbits of Hyperion and Titan exceeds the number density of interplanetary grains not more than by three orders of magnitude, so there are no chances for success for any search based on light scattering measurements. Also questionable is the possibility of *in-situ* measurements by the Cassini dust detector. However, these estimates only apply to measurements in space between Hyperion and Titan. The chances to "catch" the Hyperion dust close to the source (Hyperion) and main sink (Titan) are much better. First, it should be possible to detect "fresh", mostly submicrometer-sized ejecta from Hyperion during the Cassini flyby of this moon, at several satellite radii. Quite recently, the first successful measurements of this kind were made by Galileo spacecraft in the vicinities of the Galilean moons of Jupiter (Krüger *et al.*, 1999). A similar technique was proposed to study the dusty ejecta of the martian moons (Krivov and Jurewicz, 1999) and the famous saturnian satellite Enceladus, the likely source of the broad E ring (Spahn *et al.*, 1999). Next, it is not improbable that Cassini will detect some of the large, 10 to 100 μm-sized grains – those which remain locked in the resonance with Titan and stay in a swarm around Hyperion (see Sect. 3). Regarding the possibilities to detect dust at Titan, we do not predict an increased dust density in the vicinity of Titan, which would result from the gravitational focusing of dust by Titan. However, there must be possibilities to gain (indirect) evidence of the Hyperion dust at Titan through the chemical reactions in its atmosphere, induced by the water-containing Hyperion particles. Let us address this question in more detail.

That a majority of ice-containing dust particles kicked up from Hyperion should reach Titan is perhaps the most important conclusion which can be drawn from our study. To estimate possible consequences of this statement, we need to estimate how much dust arrives at Titan. A general scheme of dust production in the outer saturnian system (see Sect. 1) suggests three mechanisms (marked I to III in Fig. 1) that may sustain a continuous dust influx to Titan. The first one is hypervelocity impacts of interplanetary grains to Hyperion's surface. The second is hits of Hyperion by retrograde projectiles from Phoebe. Finally, the third possible source of dust for Titan is the Phoebe grains that avoid capture by both Iapetus and Hyperion and come to Titan directly. Accordingly, let us consider three dust populations at Titan: (i) the flux of Hyperion dust ejected by interplanetary impactors F_{HT}, (ii) the flux of Hyperion dust ejected by the Phoebe projectiles F_{PHT}, (iii) the flux of Phoebe particles F_{PT} (indices T, H, and P refer to Titan, Hyperion, and Phoebe, respectively). These fluxes are to be compared to the direct interplanetary flux F_T at Titan. Krivov and Banaszkiewicz (2000) find the following estimates: $F_{HT}/F_T = 1 \times 10^{-1\pm1.2}$, $F_{PHT}/F_T = 1 \times 10^{0\pm2.0}$, $F_{PT}/F_T = 9 \times 10^{-3\pm1.2}$. We see that, even at the upper limit, the direct flux of Phoebe particles at Titan makes little contribution to the dust budget and can be neglected. However, the influx of Hyperion dust to Titan is most likely comparable with (at the upper limit may be by two orders of magnitude greater than) the direct influx of IDPs at Titan. It is also evident that the bombardment of Hyperion's surface by Phoebe grains is probably more effective than that by IDPs; therefore most of the Hyperion ejecta are likely to be due to the impacts of Phoebe projectiles.

Having predicted significant contamination of Titan by the Hyperion debris, we come to a natural question: which consequences for the Titan's atmosphere can be expected? One of the intriguing problems in the photochemistry of Titan's atmosphere is the source and the abundances of oxygen-bearing compounds. Two of them, CO and CO_2 have been detected by Voyager and, quite recently, the water lines have been observed by ISO (Coustenis et al., 1998). In the most plausible scenarios CO and CO_2 are formed in chemical reactions with OH as a substrate (Lara et al., 1996). Its most likely source is photolysis of water, brought to the atmosphere by meteoroids and released from grains by ablation. English et al. (1996) have found, however, that to produce the observed abundances of CO and CO_2 the flux of pure water-ice meteoroids should exceed the measured interplanetary flux by a factor of 200. This requirement is somehow relaxed (to a factor of 20 overabundance) if one allows the grains to contain 10% of CO (like in Halley's comet). The recent ISO observations (Coustenis et al., 1998) suggest that the required flux can be decreased to the level given by English et al. (1996), therefore the local (saturnian) inventory of water containing grains is not necessarily needed as a main source. Still, several problems remain. First, if interplanetary meteoroids in the vicinity of Saturn are not composed of pure volatile materials, then an additional supply (*e.g.*, Hyperion ejecta) should be introduced. Second, the interplanetary flux shows a pronounced asymmetry when deposited on the leading and trailing hemispheres of Titan: an order of magnitude more material is delivered to the leading side. This effect is caused by the orbital motion of Titan, which leads to different relative velocities of grains approaching Titan. In the case of dust originating from Hyperion that asymmetry, according to the collision statistics we obtained, is significantly smaller: the fractions of material, deposited on the leading and trailing hemispheres, are 67% and 33% for 100 μm particles, and 56% and 44% for 10 μm ones. Also, since the encounter velocity of Hyperion ejecta that reach Titan is much smaller than that of interplanetary grains, the altitude profile of water deposition provided by the Hyperion ejecta should peak at lower altitudes than the profile calculated for interplanetary meteoroids. Further observational and modeling effort is necessary to address these issues before the Cassini mission and to distinguish between the relative contribution of interplanetary and local sources to the water production in Titan's atmosphere.

References

Banaszkiewicz, M. and Krivov, A.V. (1997) Hyperion as a dust source in the saturnian system, *Icarus* **129**, 289–303.

Burns, J., Hamilton, D.P., Mignard, F., and Soter, S. (1996) The contamination of Iapetus by Phoebe dust, in *Physics, Chemistry, and Dynamics of Interplanetary Dust (ASP Conf. Series, vol. 104)*, Gustafson, B.Å.S. and Hanner, M.S. (Eds.), ASP, San Francisco, pp. 179–182.

Coustenis, A., Salama, A., Lellouch, E., Encrenaz, T., Bjoraker, G., Samuelson, R., de Graauw, T., Feuchtgruber, H., and Kessler, M.F. (1998) Evidence for water vapor in Titan's atmosphere from ISO/SWS data, *Astron. Astrophys.* **336**, L85–L89.

English, M.A., Lara, L.M., Lorenz, R., Ratcliff, P., and Rodrigo, R. (1996) Ablation and chemistry of meteoric materials in the atmosphere of Titan, *Adv. Space Res.* **17**, (12)157–(12)160.

Farinella, P., Milani, A., Nobili, A.M., Paolicchi, P., and Zappalà, V. (1983) Hyperion – Collisional disruption of a resonant satellite, *Icarus* **54**, 353–360.

Farinella, P., Paolicchi, P., Strom, R.G., and Kargel, J.S. (1990) The fate of Hyperion's fragments, *Icarus* **83**, 186–204.

Farinella, P., Marzari, F., and Matteoli, S. (1997) The disruption of Hyperion and the origin of Titan's atmosphere, *Astron. Journal* **113**, 2312–2316.

Hamilton, D.P. and Krivov, A.V. (1996) Circumplanetary dust dynamics: Effects of solar gravity, radiation pressure, planetary oblateness, and electromagnetism, *Icarus* **123**, 503–523.

Krivov, A.V. and Banaszkiewicz, M. (2000) Unusual origin, evolution, and fates of icy ejecta from Hyperion, *Icarus*. Submitted.

Krivov, A.V. and Hamilton, D.P. (1997) Martian dust belts: Waiting for discovery, *Icarus* **128**, 335–353.

Krivov, A.V. and Jurewicz, A. (1999) The ethereal dust envelopes of the Martian moons, *Planet. Space Sci.* **47**, 45–56.

Krivov, A.V., Sokolov, L.L., and Dikarev, V.V. (1996) Dynamics of Mars-orbiting dust: Effects of light pressure and planetary oblateness, *Celest. Mech. & Dyn. Astron.* **63**,

313–339.

Krüger, H., Krivov, A.V., Hamilton, D.P., and Grün, E. (1999) Detection of an impact-generated dust cloud around Ganymede, *Nature* **399**, 558–560.

Lara, L., Lellouch, E., Lopez-Moreno, J.J., and Rodrigo, R. (1996) Vertical distribution of Titan's atmospheric neutral constituents, *J. Geophys. Res.* **101**, 23,261–23,283.

Soter, S. (1974) Brightness asymmetry of Iapetus, Paper presented at IAU Coll. No. 28, Cornell Univ., Aug. 1974.

Spahn, F., Thiessenhusen, K.U., Colwell, J.E., Srama, R., and Grün, E. (1998) Dynamics of dust ejected from Enceladus: Application to the Cassini–Enceladus encounter, *J. Geophys. Res.* **104**, 24,111–24,120.

Wilson, P.D. and Sagan, C. (1996) Spectrophotometry and organic matter on Iapetus, *Icarus* **122**, 92–106.

COLLISIONAL EFFECTS IN THE EDGEWORTH-KUIPER BELT

DONALD R. DAVIS
Planetary Science Institute
620 N. 6th Avenue, Tucson AZ 85705, USA

AND

PAOLO F. FARINELLA
Univ. di Trieste
Via Tiepolo 11, I-34131, Trieste, Italy

Abstract. The Edgeworth–Kuiper (E–K) belt, a population of small bodies a thousand times greater in number than that of the main asteroid belt orbiting outside Neptune's orbit, is undergoing collisional comminution. The discovery of these bodies confirmed strong dynamical evidence that short period comets come from a transneptunian source region. Like main belt asteroids, the size distribution of E–K belt objects has been affected by mutual impacts over Solar System history. Collisional evolution studies of the E–K belt show that bodies larger than several tens of km in diameter survive over times comparable to the age of the Solar System, hence their size distribution reflects accretionary processes in the early Solar System. At smaller sizes, however, impacts have produced a collisionally relaxed population of fragments which are predicted to have a power-law size distribution with an incremental diameter exponent near −3.5. Collisions can also inject km-sized fragments into dynamical resonances, whence they can be transported to the inner Solar System to become short period comets. However, dynamical mechanisms are needed to produce the Centaurs, a population of much larger bodies than the Jupiter family comets.

1. Introduction

More than 200 Edgeworth–Kuiper objects (EKOs) have been discovered since the first one, 1992QB$_1$, was found in 1992 (Jewitt and Luu, 1993). Based on these discoveries, the total population of EKOs is estimated to

M. Ya. Marov and H. Rickman (eds.), Collisional Processes in the Solar System, 277–285.

be more than 70,000 bodies larger than 100 km in diameter (Jewitt *et al.*, 1998). This population is widely believed to be the remnant of a more massive disk that formed in this region of the Solar System, as first proposed by Edgeworth (1949) and Kuiper (1951). Also, dynamical studies (Fernández, 1980; Duncan *et al.*, 1988; Quinn *et al.*, 1990) argued that a population of bodies must exist in the region beyond the outer edge of the known planetary region in order to explain the observed orbits of short period comets.

Based on the orbits and estimates of the total population, Stern (1995) showed that collisions occur frequently among Kuiper belt objects over Solar System history. Collisionally generated dust clouds, potentially observable in the IR, were predicted by Stern (1996). While the early disk was dynamically quiet and conducive to growing bodies as large as Pluto by accretion, a process or processes acted to stir up the orbits of this population, leading to increased impact speeds. Thus accretion terminated and collisional destruction and erosion began and continues to the present day.

In this paper, we will review our understanding of the collisional evolution of EKOs and describe the implications for the origin of Jupiter family comets and the origin of the outer Solar System. Finally, we will summarize the main outstanding questions regarding the formation of this part of the Solar System. The principal issues addressed here are:

• To what degree have collisions modified the primordial EKO size distribution over Solar System history?

• Are there today primitive bodies in the EKO population, *i.e.*, bodies which have not been altered since their formation?

• What constraints can we place on the size distribution of the EKOs at the end of an early accretionary regime?

• What is the current flux of km-sized collisional fragments that are injected into resonant "escape-routes", thus starting them on the path to becoming Jupiter family comets?

2. Collisional Evolution of the Edgeworth-Kuiper Belt

The size distribution of EKOs provides a major clue as to the collisional history of the population. The largest known EKO is, of course, Pluto; however, in the belt beyond 40 AU, the largest discovered bodies are $\sim$ 400 − 500 km diameter. Even larger bodies exist in the "scattered disk," a population that has been gravitationally scattered in large, highly eccentric orbits. Current estimates of the incremental size distribution slope (b) differ significantly. Gladman *et al.* (1998) find a mean slope of $b = -4.8\pm0.5$, over the size range 100 − 500 km. On the other hand, Jewitt *et al.* (1998) come up with a value of -3.0 ± 0.5. While the latter value is most consistent with the equilibrium slope of a collisionally relaxed population, -3.5 (Dohnanyi, 1969), there are reasons to believe that EKOs larger than 100 km are not collisionally relaxed. Hence, if the Gladman *et al.* value is correct, it may be telling us something about the accretionary size distribution.

EKOs typically have orbits with moderate eccentricities and inclinations, $< e > \sim 0.08$ and $< \sin i > \sim (0.15 - 0.20)$. However, about $10 - 15\%$ of the true population (as opposed to the discovered population) are Plutinos, *i.e.*, bodies which are trapped in a 2:3 resonance with Neptune and consequently have somewhat higher eccentricities and inclinations.

The average collision rate among EKOs is very similar to that of the main belt asteroids. The main difference between these two populations is that the average impact speed between asteroids is about a factor of five higher than that for EKOs, 5.8 km/s for asteroids vs. ~ 1.2 km/s for EKOs.

The total mass of the present E–K belt is estimated to be $\sim 0.10 - 0.30 M_\oplus$, based on estimates of the size distribution. Most of this mass is in bodies larger than 100 km diameter.

Our current understanding of how bodies would respond to collisions at hundreds of m/s is limited. A critical parameter for collisional studies is Q^*, the specific collisional energy that is needed to fracture a given sized body (dynamical strength). We will assume that EKOs are rather weak bodies (as compared with most asteroids) in terms of their dynamic impact strength. Another assumption is that EKOs break up into a size distribution of fragments moving relative to one another. The speed of fragments is critical when the target body has a significant gravity field – fragments moving slower than the local escape speed reaccumulate to form "rubble pile" structures.

Using a time dependent collisional evolution code, Farinella and Davis (1996) and Davis and Farinella (1997) confirmed Stern's (1995) estimates, and provided some further insights. In light of the large gaps in our understanding of collisional physics for EKOs described above, we varied the critical parameters of our fragmentation model to test the robustness of our conclusions to these uncertain quantities.

The results for our baseline case run are shown in Fig. 1. The initial EKO population from 30-50 AU is assumed to be a power-law distribution at sizes smaller than 300 km and falls to zero for sizes between 300 km and 500 km. The impact speeds are randomly chosen for each timestep from a distribution with a mean of 1.15 km/s and a standard deviation of 0.65 km/sec (Dell'Oro et al., 2000). As seen from Fig. 1, the population at sizes larger than ~ 100 km diameter is essentially unchanged over Solar System history, simply because at the relatively low impact speeds found in this population, even a collision between equal sized bodies cannot break up bodies larger than about $100 - 150$ km. However, at sizes smaller than

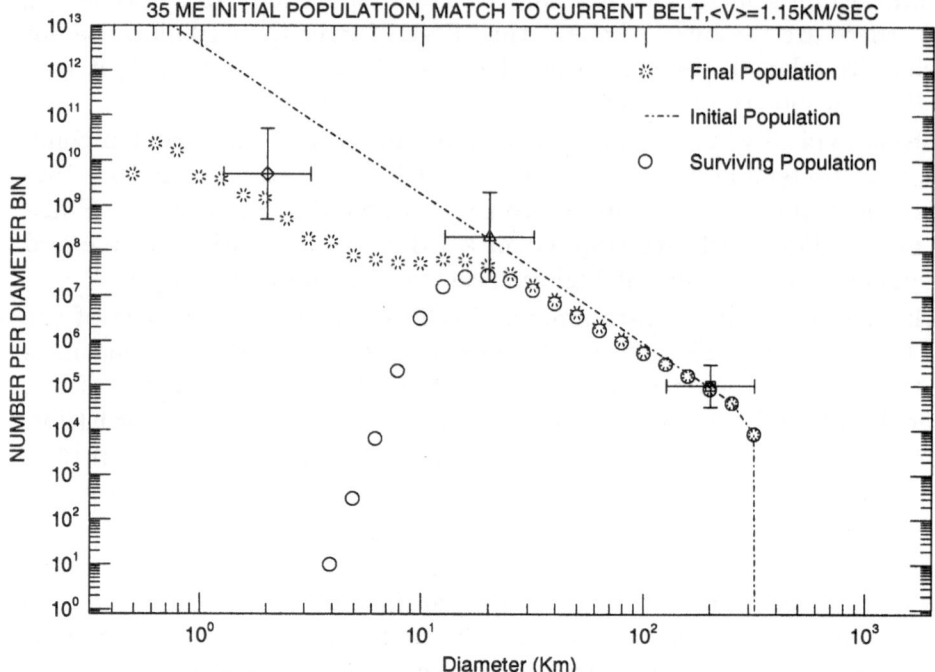

Figure 1. The collisionally evolved population of EKO after 4.5 Gy (asterisks), starting from a hypothetical initial population between 30-50 AU having a power law size distribution with an index of 3.5 for diameters < 300 km (dash-dotted line). The number of "survivors" from the original population (open circles) is also plotted. The three points with error bars represent our current estimate of the EKO size distribution. The point at $D = 200$ km is the estimate of the population larger than 100 km based on extrapolating the known discoveries. The 20 km point is from the Cochran et al/ (1995) HST detection, while the point at 2 km is the number of bodies needed to maintain the short period comet population. The error bars span a factor of ± 10 in the number of bodies, and a factor of 2 in their sizes.

100 km, there is increasing collisional depletion with decreasing size; for sizes ≈ 20 km, the population is reduced by a factor of 10 from the initial one. The slope of the small size population is very close to -3.5, the equilibrium value for a collisionally relaxed population with size-independent collisional physics (Dohnanyi, 1969).

Also shown in Fig. 1 is the number of bodies which survive from the original population – at sizes $D \approx 20$ km, about 50% of the current population consists of "survivors". The survivor fraction increases with size and for $D > 80$ km, essentially all of the bodies are survivors. Below ≈ 10 km, though, survivors are a minority of the population, and at sizes of a few km, virtually all bodies are collisionally derived fragments.

Davis and Farinella (1997) also varied the starting population to see how sensitive the results were to this initial condition. A steeper initial population at small sizes (power-law exponent of −4.0) again shows collisional depletion starting at about 100 km diameter, but depletion is much stronger (about a factor of ∼ 70) at sizes of $D \approx 1$ km. Again, though, the slope of the small size end of the distribution is −3.5. On the other hand, a shallow starting slope (−3.0) yields less collisional depletion, only about 50% at $D \approx 30$ km, and actually gives a final population larger by a factor of 2 over the starting one at sizes of 1 km diameter. Again, though, the final population is collisionally relaxed with a slope of −3.5. More radical changes in the starting population can result in a present population that is qualitatively different from those discussed above. For example, if the initial population consisted only of large bodies hundreds of km in size (*e.g.*, because the accretional process was very efficient at consolidating small bodies into large ones), then there would not have been enough collisional evolution to generate the population of small bodies that supply the short period comet population. Again using a starting population close to the current one at large sizes and following a power law for $D < 300$ km, we can estimate that the small-size slope had to be larger than about 2, otherwise the current reservoir of short period comets would be too small.

Varying the collisional parameters which are uncertain for comets (impact strength, strength scaling with size, ejecta energy partitioning coefficient, etc.) resulted in changes similar to those described above. Changes that produce collisionally weaker bodies showed a greater degree of collisional evolution, while those that produced stronger bodies had less evolution. All cases, though, led to a collisionally relaxed population at sizes smaller than about 25 km.

The main conclusions of this work are:

1. *The number of EKOs larger than ∼ 100 km diameter has not been significantly altered by collisions over the age of the Solar System.* The size distribution in this range must be the signature of the accretion process that acted to form this population. Many of these bodies, however, have likely been converted into "rubble piles", as there is a significant energy gap between the projectile energy needed to shatter the target and that required to disrupt it, *i.e.*, to disperse most of the target mass to infinity (Davis *et al.*, 1989).

2. *As shown in Fig. 1, smaller EKOs are mostly fragments undergoing a collisional cascade, with a (differential) size distribution exponent predicted to be close to the −3.5 equilibrium value (Dohnanyi, 1969).* However, their current abundance implies that there was a substantial original (accretionary) population down to a few tens of km in size, otherwise there would have been a shortage of projectiles and the current E–K belt reservoir of comets would be too small. On the

other hand, for initial values of the size distribution index ≤ -2 at diameters < 300 km, the final population is quite insensitive to initial conditions, reflecting the collisional relaxation to the -3.5 equilibrium value. If observational surveys will show that the current index has a significantly different value in the $1 - 50$ km size range, this will probably require an interpretation in terms of size-dependent collisional response parameters.

3. *As a by–product of the collisional process, about ten fragments per year 1 to 10 km in size are currently produced in the E–K belt at 40 AU.* There is an uncertainty of a factor ≈ 4, depending on the assumed collisional response parameters. With ejection speeds of $10 - 100$ m/s, similar to those inferred for asteroids, these fragments have semimajor axes about $0.1 - 1.0$ AU different than their parent bodies. This is sufficient to cause at least a few percent of these fragments ($\sim 0.2/\text{yr}$) to fall into the resonant "escape hatches" from the E–K belt (Duncan et al., 1995; Morbidelli et al., 1995) and to chaotically evolve into the planetary region of the Solar System. This is roughly in agreement with the required flux to replenish the short period comets, which have an estimated population (including extinct/dormant nuclei) of 2×10^4 bodies and a dynamical lifetime of about 3×10^5 years (Levison and Duncan, 1994).

4. *However, disruptive collisions cannot explain how large bodies such as Chiron and other sizeable members of the so-called Centaur population originate, since collisions cannot produce fragments as large as 100 km or more at such low impact speeds.* Some purely dynamical delivery mechanism is probably at work, such as slow diffusion from the vicinity of the resonance zones (*e.g.*, Morbidelli, 1997). An additional mechanism for the insertion of big Centaurs into the delivery routes is through non-disruptive collisions: a 10^{-2} projectile-to-target mass ratio is not enough to disrupt a 100-km-sized EKO, but can change its orbital velocity by several m/s, hence its semimajor axis by $\approx 0.2\% \approx 0.1$ AU. Such events may be frequent enough to provide a significant influx of Chiron-sized bodies. As for the short period comets, it remains to be seen whether the dynamical or the collisional delivery mechanism is the dominant one.

3. Implications for the Jupiter-Family Comets

The result that most short period comets are collisional fragments means that these bodies may have experienced some type of alteration in the interior of their parent bodies. At a minimum, there would be a modest compacting effect, due to the gravitational self-compression within a parent body's interior. For example, at a depth of ≈ 10 km (the median mass depth in a 100 km diameter body), the overburden pressure is about 1.5 bars (for $\rho = 1.0$ g/cm^3). Such a pressure acting over astronomical timescales could produce a "cold welding" of the material, resulting in a non-zero tensile strength of the body. This could explain why all comets do not split when they experience thermal or tidal stresses close to the Sun or Jupiter. There is some other independent evidence for a collisional processing of short period comets: (i) their irregular, triaxial shapes resemble those of fragments produced in breakup events (Jewitt and Meech, 1988); (ii) the variety of colors observed among Centaur and Kuiper belt objects can be interpreted as a result of a varying degree of collisional alteration and/or resurfacing (Luu and Jewitt, 1996).

4. Formation of the E–K Belt

The existence of Pluto and QB$_1$-sized bodies in the outer Solar System suggests that there was much more mass in this region during its formation than is found today (Stern and Colwell, 1997). They estimate that there was of order 35 $M_\oplus$ in the region $35 - 50$ AU. Davis et $al.$ (1999) found that most of this mass must have been in bodies smaller than ~ 50 km diameter, otherwise we would see many more QB$_1$-sized and larger bodies in the E–K belt than exist there today. If most of the primordial mass was in bodies smaller than 50 km, then collisional comminution could have ground down this material to dust sizes where it would have been removed by radiation forces over the age of the Solar System.

Davis et $al.$ (2000) compared the size distribution inferred from collisional models with those calculated using a multizone accretion code. This work studied accretion in the heliocentric distance range $24 - 50$ AU, which includes the region where Neptune may have formed, as well as Pluto and other E–K objects. One goal of this work was to see if the growth of Neptune can effectively terminate accretion at the right time to produce the E–K belt that we see today. Their work shows that size distributions with most of the mass at sizes < 50 km do occur after $\sim 100 - 150$ My and the total number of bodies in the 100–500 km size range is consistent with the number that we estimate to be there at present. However, the predicted size distribution in this range is significantly steeper than observed in the present belt. At later times, the slope is consistent with that inferred for the present E–K belt, but the population has grown to be much larger than what we believe to be in the E–K belt. Also, Neptune fails to form in time to stop accretion in their work. This result poses a fundamental challenge to our current understanding of the formation of the outer Solar System.

Kenyon and Luu (1999) used a single zone model to study accretion in the region from 32 − 38 AU. They also found that accretion would grow Pluto-sized bodies while preserving most of the primordial mass in small bodies, less than ∼ 20 km diameter. They found that the size distribution exponent is around $b \sim -4$ for diameters larger than ∼ 5 km, a value intermediate between the observationally inferred values of Gladman *et al.* (1998) and Jewitt *et al.* (1998). The question of the formation of Neptune or the nature of the mechanism that interrupted accretion in the E–K belt was not addressed by Kenyon and Luu.

The problem of the formation of Neptune at or near the location where it is found today is an old one in planetary science (Lissauer *et al.*, 1995). Recent work by Thommes *et al.* (2000) suggests a way out of the dilemma, namely, that Neptune (and Uranus) formed between Jupiter and Saturn and were gravitationally scattered into the outer Solar System. There, gravitational interactions with a massive population of small planetesimals stabilized and circularized their orbits. *N*-body integrations of this scenario have produced outcomes that are quite similar to our Solar System.

Countering this scenario is recent work by Owen *et al.* (1999) who found, based on Galileo probe measurements, that Jupiter's atmosphere has a higher than expected noble gas abundance. If Jupiter formed from material near where it is located today, then temperatures would have been too warm to permit trapping of noble gases. So, perhaps Jupiter formed out of material from the E–K belt, where it is cold enough to trap noble gases in planetesimals, or the primordial solar nebula was much colder than we think it was.

Clearly there is much additional work to be done to understand the formation of the outer Solar System. Comparatively, the collisional evolution of the E–K belt is relatively well understood, but putting together a fully self-consistent picture of the origin and evolution of the outer Solar System remains a challenge.

References

Cochran, A.L., Levison, H.F., Stern, S.A. and Duncan, M.J. (1995) The discovery of Halley-sized Kuiper belt objects using the Hubble Space Telescope, *Astrophys. J.* **455**, 342-346.

Davis, D.R., and Farinella, P. (1997) Collisional evolution of Edgeworth-Kuiper belt objects, *Icarus* **125**, 50-60.

Davis, D.R., Weidenschilling, S.J. and Farinella, P. (1999) Accretion of a massive Edgeworth-Kuiper belt, in *Lunar Planet. Sci. XXX*, Abstract #1883, Lunar and Planetary Institute, Houston (CD-ROM).

Davis, D.R., Weidenschilling, S.J., Farinella, P. Paolicchi, P. and Binzel, R.P. (1989) Asteroid collisional history: Effects on sizes and spins, in *Asteroids II* (Eds. R.P. Binzel, T. Gehrels, and M.S. Matthews), Univ. of Arizona, Tucson, pp. 805-826.

Davis, D.R., Farinella, P. and Weidenschilling, S.J. (2000) Accretion and collisional erosion of a massive Edgeworth-Kuiper belt: Constraints on the initial population, to be submitted to *Icarus*.

Dell'Oro, A., Marzari, F., Paolicchi, P. and Vanzani, V. (2000). Updated collisional probabilities of minor body populations, submitted to *Astron. Astrophys.*.

Dohnanyi, J.W. (1969) Collisional model of asteroids and their debris, *J. Geophys. Res.* **74**, 2531-2554.

Duncan, M., Quinn, T. and Tremaine, S. (1988) The origin of short-period comets, *Astrophys. J.* **328**, L69-73.

Duncan, M.J., Levison, H.F. and Budd, S.M. (1995) The dynamical structure of the Kuiper belt, *Astron. J.* **110**, 3073-3081.

Edgeworth, K.E. (1949) The origin and evolution of the Solar System, *Mon. Not. Roy. Astron. Soc.* **109**, 600-609.

Farinella, P., and Davis, D.R. (1996) Short period comets: Primordial bodies or collisional fragments?, *Science* **273**, 938-941.

Fernández, J.A. (1980) On the existence of a comet belt beyond Neptune, *Mon. Not. Roy. Astron. Soc.* **192**, 481-491.

Gladman, B., Kavelaars, J., Nicholson, P.D., Loredo, T.J. and Burns, J.A. (1998) Pencil-beam surveys for faint trans-Neptunian objects, *Astron. J.* **116**, 2042-2054.

Jewitt, D., Luu, J.X. and Trujillo, C. (1998) Large Kuiper Belt objects: The Mauna Kea 8K CCD survey, *Astron. J.* **115**, 2125-2135.

Jewitt, D., and Luu, J.X. (1993) Discovery of the candidate Kuiper belt object 1992 QB_1, *Nature* **362**, 730-732.

Jewitt, D., and Meech, K. (1988) Optical properties of cometary nuclei and a preliminary comparison with asteroids, *Astrophys. J.* **328**, 974-986.

Kenyon, S.J., and Luu, J.X. (1999) Accretion in the early Kuiper Belt: II. Fragmentation, *Astron. J.* **118**, 1101-1119.

Kuiper, G.P. (1951) On the origin of the solar system, in *Astrophysics: A Topical Symposium*, (J.A. Hynek, Ed.), McGraw-Hill, New York, pp. 357-424.

Levison, H.F., and Duncan, M.J. (1994) The long-term dynamical behavior of short-period comets, *Icarus* **108**, 18-36.

Lissauer, J., Pollack, J.B., Wetherill, G.W. and Stevenson, D.J. (1995) Formation of the Neptune system, in *Neptune and Triton*, Univ. of Arizona Press, Tucson, pp. 37–108.

Luu, J.X., and Jewitt, D.C. (1996) Reflection spectrum of the Kuiper belt object 1993 SC, *Astron. J.* **111**, 499-503.

Morbidelli, A. (1997) Chaotic diffusion and the origin of comets from the 2/3 resonance in the Kuiper Belt, *Icarus* **127**, 1-12.

Morbidelli, A., Thomas, F. and Moons, M. (1995) The resonant structure of the Kuiper belt and the dynamics of the first five trans-Neptunian objects, *Icarus* **118**, 322-340.

Owen, T.C., Mahaffy, P., Niemann, H.B., Atreya, S., Donahue, T., Bar-Nun, A. and de Pater, I. (1999) Low temperature condensates brought heavy elements to Jupiter, *Bull. Amer. Astron. Soc.* **31**,, 1131.

Quinn, T., Tremaine, S. and Duncan. M. (1990) Planetary perturbations and the origin of short-period comets, *Astrophys. J.* **355**, 667-679.

Stern, S.A. (1995) Collision timescales in the Kuiper disk: Model estimates and their implications, *Astron J.* **110**, 856-868.

Stern, S.A. (1996) Signature of collisions in the Kuiper disk: Model estimates and their implications, *Astron. Astrophys.* **310**, 999-1009.

Stern, S.A. and Colwell, J. (1997) Accretion of the Edgeworth-Kuiper belt: Forming 100-1000 km radius bodies at 30 AU and beyond, *Astron. J.* **114**, 841-849.

Thommes, E.W., Duncan, M.J. and Levison H.F. (2000) The formation of Uranus and Neptune in the Jupiter-Saturn region, *Nature* **402**, 635-638.

IN MEMORIAM: PAOLO FARINELLA

The career of Paolo Farinella, 47, was tragically and prematurely ended when he died on March 25, 2000, of complications following congestive heart failure. Born near Ferrara, Italy, with a congenital heart defect, Farinella had a severely restricted childhood until undergoing one of the first open heart surgeries in Italy at age eight that successfully corrected the defect. He received his education at the Scuola Normale Superiore in Pisa where he was strongly influenced by the lectures of Giuseppe Colombo. With colleagues Andrea Milani and Anna Nobili, he established the Space Mechanics Group at the University of Pisa, which rapidly became one of the outstanding research centers in planetary science in Italy. His contributions to the understanding of the origin and evolution of small bodies of the Solar System were fundamental and will be enduring. Of the more than 250 scientific papers he authored (an additional 125 or so articles were on popular science or disarmament), 200 were on the topics of small body dynamics and collisional evolution of asteroids, Trojans, comets and Kuiper Belt objects. He provided new insights and understanding of how bodies are transported from the main asteroid belt into Earth-crossing orbits where they are a source of potential impactors – the so-called potentially hazardous asteroids. Farinella and coworkers were the first to point out that the strong asteroidal resonances are actually too strong: most of the asteroids that are injected into these resonances fall into the Sun. Hence other, more subtle resonances are responsible for the delivery of most asteroids onto planet-crossing orbits.

One of his last major contributions was to show that a long known but largely neglected dynamical process, the Yarkovsky effect, was a significant mechanism for transporting to resonances to begin their delivery to Earth. This insight resolved a long-standing contradiction in the age of meteorites as calculated by their cosmic ray exposure vs. the time it takes them to reach Earth, based on resonance lifetime calculations.

Farinella enriched our understanding of the collisional evolution of small body populations, particularly mainbelt asteroids, Trojans belt objects and Edgeworth-Kuiper belt objects. Another of his principal contributions was the realization that most short period comets are probably not primordial bodies that have existed unaltered since the formation of the Solar System, but rather they are likely to be collisional fragments from a disruptional cascade in the Edgeworth-Kuiper belt. Asteroid 3248 was named Farinella in recognition of his pioneering work on dynamical and collisional evolution of small body populations.

M. Ya. Marov and H. Rickman (eds.), Collisional Processes in the Solar System, 287–288.

These are only the contributions of Farinella to planetary science; his interests reached far beyond the sphere of planetary science. Always concerned with the connections between science and society, he applied the tools and methodology of asteroid collisional evolution to a growing problem much closer to home: the increasing hazard to Earth-orbiting satellites from space debris, the swarm of small particles orbiting Earth cast aside as junk by human activities in space. He and colleagues developed the most sophisticated models to date to determine how this hazard might be expected to increase in the future. With these assessments in hand, the governments and space companies are beginning to address mitigation procedures for space debris. Another manifestation of Paolo's social conscience was a commitment to work for a major reduction in weapons of mass destruction among the nations of the world. His efforts through the Italian organization USPID (Unione Scienziati Per Il Disarmo) and Pugwash were driven by a rational perspective on the issues of arms control for the benefit of humanity, not narrow nationalistic perspectives. Paolo's excitement was boundless when Dr. J. Rotblatt was awarded the Nobel Peace Prize just before he was scheduled to visit Italy; an engagement which Rotblatt kept despite the increased demands on his time generated by the Prize.

The scientific accomplishments of Farinella were matched by his humanism and personal warmth. He was a dedicated teacher who sought to find better, more effective ways to motivate and teach students who seemed frustratingly uninterested in the fascinating material that was presented them. His lecture on the astronomical basis of Van Gogh's art was a captivating talk on the art/science connection. Paolo's ability to work with an enormous range of colleagues on a wide variety of problems demonstrated the range and vitality of his interests. His passion for justice and equality among nations, together with an interest in promoting stable and equitable economic systems that would ultimately benefit the human race, was apparent in all of his endeavors. His powerful intellect was limited by his frail physique which resulted in part from his heart condition, though his early surgery gave him nearly four more decades of life.

The memory of Paolo Farinella can be honored in a number of important ways: (a) by working to increase the supply of donor organs worldwide; (b) by supporting and contributing to rational perspectives in global cooperation and disarmament; and (c) by adhering to the scientific principles of research, strict scientific integrity and fairness, and the willingness to devote the necessary time to produce a finished scientific product before submitting it for publication. These ideals are timeless, as are Paolo's contributions to human knowledge.

Donald R. Davis

ORIGIN AND EVOLUTION OF NEAR EARTH ASTEROIDS

A. MORBIDELLI

CNRS, Observatoire de la Côte d'Azur, B.P. 4229, 06304 Nice Cedex 4, France

Abstract. Our current understanding of the origin and evolution of NEAs is the result of several research steps done essentially over the last 30 years. J.G. Williams and J. Wisdom have been the pioneer researchers who showed that some resonances may increase the eccentricity of the asteroids, thus transporting them from the main belt to terrestrial planets crossing orbits. G. Wetherill with a large number of sophisticated Monte Carlo simulations, designed a scenario for the origin and evolution of NEAs. Furthermore, Farinella and collaborators found that a typical end-state for NEAs is the collision with the Sun, and Gladman and collaborators showed, with a large number of numerical simulations, that these collisions make the dynamical lifetime of the NEAs one order of magnitude shorter than previously believed. Even more recently, Migliorini and collaborators brought attention to the fact that asteroids can leave the main belt and reach Mars–crossing orbits also under the action of numerous weak mean motion resonances and that this mechanism could account for the origin of several among the multi–kilometer NEAs. The state of the art is still in rapid evolution. It should be possible in the close future to quantify the relative importance of the different escape routes from the main belt, and to better understand the mechanisms by which the transporting resonances are resupplied of bodies.

1. Introduction

With the discovery of 433 Eros in 1898, the existence of a new population of asteroid–like bodies on orbits intersecting those of the inner planets was established. Explaining the origin of these Near Earth Asteroids (NEAs) was difficult because, at that time, it was not evident which mechanisms could have forced them to evolve from an orbit bounded by those of Mars and of Jupiter – typical of the main asteroid belt – to a planet–crossing

M. Ya. Marov and H. Rickman (eds.), Collisional Processes in the Solar System, 289–302.

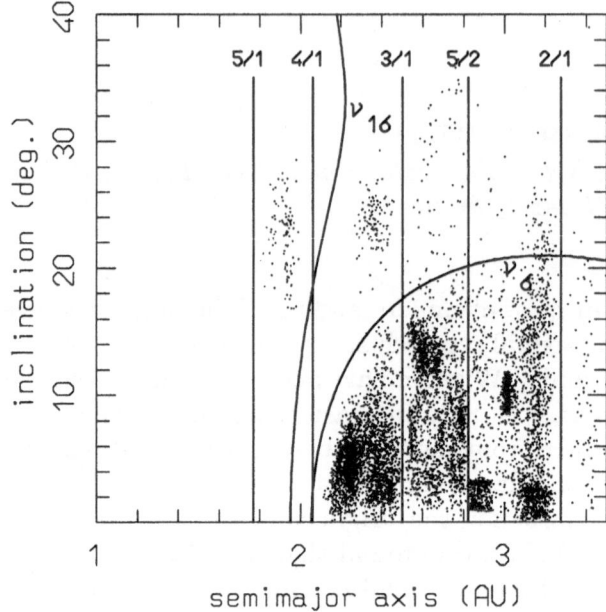

Figure 1. Orbital distribution of the numbered asteroids that do not cross the orbits of any planet. The two bold curves denote the location of the ν_6 and ν_{16} secular resonances, while the vertical lines mark the position of the main mean motion resonances with Jupiter associated to evident Kirkwood gaps.

orbit. Still in 1976, Wetherill was claiming that most of NEAs must be extinct cometary nuclei.

However, a close look to the distribution of the asteroids in the main belt shows that the belt is structured by several resonances: the existence of the Kirkwood gaps (Kirkwood, 1866) is associated with the main mean motion resonances with Jupiter (resonances between the orbital periods of the asteroid and of the planet), among which particularly evident are the 3/1, the 5/2 and the 2/1 resonances; the upper bound of the asteroid distribution, when plotted in the semimajor axis a vs. inclination i plane, corresponds with the location of the ν_6 secular resonance, which occurs when the mean precession rates of the longitudes of perihelia of the asteroid and of Saturn are equal to each other (Fig. 1). This implies that somehow the asteroids must be removed from resonant locations, and suggests that this phenomenon could be related to the origin of NEAs.

The first indication that resonances can force bodies to cross the orbits of the planets came from J.G. Williams, in a diagram reported by Wetherill

(1979), showing the amplitude of eccentricity oscillations as a function of the distance from the ν_6 resonance: at distances smaller than 0.025 AU, the amplitudes exceed 0.25, forcing the resonant bodies to cross the orbit of Mars at the top of their eccentricity oscillation. Shortly afterwards, Wisdom (1983) showed that the 3/1 mean motion resonance has a similar effect: the eccentricity of resonant bodies can have, at irregular time intervals, rapid and large oscillations whose amplitudes exceed 0.3, the threshold value to become Mars–crosser at the 3/1 location.

Following these pioneering works, several studies confirmed, both analytically and numerically, the role that resonances have in increasing asteroid eccentricities to Mars–crossing or even Earth–crossing values. For a review of the studies on mean motion resonances we recommend the paper by Moons (1997), while a compendium of the investigations on secular resonances can be found in Froeschlé and Morbidelli (1994).

The improved knowledge of resonant dynamics allowed to design a generally accepted scenario on the origin of Near Earth Asteroids, that we will call hereafter "the classical scenario".

2. The classical scenario on the origin of NEAs

G.W. Wetherill is the author who most contributed to outlining a coherent scenario for the origin and evolution of NEAs and meteoroids in a long series of papers (Wetherill, 1979, 1985, 1987, 1988). His results were reviewed, in an interpreted form, by Greenberg and Nolan in two papers (1989, 1993) that gave a well defined and concise portrait of the classical scenario (Fig. 2).

According to this scenario, collisions in the main belt continuously produce new asteroids by fragmentation of larger bodies. Some of these new asteroids are injected into the ν_6 or the 3/1 resonance by the collisions that liberated them from their parent bodies. Once inside one of these resonances, their average semimajor axes stay constant, while their eccentricities suffer large oscillations, reaching, after a typical time-scale of 1 My, values that make the asteroids intersect the orbit of Mars and/or of the Earth.

At this point, close encounters with a planet may occur. Close encounters provide an impulse velocity to the asteroid's trajectory, causing a "jump" of its semimajor axis and eccentricity by a quantity depending on the geometry of the encounter and on the mass of the planet. If the jump in semimajor axis is large enough, the asteroid is removed from the resonant location and liberated from the control of the resonance, and then evolves mainly under the sole effects of the subsequent close encounters with the planet.

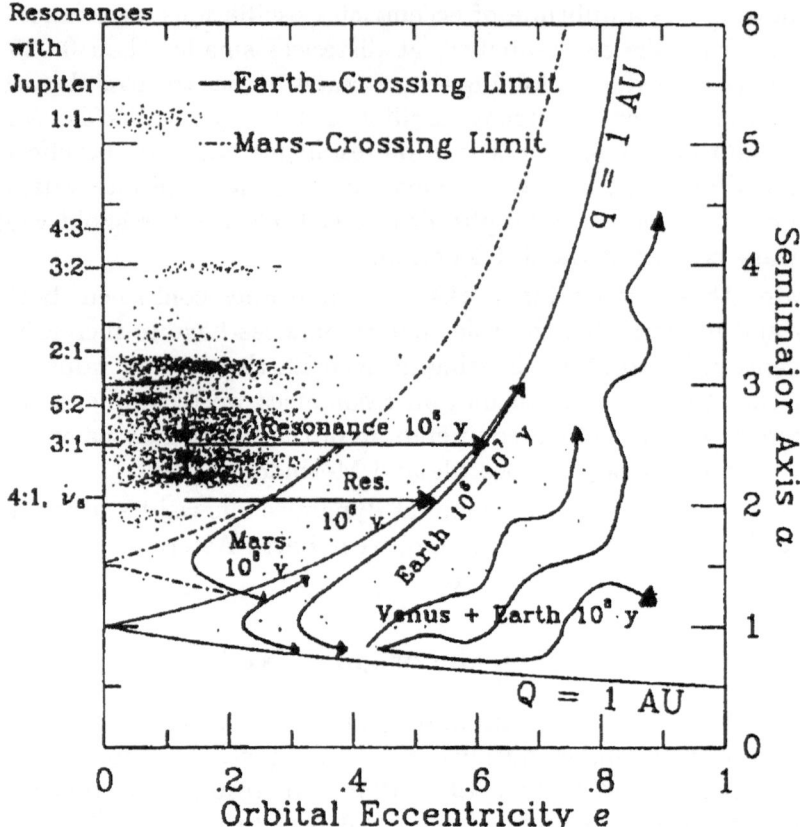

Figure 2. Schematic representation of the origin and evolution of NEAs according to the classical scenario (from Greenberg and Nolan, 1993). The dashed curves denote the set of orbits with perihelion distance q or aphelion distance Q equal to the value of the Martian semimajor axis. Solid curves denote the orbits with $q = 1$ or $Q = 1$ AU. The arrows with labels "Mars 10^8y" and "Earth 10^6–10^7y" sketch curves of constant Tisserand parameter with Mars and the Earth respectively. See text for further comments.

Repeated encounters with random geometry force the asteroid to random walk on the (a, e)–plane. This random walk, however, must follow preferential directions, approximately preserving the so–called Tisserand parameter

$$T = \frac{a_p}{a} + 2\sqrt{\frac{a(1 - e^2)}{a_p}} \cos i$$

relative to the dominating planet with semimajor axis a_p. In fact, if the planet is on a circular orbit the Tisserand parameter is equal to $3 - U^2$, where U is the norm of the relative velocity between the asteroid and the planet, which is not changed by the encounter as it is well known from the theory of scattering dynamics (Öpik, 1976). Setting the inclination to a constant value, the constancy of the Tisserand parameter defines curves on the (a, e)–plane. A few of these curves are sketched in Fig. 2.

If the asteroid has been extracted from the resonance by Mars on a solely Mars–crossing orbit, due to the small mass of the planet its random walk evolution is very slow and eventually may reach, after a typical time of 10^8 y, an Earth–crossing orbit. If, on the contrary, it has been extracted on an Earth–crossing orbit, its subsequent evolution is faster, requiring 10^6–10^7 years to reach a Jupiter–crossing or a Venus–crossing orbit. Combined encounters with the Earth and Venus break the constancy of the Tisserand parameter, since neither planet really may dominate as the "main perturber". This allows the asteroid to go all over the Earth/Venus–crossing space, on a typical time-scale of order 10^8 y. Encounters with Jupiter, conversely, quickly eject the body from the Solar System.

Undergoing this kind of evolution, a NEA survives as long as it does not collide with a planet or undergo encounters with Jupiter that eject it on an unbound orbit from the Solar System. Monte Carlo codes, which treat in a statistical way the effects of close planetary encounters (Arnold, 1965; Wetherill, 1988), predict that the median lifetime of NEAs, from their original injection into resonance, is of order several tens of My (Gladman, personal communication).

3. The Solar sink

The first indication that some qualitatively important dynamical features are missing from the classical scenario came from Farinella *et al.* (1994), who showed that, among the bodies in the ν_6, 3/1 or 5/2 resonances, the collision with the Sun is a fairly common fate. The collision with the Sun happens because the eccentricity increases up to values close to unity, so that the perihelion distance becomes smaller than the solar radius.

The reasons for which the main resonances pump the eccentricity of resonant bodies to unity were quickly understood. The ν_6 resonance location is almost independent of the eccentricity (Williams and Faulkner, 1981), so that the eccentricity may have an infinite regular growth (Morbidelli, 1993). Inside the 3/1 and 5/2 resonances, conversely, secular resonances are present and overlap each other, making most of the resonant phase space chaotic. This allows the eccentricity to evolve in an irregular manner without upper bound (Moons and Morbidelli, 1995).

Farinella *et al.* also showed that NEAs with very large orbital eccentricity – like the Taurid asteroids – may also easily collide with the Sun, despite not being inside any notable resonance. Actually, they may be forced to collide with the Sun by non–resonant secular oscillations of the eccen-

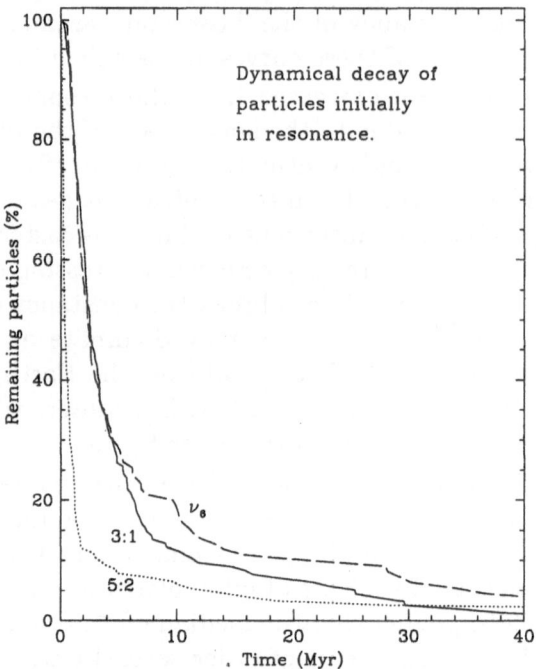

Figure 3.   Decay of populations of particles initially placed in the 3/1, 5/2 and ν_6
resonances (Gladman *et al.*, 1997). The median dynamical lifetime is of order 2 My for
the 3/1 and ν_6 resonances and less than 1 My for the 5/2 resonance.

tricity, which are particularly large when all secular arguments are in phase
(Levison and Duncan, 1994; Valsecchi *et al.*, 1995).

The Farinella *et al.* simulations, however, were based on too few bodies
to conclude on the statistical importance of solar collisions in the overall
scenario of the origin and evolution of NEAs. The improvements in computer technology and, in particular, the availability of a new integration
algorithm (Wisdom and Holman, 1991; Levison and Duncan, 1994) a few
years later allowed Gladman *et al.* (1997) to numerically simulate the dynamical evolution of several hundred test particles, initially placed into the
ν_6, 3/1 and 5/2 resonances. They found that the median lifetime of the
simulated particles is about 2 My, while only ∼10% of them survive longer
than ≃10 My (Fig. 3).

This happens precisely because the resonances tend to pump the eccentricity of almost all the resonant bodies to unity on a My time-scale. As a
result of the rapid growth of *e*, Mars is able to extract only a few percent of
resonant bodies before an Earth-crossing orbit is achieved. Of all the simulated particles extracted by Mars, none was transported to Earth–crossing

orbits by successive Martian encounters, conversely to what was expected in the classical scenario. The reason for this is that the 'kicks' provided by Martian encounters are so small that the particles cannot jump over the resonances and (after a typical time ranging from 1 to a few 10 My) always find themselves again injected into some resonant mechanism which takes them to Earth-crossing orbits.

In the Gladman *et al.* simulation, Earth and Venus are found to be much more efficient than Mars in extracting bodies from resonances. However, the "mortality" of the extracted particles is still very high. Those which are driven by close encounters to $a > 2.5$ AU are usually ejected by Jupiter on hyperbolic orbits. Many others are injected again into the 5/2, 3/1 or ν_6 resonances and subsequently are forced to collide with the Sun. Only the bodies that reach semimajor axes <1.8 AU may live much longer than the median lifetime because, in this region, although many resonances exist and influence the dynamics (Michel and Froeschlé, 1997; Michel, 1997), no statistically significant dynamical mechanisms have been found which pump the eccentricities up to Sun–grazing values. Therefore, these bodies die only by colliding with a planet (rare), or after being driven back to $a > 1.8$ AU (after a typical 1–10 My journey) and then usually being pushed by a resonance into the Sun.

Finally, the Gladman *et al.* simulations show that particles extracted from the resonances do not evolve by closely following lines of constant Tisserand parameter, as described by the classical scenario. In reality the dynamics is much more complicated: there are many high-order resonances which force the orbits to evolve transversally with respect to the curves of constant Tisserand parameter (see Michel *et al.*, 1996), so that the Tisserand parameter is very poorly conserved on a My or longer time scale and, in practice, extracted particles are quickly spread all over the Earth- and Venus-crossing region.

Despite the fact that collisions with the Sun quantitatively change the classical scenario regarding dynamical paths and typical lifetime, the basic concept that NEAs are asteroid fragments pushed to Earth–crossing orbits after being injected into the 3/1 or ν_6 resonances could still be considered valid. The resonant bodies' median lifetime of 2 My simply implies that, to sustain the NEA population in steady state, the number of asteroids injected into resonance per My needs to be roughly of order of 1/4 of the total number of NEAs. This is plausible for small to km–sized bodies. For instance, 2000 NEAs are estimated to exist with diameter of order of 1 km (Morrison, 1992), while the number of 1–km bodies injected into the 3/1 or ν_6 resonances per My is estimated to be ~400 by Menichella *et al.* (1996). Moreover, the orbital distribution of NEAs seems to be consistent with the one expected on the basis of the Gladman *et al.* simulations, once

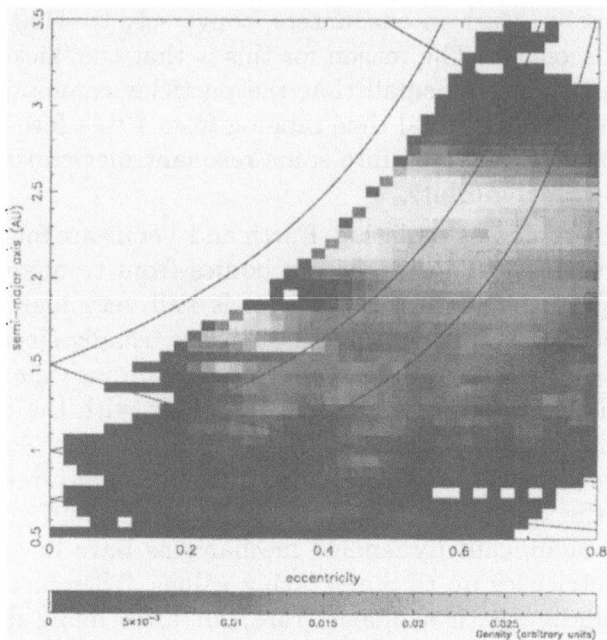

Figure 4.   Expected unbiased orbital distribution of NEAs ($q < 1.3$ AU) coming from the 3/1 and ν_6 resonances, assuming that 5 times more particles are injected into the 3/1 than in the ν_6 resonance per unit time. The grey scale denotes the relative density of NEAs on the (e, a)-plane.

observational biases are quantitatively taken into account (Bottke, Jedicke and Morbidelli, work in progress). To have a decent fit between the expected and observed orbital distributions of NEAs, the number of bodies injected into the 3/1 resonance per unit time should be about 5 times larger than the number of bodies injected into the ν_6 resonance (Fig. 4). This ratio is close to the one found by Morbidelli and Gladman (1998) in order to explain the observed orbital distribution of fireballs of chondritic origin, and is reasonable from the point of view of the location of the source asteroids (Farinella *et al.*, 1993).

However, the short dynamical lifetime of 3/1 and ν_6 resonant particles produces inconsistencies for the classical scenario concerning the origin of multi–kilometer NEAs. About 10 bodies larger than 5 km exist on Earth–crossing orbits. To sustain this population, 2–3 bodies of this size should be injected into 3/1 or ν_6 resonance per My. However, bodies of this size can be injected into resonance only during very energetic and rare break–up events, such as those leading to the formation of asteroid families (Menichella *et al.*, 1996; Zappalà *et al.*, 1998).

4. A new scenario on the origin of large NEAs

A hint for a better understanding of the origin of multi–kilometer Earth–crossers comes from the observation that Mars–crossing asteroids of similar size are much more numerous. Taking into account in the Mars–crossing population also the bodies that will intersect the orbit of Mars within the next 300,000 y (namely during their next secular eccentricity cycle), the number of Mars–crossers with diameter larger than 5 km is $\sim$350, $i.e.$, 35 times larger than the number of Earth–crossers of comparable size (Migliorini $et\ al.$, 1998). Numerical integrations done by Migliorini $et\ al.$ show that Mars–crossers evolve to Earth–crossing orbits on a typical (median) time-scale of 20–25 My; this suggests that Mars–crossing asteroids might constitute the intermediate reservoir of Earth–crossers, but moves the fundamental question from the origin of multi–kilometer Earth–crossers to the origin of multi–kilometer Mars–crossers.

The orbital distribution of Mars–crossing asteroids shows no concentration around the 3/1 and ν_6 resonances (Fig. 5). It reveals the existence of four main groups – denoted by Migliorini $et\ al.$ as MB, MB2, HU and PH – with values of semimajor axis a and inclination i (compare with Fig. 1) similar to those of four populations of non–planet–crosser asteroids: the main belt below the ν_6 resonance, the main belt beyond the 3/1 resonance and above the ν_6, the Hungarias and the Phocaeas. This similarity suggests that these populations continuously lose objects to the Mars–crossing region, sustaining the MB, MB2, HU and PH groups. Only 2% of Mars–crossers larger than 5 km have a and i very different from those of non–planet–crossing asteroids and therefore must have evolved relative to the orbit that they had when they first crossed the orbit of Mars: for this reason they have been denoted as EV.

In order to better understand the mechanism by which the main belt sustains the MB Mars–crossing population, Migliorini $et\ al.$ numerically integrated a sample of 412 asteroids with osculating perihelion distance smaller than 1.8 AU, semimajor axis in the range 2.1–2.5 AU, inclination smaller than 15 degrees and which are not Mars–crossers in the first 300,000 y. Figure 6 shows the evolution of the proper semimajor axis and eccentricity of the integrated bodies. Very few asteroids exhibit regular dynamics (those which appear as dots in Fig. 6); the vast majority exhibit macroscopic diffusion in eccentricity – that is, a relevant change of proper eccentricity – in agreement with the result of Morbidelli and Nesvorný (1998) that most of the bodies in the inner belt are chaotic, mainly due to the dense location of mean motion resonances with Mars.

In the Migliorini $et\ al.$ simulations, 17% of the integrated bodies become MB Mars–crossers in the first 25 My by diffusing to larger eccentricity. Scal-

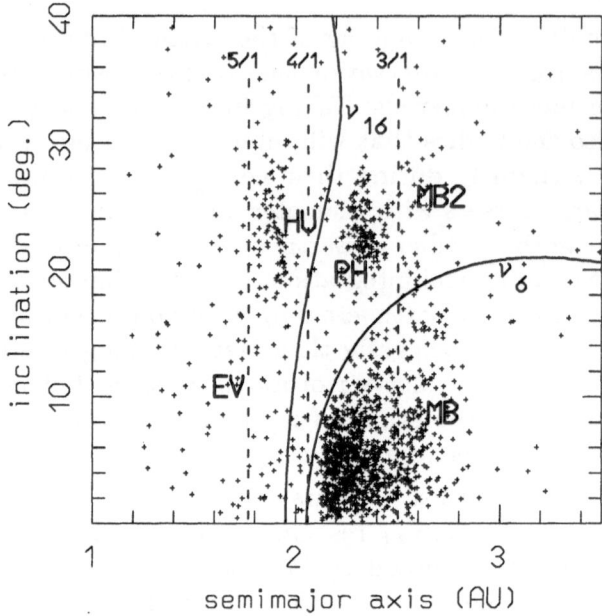

Figure 5.   Orbital distribution of Mars–crossers. Note the similarity with the (a, i) distribution of non planet–crossing main belt asteroids (Fig. 1).

ing to the number of asteroids larger than 5 km existing in the population sampled by the integrated bodies, this implies that 44 new multi–kilometer asteroids should become MB Mars–crossers in the next 25 My. This number could probably be doubled, if one took into account that the chaotic process which leads to the origin of Mars–crossers concerns basically all the asteroids with proper perihelion distance smaller than 1.92 AU (Fig. 6) and that, according to the 1994 update of the proper elements catalog (see Milani and Knežević, 1994), about 1000 asteroids larger than 5 km in the inner belt have proper perihelion distances smaller than this threshold, including parts of the Flora and the Nysa families. On the other hand, in the same time interval (25 My), the MB Mars–crossing population should lose ~ 90 bodies (50% of the total population), which evolve to the Earth–crossing region and subsequently collide with the Sun or another planet or are ejected on an hyperbolic orbit. Therefore, the diffusion process should be sufficient to keep the MB Mars–crosser population in a steady state, at least for the next 25 My.

If chaotic diffusion is able to sustain the MB and HU populations in

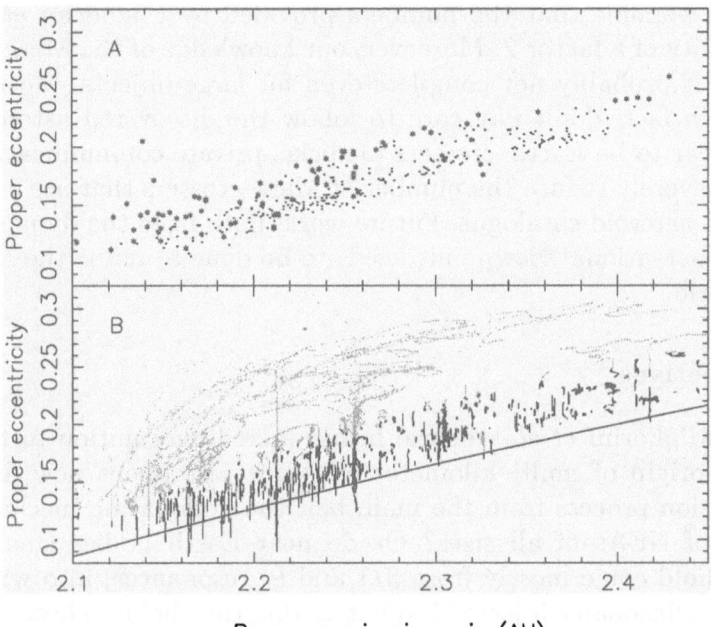

Figure 6. Origin of MB Mars–crossers from the main belt. (A) the initial proper semi-major axis and eccentricity of the integrated asteroids. Asterisks denote the bodies that will become Mars–crossers within the integration time span. (B) The trace left by their time evolution. Proper elements are constant for regular asteroids, whose orbital elements have quasi–periodic evolution, but change over time for those evolving chaotically. If all asteroids were regular, panel B and panel A would be identical, so that the comparison between the two plots allows to appreciate that most of the asteroids chaotically migrate in the proper element plane. Before becoming Mars–crossers (black), asteroids migrate in proper eccentricity, keeping the proper semimajor axis close to its initial value. During the Mars–crossing regime (grey) asteroids migrate also in proper semimajor axis, as a result of close encounters with Mars which force them to evolve along curves of quasi–constant Tisserand invariant. The solid curve denotes proper perihelion distance equal to 1.92 AU (from Migliorini *et al.*, 1998).

steady state, by liberating to the Mars–crossing region a sufficiently large number of asteroids from the main belt and the Hungaria region, then the Migliorini *et al.* simulations show also that the number of asteroids larger than 5 km that should be expected on Earth–crossing orbits and on EV Mars–crossing orbits is very close to that presently observed (10 and 7, respectively). Therefore, the new scenario for the origin of multi–kilometer NEAs appears to be quantitatively able to supply both the Mars–crossing and the Earth–crossing asteroids.

It should be qualified, however, that the Migliorini *et al.* simulations concerned only approximately 1/3 of the known total Mars–crossing population; the integration of the remaining bodies is presently undergoing. It is therefore possible that the numbers provided by Migliorini *et al.* have an uncertainty of a factor 2. Moreover, our knowledge of the Mars–crossing population is probably not complete even for large objects. Most surveys for NEAs, in fact, don't pay care to follow the discovered asteroids that do not appear to be Earth–crossers (Jedicke, private communication), and this could severely reduce the number of Mars–crossers that are eventually listed in the asteroid catalogue. Future work, both from the dynamical and from the observational viewpoint, needs to be done to refine the Migliorini *et al.* scenario.

5. Perspectives

While the Migliorini *et al.* scenario brings a credible solution to the problem of the origin of multi–kilometer NEAs, it also opens new dilemmas. Is the diffusion process from the main belt the dominating mechanism for the origin of NEAs of all sizes? Or do near–Earth bodies smaller than some threshold come mostly from 3/1 and ν_6 resonances, into which they have been collisionally injected? What is this threshold? These questions are intimately related to the size distribution of NEAs (extremely poorly known), because the size distribution of NEAs should be main belt–like in the size–range where the diffusion process is the dominating mechanism, while it might become steeper at sizes such that the contribution of bodies collisionally injected into the main resonances becomes substantial.

A significant advancement in this direction may be achieved by quantitatively fitting the observed orbital distribution of Earth–crossers by the simulated orbital distributions of bodies coming out from ν_6 resonance, 3/1 resonance and Mars–crossing intermediate reservoir. This however requires a reliable knowledge of the observational biases. In principle, this procedure could allow the determination of the relative weight of the ν_6, 3/1 and Mars–crossing sources. On the other hand, an improved knowledge of the collisional mechanisms in the main belt could refine the estimates of the number of bodies injected into the main resonances, as a function of body size.

Another new open question raised by the Migliorini *et al.* results is that the escape rate of asteroids from the main belt seems to decrease, after the first 10 My, to about 10% of the population per 100 My. This escape rate would be insufficient to keep the present Mars–crossing population in steady state on a 10^8 y time-scale. In the Migliorini *et al.* simulation this happens because the bodies in the main diffusion tracks escape in majority in the first 10 My, so that these tracks result depleted of objects; subsequently, only the bodies in the "diffusion background" contribute to sustain the Mars–crossing population, but can do so only at a much lower rate. However, in reality the main diffusion tracks are not associated with gaps in the distribution of asteroids; this indicates that they must be resupplied with new objects on a ~10 My time–scale by some process(es) that are not taken into account in the Migliorini et al. simulation. The important processes that could bring new bodies to the considered large–eccentricity parts of the main diffusion tracks could be *(i)* diffusion from the low eccentricity portion of the belt, *(ii)* injection into resonance by collisions and/or encounters among asteroids, *(iii)* migration in semimajor axis due to some non–conservative force, such as that given by the Yarkovsky thermal re–emission effect (Burns et al., 1979; Rubincam, 1995; Farinella et al., 1998) which could allow for a 0.01 AU mobility of multi–kilometer asteroids over their collisional lifetime (Farinella and Vokrouhlický, 1998). Until a quantitative analysis of these processes is done, a realistic understanding of the long–term evolution of the asteroid belt will not be achieved, limiting in turn our understanding of the NEA origin process.

The picture of the origin of NEAs is getting more complex, but this is because it is getting more realistic.

References

Arnold, J.R. (1965) The origin of meteorites as small bodies. II. The model, *Ap. J* **141**, 1536–1547.

Burns, J.A., Lamy, P.H. and Soter, S. (1979) Radiation forces on small particles in the solar system, *Icarus* **40**, 1–48.

Farinella, P., Gonczi, R., Froeschlé, Ch. and Froeschlé, C. (1993) The injection of asteroid fragments into resonance, *Icarus* **101**, 174–187.

Farinella, P., Froeschlé, Ch., Froeschlé, C., Gonczi, R., Hahn, G., Morbidelli, A. and Valsecchi, G.B. (1994) Asteroids falling onto the Sun, *Nature* **371**, 315–317.

Farinella, P., Vokrouhlický, D. and Hartmann, W.K. (1998) Meteorite delivery via Yarkovsky orbital drift, *Icarus* **132**, 378–387.

Farinella, P. and Vokrouhlický, D. (1998) Semimajor axis mobility of asteroidal fragments, *Science* **283**, 1507–1510.

Froeschlé, Ch. and Morbidelli, A. (1994) The secular resonances in the solar system, in *Asteroids, Comets, and Meteors, 1993* (A. Milani, M. Di Martino, A. Cellino, Eds.) Kluwer: Boston, pp. 189–204.

Gladman, B., Migliorini, F., Morbidelli, A., Zappalà, V., Michel, P., Cellino, A., Froeschlé, Ch., Levison, H., Bailey, M. and Duncan, M. (1997) Dynamical lifetimes of objects injected into asteroid main belt resonances, *Science* **277**, 197–201.

Greenberg, R. and Nolan, M. (1989) Delivery of asteroids and meteorites to the inner solar system, in Asteroids II, eds. R.P. Binzel, T. Gehrels and M.S. Matthews (Tucson: Univ. Arizona Press), pp. 778–804.

Greenberg, R. and Nolan, M. (1993) Dynamical relationships of near–Earth asteroids to main–belt asteroids, in *Resources of Near–Earth space*, J. Lewis et al. eds. (Tucson: Univ. Arizona Press), pp. 473–492.

Kirkwood, D. (1866) in *Proceedings of the American Association for the Advancement of Science for 1866*.

Levison, H.F. and Duncan, M. (1994) The long term dynamical behaviour of short period comets, *Icarus* **108**, 18–36.

Menichella M., Paolicchi, P. and Farinella, P. (1996) The main belt as a source of Near

Earth–Asteroids, *Earth, Moon and Planets* **72**, 133–149.

Michel, P., Froeschlé, Ch. and Farinella, P. (1996) Dynamical evolution of NEAs: close encounters, secular perturbations and resonances, *Earth Moon and Planets* **72**, 151–164.

Michel, P. and Froeschlé, Ch. (1997) The location of secular resonances for semimajor axes smaller than 2 AU, *Icarus* **128**, 230–240.

Michel, P. (1997) Effects of linear secular resonances in the region of semimajor axes smaller than 2 AU, *Icarus* **129**, 348–366.

Migliorini, F., Michel, P. Morbidelli, A. Nesvorný, D. and Zappalà, V. (1998) Origin of Earth–crossing asteroids: the new scenario, *Science* **281**, 2022–2024.

Milani, A. and Knežević, Z. (1994) Asteroid proper elements and the chaotic structure of the asteroid main belt, *Icarus* **107**, 219–254.

Moons, M. (1997) Review of the dynamics in the Kirkwood gaps, *Cel. Mech.* **65**, 175–204.

Moons, M. and Morbidelli, A. (1995) Secular resonances inside mean-motion commensurabilities: the 4/1, 3/1, 5/2 and 7/3 cases, *Icarus* **114**, 33–50.

Morbidelli, A. (1993) Asteroid secular resonant proper elements, *Icarus* **105**, 48–66.

Morbidelli, A. and Gladman, B. (1998) Orbital and temporal distribution of meteorites originating in the asteroid belt, *Meteoritics and Planet. Sci.* **33**, 999–1016.

Morbidelli, A. and Nesvorný, D. (1998) Numerous weak resonances drive asteroids towards terrestrial planets oribts, *Icarus* **139**, 295-308.

Morrison, D. Ed. (1992) *The Spaceguard Survey. Report of the NASA near-Earth object detection workshop.*, NASA, Washington, D.C.

Öpik, E.J. (1976) *Interplanetary Encounters*, Elsevier Press, New York.

Rubincam, D.P. (1995) Asteroid orbit evolution due to thermal drag, *JGR* **100**, 1585–1594.

Valsecchi, G.B., Morbidelli, A., Gonczi, R., Farinella, P., Froeschlé, Ch. and Froeschlé, C. (1995) The dynamics of objects in orbits resembling that of P/Encke, *Icarus* **117**, 45–61.

Wetherill, G.W. (1976) Where do meteorites come from? A re–evaluation of the Earth–crossing Apollo objects as sources of chondritic meteorites, *Geochimica et Cosmochimica Acta* **40**, 17–1317.

Wetherill, G.W. (1979) Steady–state populations of Apollo–Amor objects, *Icarus* **37**, 96–112.

Wetherill, G.W. (1985) Asteroidal sources of ordinary chondrites, *Meteoritics* **20**, 1–22.

Wetherill, G.W. (1987) Dynamic relationship between asteroids, meteorites and Apollo–Amor objects, *Phil. Trans. Royal Soc. London* **323**, 323–337.

Wetherill, G.W. (1988) Where do the Apollo objects come from?, *Icarus* **76**, 1–18.

Williams, J. G. and Faulkner, J. (1981) The position of secular resonance surfaces, *Icarus* **46**, 390–399.

Wisdom, J. (1983) Chaotic behavior and the origin of the 3/1 Kirkwood gap, *Icarus* **56**, 51–74.

Wisdom, J. and Holman, M. (1991) Symplectic maps for the N–body problem, *Astron. J.* **102**, 1528–1538.

Zappalà, V., Cellino, A., Gladman, B., Manley, S. and Migliorini, F. (1998) Note: asteroid showers on Earth after family break–up events, *Icarus* **134**, 176–179.

FORMATION OF ASTEROID FAMILIES AND DELIVERY OF NEO SHOWERS FROM THE MAIN BELT

VINCENZO ZAPPALÀ AND ALBERTO CELLINO
Osservatorio Astronomico di Torino
strada Osservatorio 20
I-10025 Pino Torinese (TO), Italy

Abstract. Dynamical families are the observable outcomes of collisional events that occurred in the asteroid Main Belt. They are critically important in many respects. First, they are major sources of information about the physics governing the processes of catastrophic break-up. Second, families give some unique opportunities to understand the structures and internal compositions of asteroidal bodies. Third, they provide essential constraints on the plausible collisional history of the Main Belt. Last but not least, family-forming events have been proven to play an important role as sources of near-Earth objects (NEOs). In particular, injection of very large numbers of objects into the region of the terrestrial planets, over relatively short timescales, eventually occurred when some of the major families presently known were formed. These asteroid showers are expected to have affected the cratering history of the inner planets, and to have temporarily increased the NEO population very much above the supposedly steady-state inventory.

1. Introduction. Collisional events in the asteroid Main Belt

Whatever process was responsible for the early depletion of over 99% of the solid matter originally located in the region of the Solar System where the present asteroid Main Belt is located (see, *e.g.*, Wetherill, 1992), it is widely accepted that catastrophic collisions have been the major physical process that has governed the evolution of the asteroid population since the early epoch of planetary accretion.

In particular, it is believed that both the size and spin rate distributions of present-day asteroids are the evolutive outcome of a complex process of

303

M. Ya. Marov and H. Rickman (eds.), Collisional Processes in the Solar System, 303–321.
© 2001 *Kluwer Academic Publishers. Printed in the Netherlands.*

ongoing collisional evolution. This process has been extensively analyzed by means of numerical simulations (Davis *et al.*, 1989, 1994; Farinella *et al.*, 1992). The results indicate that only the very few asteroids larger than about 4-500 km can be considered as survivors of the early population of planetesimals accreted in the asteroid belt, while most of the objects presently existing should be the final outcomes of a complex collisional history.

Although some problems are still open in this scenario (see, *e.g.*, Campo Bagatin *et al.*, 1994) it is presently widely believed that the initial total mass present in the Main Belt at the beginning of the present regime of disruptive collisions (with relative velocities of the order of 5 km/s, see Farinella and Davis, 1992; Bottke *et al.*, 1994; Dell'Oro and Paolicchi, 1997) was only a few times the present one. It is interesting to note that the present size distribution of the Main Belt population can be represented by means of a power-law, as predicted by classical models of a collisionally evolved population (Dohnanyi, 1969, 1971). The typical exponent of the power law describing the size distribution, however, does not fit the classical predictions, but it appears to be much shallower (Cellino *et al.*, 1991; Jedicke and Metcalfe, 1998).

The fact that the overall physical properties of the asteroid population are due to a long-term collisional evolution is also supported by the observational evidence. In particular, it has long been known that some proofs of the occurrence of individual, energetic events of collisional shattering and fragment dispersion can be identified in the belt. These are the cases of the so-called dynamical families, which can be interpreted as the direct, observable evidence of the occurrence of such events. Of course, asteroid families are not among the most energetic collisional processes which have played an essential role in the evolution of planetary bodies in our Solar System (like the probable collisional origin of Earth's Moon, or the collision responsible of the anomalous tilt of Uranus' obliquity angle). On the other hand, families are extremely important for understanding many properties of the asteroid population, and to constrain its overall collisional history. More in general, families are physically interesting because they are examples of collisional processes involving objects with masses in the zone of transition between fully gravity-dominated bodies and bodies mostly shaped by solid-state forces.

In this paper, we review what is presently known about physical properties of families. In particular, we will concentrate on properties, like the size and velocity distributions of the fragments produced by these events, which are directly relevant from the point of view of the origin of NEOs and meteorites. In this way, we will show that family-forming events have been important landmarks during the history of the NEO population, strongly

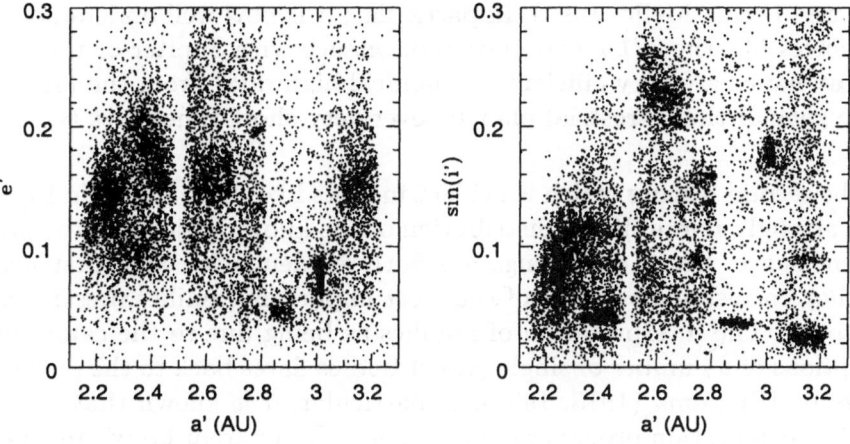

Figure 1. Distribution of the set of more than 12,000 asteroids used by Zappalà *et al.*
(1995) to identify asteroid families, in the $a' - e'$ and $a' - \sin i'$ planes, a', e' and i' being
the proper semi-major axis, eccentricity and inclination, respectively.

affecting the NEO inventory and the cratering rate of terrestrial planets
over relatively short timescales.

2. Asteroid families

Figure 1 shows the distribution of a set of more than 12,000 asteroids
for which orbital *proper* elements were computed by Milani and Knežević
(1994) in the (proper) eccentricity *versus* semi-major axis and sinus of inclination *versus* semi-major axis planes, respectively.

From these plots, it is easy to see that the distribution of the orbital
proper elements for the asteroids in the Main Belt is far from uniform.
Here we remind the reader that the proper elements are quasi-constants of
the orbital motion, and can be considered, roughly speaking, as a kind of
average of the secular variations of the mean osculating elements due to
planetary perturbations. The stability of proper elements has been extensively tested by means of numerical integrations of the orbits (Milani and
Knežević, 1994).

Two main features are clearly recognizable in the plots. The first, is the
presence of some well defined depleted regions corresponding to the major
Kirkwood gaps and (mainly visible in the inner region of the $a' - \sin i'$ plot)
the ν_6 secular resonance. The depletion of the Kirkwood gaps (associated
with some of the main mean-motion resonances with Jupiter) and of the
ν_6 secular resonance have been extensively studied in recent years by many
researchers (see, *e.g.*, the reviews on these subjects by Knežević and Milani,

1994, and Ferraz Mello, 1994). In particular, it is now well understood that resonances like the 3/1 mean-motion resonance with Jupiter or the ν_6 secular resonance are very efficient "dynamical highways" from the Main Belt to the zone of the terrestrial planets, over very short timescales (Gladman et al., 1997).

The second feature which is evident in plots like those shown in Figure 1 is the strongly non-uniform distribution of the known objects in the proper element space. Some clusterings are fairly evident even at a first glance, and they correspond to some of the most important families present in the belt today. The interpretation of families as being due to the catastrophic disruptions of a number of single parent bodies dates back to the pioneering papers by Hirayama (1918, 1933). In particular, it is known that the collisional fragmentation process can be imagined as a parent body breaking up into a large number of fragments, each achieving an abrupt velocity change $\delta \vec{v}$ with respect to the original orbital velocity vector of the parent at the instant of break-up. When $\delta \vec{v}$ is small with respect to the orbital velocity of the parent body (as usually occurs in practice dealing with asteroid break-up phenomena) the corresponding variations δa, δe and δi of the orbital elements of the fragments with respect to the original elements of the parent are given by the Gauss equations. In practice, when the ejection velocities are not extremely high, what happens is that the new orbital elements are fairly close to those of the original parent body. These differences are essentially conserved when working in terms of proper elements, and what is expected is the occurrence of a clustering of objects in the space of proper elements. In this way, families form and can be recognized when analyzing the distribution of the asteroids in the proper element space.

This task is not trivial, however, since refined techniques of statistical identifications are needed to discriminate among real families and random fluctuations of the number density of objects in some region, without any special physical significance. In this respect, major advances have been achieved during the last decade, and there is presently a large agreement on the existence of about 20 families not questionable on statistical grounds, while a larger number of smaller groupings have been also identified, with lesser confidence (Zappalà et al., 1990, 1994, 1995; Bendjoya et al., 1991; Bendjoya, 1993). The unambiguous identification of a set of well defined families has opened new perspectives to the studies aimed at understanding the physics of catastrophic collision processes.

In particular, it is known that the outcomes of catastrophic impacts depend upon several physical parameters characterizing the colliding bodies, including mass, shape, composition, internal structure, spin rate and, of course, relative velocity. At present, attempts at developing a global model able to predict in a quantitative way the final outcome, starting from a well defined set of initial conditions, have not yet been fully successful. The main constraints for the models have come essentially from the results of a large number of experiments carried out in the laboratory. These data have been interpreted and compared with the predictions of different theoretical models. Both complex hydrocodes (Melosh et al., 1992; Benz and Asphaug, 1994; Benz et al., 1994; Asphaug et al., 1998) and simpler semi-analytical methods (Paolicchi et al., 1989, 1996; Verlicchi et al., 1994) have been developed.

A major problem comes from the difficulty in developing appropriate scaling laws needed to extrapolate laboratory experiments obtained for cm-sized objects to much larger, gravity dominated, bodies like asteroids. To understand this problem it should be taken into account that several physical parameters, playing a major role during the fragmentation-dispersion process, are poorly known. Qualitatively, it is known that a critical parameter is the *specific impact energy*, given by the ratio between the kinetic energy of the projectile and the mass of the target body. Depending on this parameter, the outcomes vary from cratering, to partial or total fragmentation. In the case of relatively large bodies like asteroids, gravitational binding energy also plays an important role, and strongly influences the final outcomes. In particular, since gravitational binding energy increases for increasing parent body mass, large asteroids can be hit by projectiles large enough to break them apart, yet the energy can still be insufficient to cause the fragments to disperse. The fragments thus fall back together under their mutual attraction, leading to the formation of a gravitationally bound "rubble-pile" (Farinella et al., 1981, 1982; Davis et al., 1989). On the other hand, when a significant fraction of the fragments are dispersed, the formation of a dynamical family can be expected. For increasingly smaller parent bodies catastrophic impacts with a total dispersion of the fragments become more frequent, since the required impact energy tends to decrease. Formation of dynamical families is still possible, but the fragment ejection velocities can eventually become so high, and the fragment sizes can become so small, to make it more difficult to identify the fragments as a statistically reliable clustering.

In principle, the study of asteroid families has a crucial importance for better understanding the physical processes of fragmentation. In particular, families provide a unique opportunity for analyzing the outcomes of catastrophic disruption events involving bodies as large as tens or hundreds of kilometers in size. Therefore, it is important to compare the observational evidence on families with the predictions coming from different theoretical models of catastrophic break-up events. In this way, it should be possible to put some more direct constraints on the various unknown parameters characterizing different theoretical models. For this reason, much effort has been devoted in recent years in order to extract as much physical information as possible from the observable properties of families.

3. Physical properties of asteroid families

Here, we briefly sketch a summary of some recent advances in understanding the physical properties of asteroid families. The emphasis is put on results having a direct relevance from the point of view of the origin of family-born NEOs and asteroid showers, which will constitute the subject of the next Section.

3.1. SPECTROSCOPIC PROPERTIES

Families provide a unique opportunity to get information on the inner layers of asteroidal bodies. The reason is that family members are collisional fragments, and many of them originated from the parent body's interior. As a consequence, an intense observational activity, mainly using spectroscopic techniques, has been devoted to family members in recent years, in order to characterize the plausible mineralogic composition of these bodies. At the same time, spectroscopy is also a very important complement to the purely statistical procedures of family identification. The reason is that reflectance spectra can lead to the identification of random interlopers within families. These objects share by chance the same orbital properties as "true" family members, but they were not produced by the disruption of the family's parent body. In some cases they can be discriminated, because they can be characterized by compositional properties incompatible with those of the family. For instance, an S–type object belonging to a family of C–type members should be considered as a very likely interloper. The presence of interlopers and their plausible number is easily predictable also in purely statistical terms (Migliorini *et al.*, 1995), but it is clear that direct observations allow us to identify not only how many nominal members are actual interlopers, but also which ones. Moreover, it is also true that spectra can be used to enlarge the nominal membership of some families, in the cases in which some peculiar spectral feature characterizing a given family can be identified. In this way, objects sharing the same feature, but located beyond the nominal family borders, can be added to the list of plausible members.

A good example of the latter effect has been given by spectroscopic observations of the family associated with the large asteroid (4) Vesta (Binzel and Xu, 1993). In addition to a spectacular confirmation of the collisional nature of this family, it has been possible to discover a number of genetically related objects in addition to those of the nominal member list. This was possible in this particular case, due to the peculiar spectral properties of Vesta (a unique example of an asteroid showing spectral properties diagnostic of a basaltic surface composition, similar to HED meteorites). The observations have also convincingly shown that large ejection velocities of the fragments, of the order of several hundred meters per second are possible in these events. As a consequence, it is presently recognized that Vesta is the original source of both HED meteorites and V–type NEOs.

Other families have been extensively observed, including all the most important ones like those of Koronis (Binzel *et al.*, 1993), Eos (Doressoundiram *et al.*, 1998a; Zappalà *et al.*, 2000), Eunomia (Lazzaro *et al.*, 1999), Veritas (Di Martino *et al.*, 1997), Hoffmeister (Migliorini *et al.*, 1996), Nysa (Doressoundiram *et al.*, 1998b), Flora (Florczac *et al.*, 1998), Maria (Zappalà *et al.*, 1997). The results have not shown yet any case of very marked mineralogic heterogeneity, although some observable ranges of variation of the spectral properties do exist in some cases, indicating either a mild differentiation of the parent body, or some kind of ongoing space weathering processes affecting the surface layers of family members.

Some important results have been obtained in a few cases (Eos, Maria) by combining spectroscopic data and physical models of the original break-up events from which the families originated. As we will see later, in these cases the results are interesting from the point of view of the sources of well defined classes of meteorites, like CV/CO in the case of Eos, or ordinary chondrites in the case of Maria.

3.2. SIZE DISTRIBUTIONS

Size distributions of families are very important in many respects. First, they are directly related to the actual mechanisms of collisional break-up, thus they provide some essential data for any attempt at modelling the physics of these events. Moreover, size distributions determine how many fragments of different sizes are produced by family-forming events. As a consequence, they indicate how the asteroid inventory typically changes after such events. Coupled with information on typical ejection velocities (see later) family size distributions can be used to assess how many fragments of any given size can be expected to be eventually injected into neighbouring resonances. In this way, it is possible to predict the efficiency of family-forming events as sources of NEOs.

It has long been known that family size distributions can be fitted by power laws (Cellino *et al.*, 1991). In recent years it has been recognized that the slopes of these power-laws are generally steep, well beyond the typical values characterizing the background population of non-family asteroids in different regions of the Main Belt (Cellino and Zappalà, 1999; Tanga *et al.*, 1999). If the steep size distributions of asteroid families can be proved to

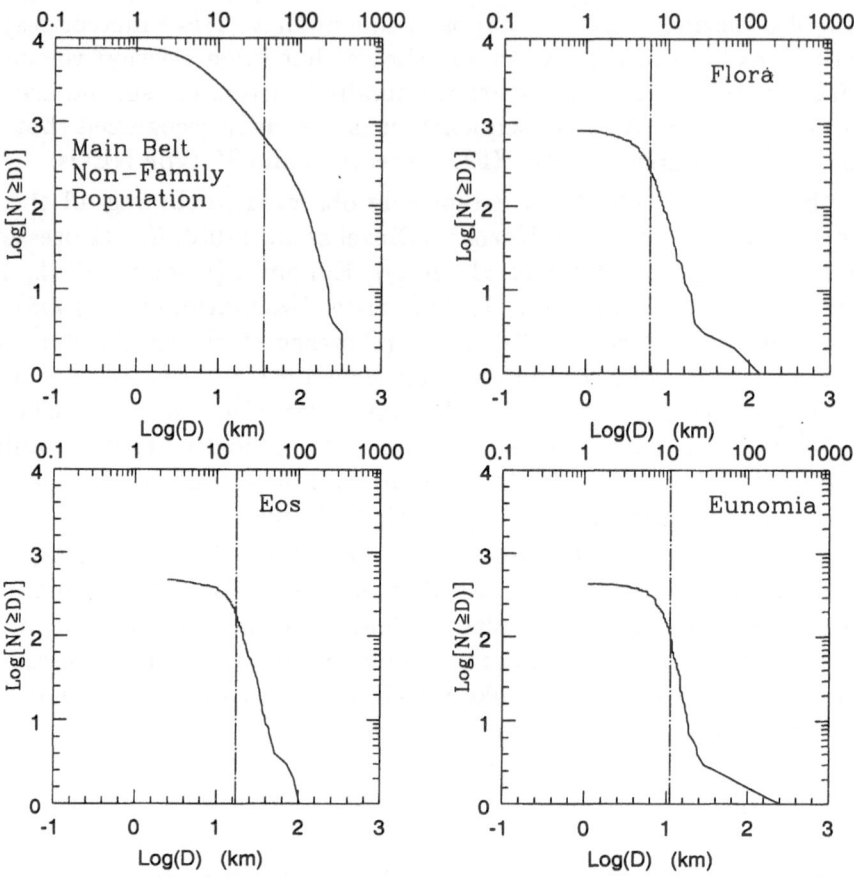

Figure 2. Comparison between the size distributions of some families and that of the whole present population of non-family asteroids. Due to the fact that family memberships date back to 1995, it is probable that the background population shown at top left is contaminated by the presence of many family objects (see text).

hold down to small sizes, of the order of 1 km (as presently suggested by the data at our disposal, see Tanga *et al.*, 1999) it is necessary to conclude that family members account for a vast fraction of the whole asteroid population at sizes of this order (Zappalà and Cellino, 1996). The situation is not fully clear, however, since unambiguous family signatures have not been identified by searches aimed at giving an estimate of the de-biased inventory of Main Belt asteroids (Jedicke and Metcalfe, 1998).

In Fig. 2, we compare the size distribution of the whole population of background, non-family asteroids presently known (ASTORB.DAT file by

E.L.M. Bowell), with the size distributions of some of the most important families identified in the belt. In these plots, sizes were taken from the IRAS Minor Planet Survey (IMPS) catalog (Tedesco, 1992), when available, or they were computed on the basis of the known values of absolute magnitude H, and assuming an albedo value. For each family, the adopted albedo value corresponds to the average value of IRAS-observed members. For the background population, albedos were randomly generated, according to the debiased albedo distribution in different regions of the Main Belt (Tedesco, 1999, private communication). Vertical lines indicate in each case the diameter of completeness, that is the limit in size above which the inventories of the objects should be complete, depending on heliocentric distance, albedo and assuming that all asteroids brighter than $V = 15.5$ mag at mean opposition have been discovered (Zappalà and Cellino, 1996; Jedicke and Metcalfe, 1998). The background population was obtained by removing from the whole present asteroid inventory the objects listed as family members by Zappalà et al. (1995), according to the Hierarchical Clustering Technique (HCM). It should be noted that the resulting background population is probably severely contaminated by the presence of actual family members which were not included in the family lists, because they were still undiscovered or did not have sufficiently good orbits to compute proper elements in the 1995 family search. Even bearing this caveat in mind, a clear evidence for a much steeper increase of the number of objects at smaller sizes for the sample of families, with respect to the background population, is apparent. In particular, it can be seen that an extrapolation down to 1 km of the size distributions, following the trends exhibited at sizes close to the completeness limits, leads to the conclusion that some families alone should contain a number of objects equal or even larger than the background population in the whole belt. We have already noted that the difference in slope between the background population and the families shown in the figure could be even larger, actually, since the background plausibly includes also a large fraction of family members not yet identified as such by the family identification techniques. A major example of this effect should be given by the large Flora family in the inner belt, as suggested by a statistical analysis carried out by Migliorini et al. (1995).

In some cases the exponents of the size distributions for several families turn out to be beyond the limit giving an infinite mass of the parent body (-3). In this sense, we are aware that at some value of size, the distributions must become less steep. According to a recent analysis by Tanga et al. (1999), however, it is probable that the steep slopes hold down to values around 1 km. These steep slopes have been explained as being due to geometric constraints coming from the finite volume of the parent body, which influences the process of fragmentation (Tanga et al., 1999).

For the above reasons, it seems presently very plausible that family-forming events produce huge amounts of fragments at small sizes, of the order of 1 km, strongly affecting the overall inventory of the whole Main Belt population. It has been shown that the subsequent collisional evolution of the newly formed family objects acts in such a way as to smooth down with time the original steepness of the size distribution (Marzari et al., 1995), though over relatively long time scales (depending on size, but typically of the order of 10^8–10^9 years). The typical exponents of the size distributions are expected to slowly converge toward the classical limit of 2.5 predicted by Dohnanyi for a collisionally evolved population. Interestingly enough, however, the normal, background population shows a size distribution exponent significantly smaller than the Dohnanyi limit (Cellino et al., 1991; Jedicke and Metcalfe, 1998). In this respect, families should continue to dominate the asteroid population at small sizes, in any case.

3.3. EJECTION VELOCITY FIELDS

The original ejection velocities of the fragments from family-forming events can be estimated from the knowledge of the relative differences in proper orbital elements, by means of the well known Gauss equations. A problem which has long prevented any advance in this field has been the presence in the Gauss equations of a couple of unknown angles, namely the true anomaly and the argument of perihelion of the parent body at the instant of break-up. This difficulty has been overcome recently by Zappalà et al. (1996), who have shown that in any case a reliable reconstruction of the original ejection velocity fields (and an estimate of the unknown angles) is possible when a sufficiently large number of family members are available.

This allows us to compare the resulting velocity fields with the evidence coming from laboratory experiments and with the predictions of theoretical models. The results indicate that in most cases the velocity fields characterizing family-forming events were not very exotic, but they shared some properties of axial symmetry which have also been found in laboratory experiments (Zappalà et al., 1996). On the other hand, a major difference is a large discrepancy in the absolute scale of the velocity fields. Family members have been ejected at very high velocities, implying that the so-called *impact strength* of the parent bodies was much higher than current predictions. The reason is that fragment ejection with so high speeds would imply that the impact energy should have been largely sufficient to completely pulverize the parent bodies, if they did not have very high values of impact strength (the parameter describing the resistance of a body to shattering). This is a well known open problem, which still deserves careful analysis.

More recently, Cellino *et al.* (1999) have undertaken a systematic analysis of the size – ejection velocity relationship for a large number of families. The results confirm that, as already suggested by simple analysis of the trends of semi-major axis histograms as a function of size (Morbidelli *et al.*, 1995), a well defined relationship exists between size and ejection velocity. In particular, it turns out that velocities tend to increase for decreasing sizes. This does not correspond to a simple energy equipartition principle as already suggested by Wiesel (1978). According to the data at our disposal, what happens is that for any size there is an interval of possible velocities, with a maximum allowed value which increases for decreasing sizes, according to the following relationship:

$$\log \left(\frac{d}{D} \right) = -C \log V - K'$$

where d and D are the diameters of the generic fragment and of the original parent body, respectively, K' depends on some basic parameters characterizing the impact, and C is a constant with a value roughly between 2/3 and 1 according to Cellino *et al.* (1999).

The above results are important, because, coupled with the knowledge of the size distribution, they make it possible to develop new refined models of family-forming events. These models can be used both for an updated numerical simulation of the overall collisional evolution of the asteroid belt, as well as for an extensive analysis of the effectiveness of known families as sources of NEOs, through injection of fragments into neighbouring resonances.

4. Families as sources of NEOs

The close proximity of several of the most important families to some of the main resonances (Kirkwood gaps and ν_6 secular resonance) in the Main Belt has stimulated some authors to try to quantify the amount and size distribution of family members injected into these resonances. Both immediate injection that occurred just after family formation, and slower injection due to diffusive and collisional processes acting on longer timescales should be taken into account. Resonances are associated with zones of chaotic orbital behaviour, leading to large eccentricity increase, and consequently to close approaches with planets. Planetary approaches are able to extract the objects from the resonance in which they are located, and lead to either ejection from the Solar System (mainly when the perturbing planet is Jupiter) or to temporary capture in the region of the terrestrial planets, where the NEO population is located. Actually, this kind of evolution has long been thought to be the continuous source of NEOs, since these objects have short dynamical lifetimes, and require a steady source of replenishment.

A cornerstone paper was published in 1995 by Morbidelli *et al.*, who developed refined methods to describe the dynamics of objects located close to the edges of the main mean-motion resonances with Jupiter. In this paper, it was shown unambiguously that several of the most important and populous families are sharply cut by the edges of neighbouring resonances. This has two main consequences: first, it is evident that the original events of family formation led in many cases to massive injection of objects into resonances. Second, these events produced and placed large numbers of objects very close to the borders of resonances. As a consequence, these families should steadily inject material into these resonances, as a consequence of collisional and diffusive processes (including possibly the Yarkovsky effect, see Farinella *et al.*, 1998; Farinella and Vokrouhlický, 1999). Morbidelli *et al.* (1995) also made some crude estimates of the numbers and sizes of the bodies originally injected by families into neighbouring resonances, based on the semi-major axis histograms of family members in different size ranges. This led to the conclusion that in several cases families were able to inject large amounts of fragments, even at fairly large sizes, of the order of some kilometers.

A more detailed analysis of the orbital evolution and dynamical lifetimes of family fragments injected into the main Kirkwood gaps and the ν_6 secular resonance was performed by Gladman *et al.* (1997). The main result of this analysis was that the dynamical lifetimes of the injected objects were very short, of the order of a few million years. This result opened some new problems, since some new, slower sources were needed to account for the existence of the observed number of NEOs without implying an unreasonably high rate of injection into the 3/1 and ν_6 resonances, and to reconcile the dynamical lifetimes of the objects with the cosmic-ray exposure ages of meteorites (Caffee *et al.*, 1988; Marti and Graf, 1992). This alternative source was then identified by Migliorini *et al.* (1998) with the population of Mars-crosser asteroids in the inner region of the Main Belt. In the presently accepted scenario (see also Morbidelli, this volume), the inner belt is slowly "evaporating", through the action of exterior mean-motion resonances with Mars like 4/7 and 7/13, high-order resonances with Jupiter (7/2, 10/3), and three-body mean motion resonances (Nesvorný and Morbidelli, 1998). According to Migliorini *et al.* (1998), 10% of the Mars-crossing population in the inner belt is removed every 100 Myrs. Taking into account collisions, which can generate new multi-kilometer sized bodies at an appropriate rate, the Mars-crossing population should be able to provide a continuous supply of NEOs with a rate which is consistent with the present population, supposed to be in a steady-state. It should be noted that the Mars-crosser population includes also fairly sizeable asteroids, some tens of kilometers in diameter; thus it can be the source of the largest NEOs, which are difficult

TABLE 1. For different families, the Table gives the involved resonance, the plausible range of 1-km fragments injected into it, the consequent range of estimated impacts with the Earth, and the duration of the shower in Myr. (From Zappalà et al., 1998)

Family	Resonance	$N_{1\,km}$	$N_{Impacts}$	Duration (Myr)
Flora	ν_6	250 – 700	4 – 11	30
Vesta	3/1	10 – 1700	0 – 1	10
Eunomia	3/1	7500 – 85000	12 – 135	10
Eunomia	8/3	2200 – 25000	0 – 4	15
Gefion	5/2	10000 – 180000	2 – 30	5
Dora	5/2	16000 – 100000	2 – 14	5
Koronis	5/2	3000 – 10000	0 – 2	5
Eos	9/4	32000 – 140000	2 – 10	140
Themis	2/1	140000 – 380000	3 – 7	90

to produce in other ways.

At the same time, the above results do not detract from the importance of families. In particular, Zappalà et al. (1998) carried out a refined analysis in order to predict in a more quantitative way the number and size distribution of the objects injected into resonances by some of the most important family-forming events. This was done by taking into account the observed size distributions of these families, and by making some (even slightly pessimistic, according to Cellino et al., 1999) estimates of the fragment ejection velocities in each case. The results were striking, since it was shown that in several cases families injected into neighbouring resonances many thousands, or tens of thousands, fragments with size of the order of 1 km (see Table 1). At least in some cases the involved resonances are in the inner belt (up to the 3/1 Kirkwood gap at 2.5 AU), and most of the produced objects should end their existence in the inner Solar System, by a collision with one of the terrestrial planets, or, mostly, with the Sun itself (Farinella et al., 1994). The duration of these "asteroid showers" were found to be fairly short, of the order of 10 Myrs, with the exception of the longer duration (of the order of 10^8 years) for the evolution of fragments injected by the Eos family into the 9/4 mean-motion resonance. The results show that family-forming events can produce temporarily a complete revolution of the NEO inventory. For instance, the Eunomia family alone was probably able to produce more than 5 times the presently estimated (Rabinowitz et al., 1994) NEO population with size of 1 km. These events led therefore also to temporary increases of the cratering rate for the terrestrial planets.

While the limited durations of the showers should prevent us from being able to identify the role of the showers into the surfaces of bodies like our Moon (Zappalà *et al.*, 1998), we cannot exclude that an enhancement of the cratering rate by objects 1 km in size might have played some role in the past history of Earth's biosphere.

Another important result of the above analysis was a confirmation of the fact that families are not very efficient sources of large NEOs, with sizes larger than a couple of kilometers. This strenghtens the need for an alternative source for NEOs, as quoted above. More in general, however, the above results indicate that the assumption of a steady-state NEO population cannot be assumed to be always valid, at least at small sizes.

Recently, some interesting results have been obtained in the cases of the Maria and Eos families, by complementing a reconstruction of the original velocity fields of fragments with spectroscopic observations. For Maria, the interesting point is that the spectra are compatible with an ordinary chondrite composition. Since the family has produced a large number of fragments just on the edge of the 3/1 resonance, a steady collisional evolution of these objects can eventually produce a large amount of meteorites; thus this family could be considered as an important potential source of ordinary chondrite material (Zappalà *et al.*, 1997).

In the case of Eos, recent spectroscopic observations of objects which are presently observed inside the 9/4 mean-motion resonance indicate that these bodies share the same spectroscopic properties as the family. This should be the first case in which objects are observed *en route* during the dynamical process of leaving the Main Belt (Zappalà *et al.*, 2000). Interestingly enough, the Eos members exhibit a spectral trend which is not common in the asteroid population, and fairly resembles that of CV/CO meteorites. Due to the fairly slow dynamical evolution of bodies injected into the 9/4 resonance, it has been shown that these objects may have a chance to be captured by Mars before being ejected from the Solar System by jovian perturbations. For this reason, a genetic relationship between the Eos family and CV/CO meteorites seems not in contradiction with dynamics (Zappalà *et al.*, 2000).

5. Summary and future developments

As we have seen in the previous Sections, inter-asteroid collisions can be considered as the main physical processes able to produce fragments in the Main Belt which can be sooner or later transferred through dynamical mechanisms and also possibly non-gravitational processes like the Yarkovsky effect (Vokrouhlický and Farinella, 1998; Farinella *et al.*, 1998; Farinella and Vokrouhlický, 1999), into the inner Solar System. Both very fast dynamical tracks, mainly associated with the 3/1 and the ν_6 resonances (Gladman *et al.*, 1997), and slower diffusive processes in the Mars-crosser region in the inner belt (Migliorini *et al.*, 1998) have been extensively studied in recent times.

In principle, collisional processes can produce different kinds of outcomes. When the impact energy is very low and/or the size of the impacted body is sufficiently large to provide a substantial gravitational field, only a few fragments are produced and ejected from the parent body. These events do not produce very evident outcomes and cannot be generally identified. As opposite, there are events producing a substantial dispersion of fragments. In these cases, a wide range of outcomes is possible, ranging from simple cratering with relatively small ejecta leaving the target, to a total disruption and dispersion of the parent body. In principle, an event becomes recognizable as a family, if it produces an observable clustering of objects sharing similar orbital elements. In this respect, it is clear that a critical role is played by the local density of the background and by observational detection limits. In particular, the denser the background, the smaller the possibility to statistically identify significant clusterings, and the higher the probability to consider as family members also a large fraction of random interlopers. We have seen that smaller (and therefore fainter) fragments tend to achieve larger velocities (Cellino *et al.*, 1999). As a consequence, it is obvious that the external layers of any given family tend to smoothly merge with the background, and statistical methods can hardly identify these objects and include them in the family membership lists.

As a conclusion, the groupings commonly accepted as families are probably the inner cores of much larger structures. In the meantime, many outcomes of collisional events escape identification completely, either because they occurred far in the past, and a large part of the fragments were subsequently erased due to subsequent collisional evolution, or because they involved small parent bodies, and cannot be identified due to the limited number of members brighter than the detection limit. Yet another possibility is that fragments were ejected with velocities so high as to prevent the formation of an observable family.

For all the above reasons, it is very hard to identify but a little fraction of the single collisional events which actually shaped in the past the present asteroid population. The presently identified families are the outcomes of only a minor fraction of the collisions which occurred in the Main Belt during the history of the Solar System. Both the Main Belt population at small sizes and the whole NEO population consist of debris from a complex collisional history. What seems important now is to use the number of "classical" families presently identified for placing some constraints on the rate of collisional evolution of the Main Belt, and on the identity of the parent bodies of present-day NEOs. In particular, in spite of the fact that most collisional events do not produce outcomes observable as a family, and thus they can contribute to a continuous supply of small-size NEOs, there are also indications that the production of large NEOs, with sizes larger than a couple of kilometers, can hardly be obtained from single collisional events in the inner belt without forming at the same time an observable family.

In other words, we should test the possibility that, taking also into account the completeness of the asteroid inventory as a function of apparent brightess (Zappalà and Cellino, 1996; Jedicke and Metcalfe, 1998) the production of NEOs with sizes of some kilometers might not occur in the inner belt but in events producing a sufficient number of objects to be identified as an observable, individual family. This possibility is qualitatively suggested by the most recent findings about size and velocity distributions of fragments in family-forming events (Tanga *et al.* 1999; Cellino *et al.* 1999). If this can be proved to be true, we should conclude that the largest NEOs are not produced by single collisional events injecting fragments into the very fast-track 3/1 and ν_6 resonances, but they come mainly from the Mars-crosser region in the inner belt, by slow diffusive mechanisms (Migliorini *et al.*, 1998). A full quantitative analysis of this issue has still to be done, but we are convinced that it should be very useful to put some firm constraints on the plausible origin of the largest NEOs and the rate of collisional evolution of the whole asteroid population.

In principle, finally, coupling statistical methods with a systematic spectroscopic characterization of each individual object of the belt could help to improve our ability to identify debris sharing a common origin from the same parent body. Preliminary results of a very extensive observational campaign by S.J. Bus and co-workers, covering the asteroid population in a limited range of semi-major axes mainly around 2.7 AU (Bus, 1999) have already shown the great power of this approach. However, it is very hard to imagine extending this research to a significant fraction of the smallest asteroids of the Main Belt in the near future.

References

Asphaug, E., Ostro, S.J., Hudson, R.S., Scheeres, D.J., and Benz, W. (1998) Disruption of kilometre-sized asteroids by energetic collisions, *Nature* **393**, 437–440.

Bendjoya, P. (1993) A classification of 6479 asteroids into families by means of the wavelet clustering method, *Astron. Astrophys. Suppl. Ser.* **102**, 25–55.

Bendjoya, P., Slezak, E., and Froeschlé, C. (1991) The wavelet transform: a new tool for asteroid family determination, *Astron. Astrophys.* **272**, 651–670.

Benz, W., and Asphaug, E. (1994) Impact simulations with fracture: I. Method and tests, *Icarus* **107**, 98–116.

Benz, W., Asphaug, E., and Ryan, E.V. (1994) Numerical simulations of catastrophic disruption: recent results, *Planet. Space Sci.* **42**, 1053–1066.

Binzel, R.P., and Xu, S. (1993) Chips off asteroid 4 Vesta: evidence for the parent body of basaltic achondrite meteorites, *Science* **260**, 186–191.

Binzel, R.P., Xu, S., and Bus, S.J. (1993) Spectral varitions within the Koronis family: possible implications for the surface colors of asteroid 243 Ida, *Icarus* **106**, 608–611

Bottke, W.F., Nolan, M.C., Greenberg, R., and Kolvoord, R.A. (1994) Velocity distributions among colliding asteroids, *Icarus* **107**, 255–268

Bus, S.J. (1999) Compositional Structure in the Asteroid Belt: Results of a Spectroscopic Survey. Ph.D. Thesis, M.I.T.

Caffee, M.W., Goswami, J.N., Hohenberg, C.M., Marti, K., and Reedy, R.C. (1988) Irradiation records in meteorites, in *Meteorites and the early Solar System* (J.F. Kerridge and M.S. Matthews, Eds.), Univ. of Arizona Press, Tucson, pp. 205–245.

Campo Bagatin, A., Cellino, A., Davis, D.R., Farinella, P., and Paolicchi, P. (1994) Wavy size distributions for collisional systems with a small-size cutoff, *Planet. Space Sci.* **42**, 1079–1092.

Cellino, A., and Zappalà, V. (1999) Structure and inventory of the asteroid main belt population, in *Asteroids, Comets, Meteors 1996* (A.C. Levasseur-Regourd and M. Fulchignoni, Eds.), in press

Cellino, A., Zappalà, V., and Farinella, P. (1991) The size distribution of main-belt asteroids from IRAS data, *Mon. Not. R. Astr. Soc.* **253**, 561–574.

Cellino, A., Michel, P., Tanga, P., Zappalà, V., Paolicchi, P., and Dell'Oro, A. (1999) The velocity – size relationship for members of asteroid families and implications for the physics of catastrophic collisions, *Icarus* **141**, 79–95.

Davis, D.R., Farinella, P., Paolicchi, P., Weidenschilling, S.J., and Binzel, R.P. (1989) Asteroid collisional history: effects on sizes and spins, in *Asteroids II* (R.P. Binzel, T. Gehrels and M.S. Matthews, Eds.), Univ. of Arizona Press, Tucson, pp. 805–826.

Davis, D.R., Ryan, E.V., and Farinella, P. (1994) Asteroid collisional evolution: results from current scaling algorithms, *Planet. Space Sci.* **42**, 599–610.

Dell'Oro, A., and Paolicchi, P. (1997) A new way to estimate the distribution of encounter velocity among the asteroids, *Planet. Space Sci.* **45**, 779–788.

Di Martino, M., Migliorini, F., Zappalà, V., Manara, A., and Barbieri, C. (1997) Veritas asteroid family: remarkable spectral differences inside a primitive parent body, *Icarus* **127**, 112–120.

Dohnanyi, J.S. (1969) Collisional model of asteroids and their debris, *J. Geophys. Res.* **74**, 2531–2554.

Dohnanyi, J.S. (1971) Fragmentation and distribution of asteroids, in *Physical studies of minor planets* (T. Gehrels, Ed.), NASA SP-267, Washington, DC, pp. 263–295.

Doressoundiram, A., Barucci, M.A., and Fulchignoni M. (1998) Eos family: A spectroscopic study, *Icarus* **131**, 15–31.

Doressoundiram, A., Cellino, A., Di Martino, M., Migliorini, F., and Zappalà, V. (1998) The puzzling case of the Nysa-Polana family finally solved?, *B.A.A.S.* **30**, 505

Farinella, P., and Davis, D.R. (1992) Collision rates and impact velocities in the main asteroid belt, *Icarus* **97**, 111–123.

Farinella, P., and Vokrouhlický, D. (1999) Semimajor axis mobility of asteroidal fragments, *Science* **283**, 1507–1510.

Farinella, P., Paolicchi, P., Tedesco, E.F., and Zappalà, V. (1981) Triaxial eliquibrium ellipsoids among the asteroids?, *Icarus* **46**, 113–123.

Farinella, P., Paolicchi, P., and Zappalà, V. (1982) The asteroids as outcomes of catastrophic collisions, *Icarus* **52**, 409–433.

Farinella, P., Davis, D.R., Cellino, A., and Zappalà, V. (1992) Asteroid collisional evolution: an integrated model for the evolution of asteroid rotation rates, *Astron. Astrophys.* **253**, 604–614.

Farinella, P., Froeschlé, Ch., Froeschlé, C., Gonczi, R., Hahn, G., Morbidelli, A., and Valsecchi, G. (1994) Asteroids falling into the Sun, *Nature* **371**, 314–317.

Farinella, P., Vokrouhlický, D., and Hartmann, W.K. (1998) Meteorite delivery *via* Yarkovsky orbital drift, *Icarus* **132**, 378–387.

Ferraz Mello, S. (1994) Kierkwood gaps and resonant groups, in *Asteroids, Comets, Meteors 1993* (A. Milani, M. Di Martino and A. Cellino, Eds.), Kluwer, Dordrecht, pp. 175–188.

Florczac, M., Barucci, M.A., Doressoundiram, A., Lazzaro, D., Angeli, C.A., and Dotto, E. (1998) A visible spectroscopic survey of the Flora clan, *Icarus* **133**, 233–246.

Gladman, B.J., Migliorini, F., Morbidelli, A., Zappalà, V., Michel, P., Cellino, A., Froeschlé, Ch., Levison, H.F., Bailey, M., and Duncan, M. (1997) Dynamical lifetimes of objects injected into asteroid belt resonances, *Science* **277**, 197–201.

Hirayama, K. (1918) Groups of asteroids probably of common origin, *Proc. Phys.-Math. Soc. Japan* **9**, 354–361.

Hirayama, K. (1933) Present state of the families of asteroids, *Proc. Imp. Acad. Tokyo* **9**, 482–485.

Jedicke, R., and Metcalfe, T.S. (1998) The orbital and absolute magnitude distributions of main belt asteroids, *Icarus* **131**, 245–260.

Knežević, Z., and Milani, A. (1994) Asteroid proper elements: the big picture, in *Asteroids, Comets, Meteors 1993* (A. Milani, M. Di Martino and A. Cellino, Eds.), Kluwer, Dordrecht, pp. 143–158.

Lazzaro, D., Mothé-Diniz, T., Carvano, J.M., Angeli, C., Betzler, A.S., Florczac, M., Cellino, A., Di Martino, M., Doressoundiram, A., Barucci, M.A., Dotto, E., and Bendjoya, P. (1999) The Eunomia family: a visible spectroscopic survey. Submitted to *Icarus*

Marti, K., and Graf, T. (1992) Cosmic-ray exposure history of ordinary chondrites, *Ann. Rev. Earth Planet. Sci.* **20**, 221–243

Marzari, F., Davis, D.R., and Vanzani, V. (1995) Collisional evolution of asteroid families, *Icarus* **113**, 168–187.

Melosh, H.J., Ryan, E.V., and Asphaug, E. (1992) Dynamic fragmentation in impacts: hydrocode simulation of laboratory impacts, *J. Geophys. Res.* **97**, 14,735–14,759.

Migliorini, F., Zappalà, V., Vio, R., and Cellino, A. (1995) Interlopers within asteroid families, *Icarus* **118**, 271–291.

Migliorini, F., Manara, A., Di Martino, M., and Farinella, P. (1996) The Hoffmeister family: inferences from physical data, *Astron. Astrophys.* **310**, 681–685.

Migliorini, F., Michel, P., Morbidelli, A., Nesvorný, D., and Zappalà, V. (1998) Origin of multi-kilometer Earth- and Mars-crossing asteroids: a quantitative simulation, *Science* **281**, 2022–2024.

Milani, A., and Knežević, Z. (1994) Asteroid proper elements and the dynamical structure of the asteroid belt, *Icarus* **107**, 219–254.

Morbidelli, A., Zappalà, V., Moons, M., Cellino, A., and Gonczi, R. (1995) Asteroid families close to mean-motion resonances: dynamical effects and physical implications, *Icarus* **118**, 132–154.

Nesvorný, D., and Morbidelli, A. (1998) Three-body mean motion resonances and the chaotic structure of the asteroid belt, *Astron. J.* **116**, 3029–3037.

Paolicchi, P., Cellino, A., Farinella, P., and Zappalà, V. (1989) A semiempirical model of catastrophic breakup processes, *Icarus* **77**, 187–212.

Paolicchi, P., Verlicchi, A., and Cellino, A. (1996) An improved semi-empirical model of catastrophic impact processes. I. Theory and laboratory experiments, *Icarus* **121**, 126–157.

Rabinowitx, D., Bowell, E., Shoemaker, E., and Muinonen, K. (1994) The population of Earth-crossing asteroids, in *Hazards due to comets and asteroids* (T. Gehrels, Ed.), Univ. of Arizona Press, Tucson, pp. 285–312.

Tanga, P., Cellino, A., Michel, P., Zappalà, V., Paolicchi, P., and Dell'Oro, A. (1999) On the size distribution of asteroid families: the role of geometry, *Icarus* **141**, 65–78.

Tedesco, E.F., Ed. (1992) *The IRAS Minor Planet Survey*, Phillips Laboratory Technical Report No. PL-TR-92-2049, Hanscom Air Force Base, MA

Verlicchi, A., La Spina, A., Paolicchi, P., and Cellino, A. (1994) The interpretation of laboratory experiments in the framework of an improved semi-empirical model, *Planet*

Space Sci. **42**, 1031–1041.

Vokrouhlický, D, and Farinella, P. (1998) Orbital evolution of asteroidal fragments into the ν_6 resonance *via* Yarkovsky effects, *Astron. Astrophys.* **335**, 351–362.

Wetherill, G.W. (1992) An alternative model for the formation of the asteroids, *Icarus* **100**, 307–325.

Wiesel, W. (1978) Fragmentation of asteroids and artificial satellites in orbit, *Icarus* **34**, 99–116.

Zappalà, V., and Cellino, A. (1996) Main Belt asteroids: present and future inventory, in *Completing the inventory of the Solar System* (T.W. Rettig and J.M. Hahn, Eds.), A.S.P. Conf. Series, Vol. 107, pp. 29–44.

Zappalà, V., Cellino, A., Farinella, P., and Knežević, Z. (1990) Asteroid families. I. Identification by hierarchical clustering and reliability assessment, *Astron. J.* **100**, 2030–2046.

Zappalà, V., Cellino, A., Farinella, P., and Milani, A. (1994) Asteroid families: extension to unnumbered multi-opposition asteroids, *Astron. J.* **107**, 772–801.

Zappalà, V., Bendjoya, P., Cellino, A., Farinella, P., and Froeschlé, C. (1995) Asteroid families: search of a 12,487 asteroid sample using two different clustering techniques, *Icarus* **116**, 291–314.

Zappalà, V., Cellino, A., Dell'Oro, A., Migliorini, F., and Paolicchi, P. (1996) Reconstructing the original ejection velocity fields of asteroid families, *Icarus* **124**, 156–180.

Zappalà, V., Cellino, A., Di Martino, M., Migliorini, F., and Paolicchi, P. (1997) Maria's family: physical structure and implications for the origin of giant NEAs, *Icarus* **129**, 1–20.

Zappalà, V., Cellino, A., Gladman, B.J., Manley, S., and Migliorini, F. (1998) Asteroid showers on Earth after family breakup events, *Icarus* **134**, 176–179.

Zappalà, V., Bendjoya, P., Cellino, A., Di Martino, M., Doressoundiram, A., Manara, A., and Migliorini, F. (2000) Fugitives from Eos family: first spectroscopic confirmation, *Icarus* **145**, 4–11.

NEAR-EARTH ASTEROID SURVEYS

ALAN W. HARRIS
Jet Propulsion Laboratory
Pasadena, CA 91109 USA

Abstract. To date, more than 700 Near-Earth Asteroids have been discovered; of those, ~290 are estimated to be larger than 1 km in diameter. We estimate the total population of NEAs > 1 km to be ~1600. Present surveys are covering about half of the available sky each month to a limiting magnitude of ~19.0, resulting in about 20 discoveries per month of all sizes, including about 6 larger than 1 km diameter. This is about a factor of eight less than the discovery rate required to accomplish the *Spaceguard Goal* of discovering 90% of all NEAs larger than 1 km in diameter in ten years. We estimate that this goal will require an all-sky survey to a limiting magnitude of 20.5.

1. Introduction

A Near-Earth Asteroid (NEA) is defined to be any asteroid with a perihelion distance under 1.3 AU. By this definition, the first NEA discovered was (433) Eros, in 1898. The first truly Earth-Crossing Asteroid (ECA) discovered, with $q < 1.0$ AU, was (1862) Apollo, in 1932. Until 1973, discoveries were largely serendipitous. Beginning in that year, Shoemaker and Helin began a systematic survey specifically directed toward discovering NEAs, using the Schmidt telescopes on Mount Palomar. Figure 1 is a logarithmic plot of the cumulative number of NEAs discovered vs. time. The slope is remarkably constant up until 1973, and is again fairly constant at a steeper slope from 1973 to 1989, while the discovery rate was dominated by the Palomar photographic surveys. The Spacewatch Camera on Kitt Peak, using a CCD detector, began making discoveries in 1989. Although that camera has a much narrower field of view than the photographic systems, the high quantum efficiency of the CCD and the automated computer detection methods allowed that system to dominate the discovery rate until

323

M. Ya. Marov and H. Rickman (eds.), Collisional Processes in the Solar System, 323–332.
© *2001 Kluwer Academic Publishers. Printed in the Netherlands.*

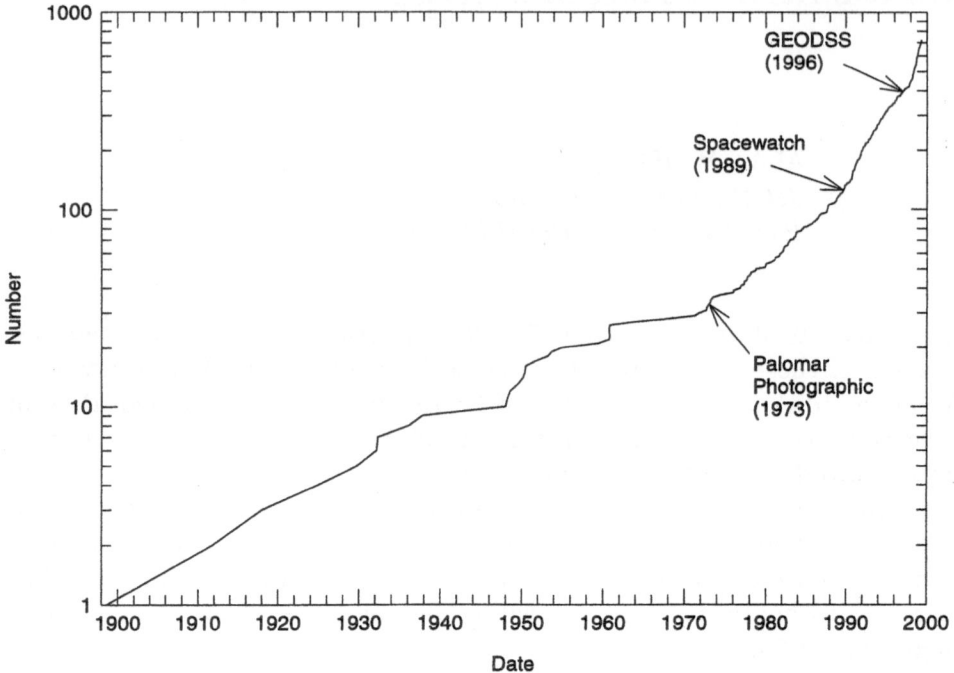

Figure 1.   Cumulative number of NEAs discovered vs. time.

1996, when wide-field CCD survey systems began sweeping the sky, covering a fair fraction of all sky visible from a single site each month. The most
successful of these systems, LINEAR, is based on a Lincoln Laboratories
CCD chip operated on a 1 m f/2.15 telescope in Socorro, NM, of the design used for the USAF Groundbased Electro-Optical Deep-Space Surveil-
lance (GEODSS) system. A second system, NEAT, using a JPL-developed
CCD camera, is used on a GEODSS telescope in Maui. Two other wide-
field CCD systems are installed in Schmidt telescopes of roughly 0.5 m
aperture: the LONEOS system operated by Lowell Observatory, and the
Catalina Sky Survey system operated by the University of Arizona. The
Spacewatch Camera is also operated by University of Arizona, and will
soon be joined by a second camera of 1.8 m aperture. The GEODSS-based
LINEAR system currently dominates the discovery rate. At the time of this
writing, over 700 NEAs have been discovered, and the rate of discoveries
(of all sizes of objects) is about 20 per month.

2. NEA Population

With over 700 objects discovered, we can now estimate the number of the
largest NEAs based on the assumption that we have discovered all of the

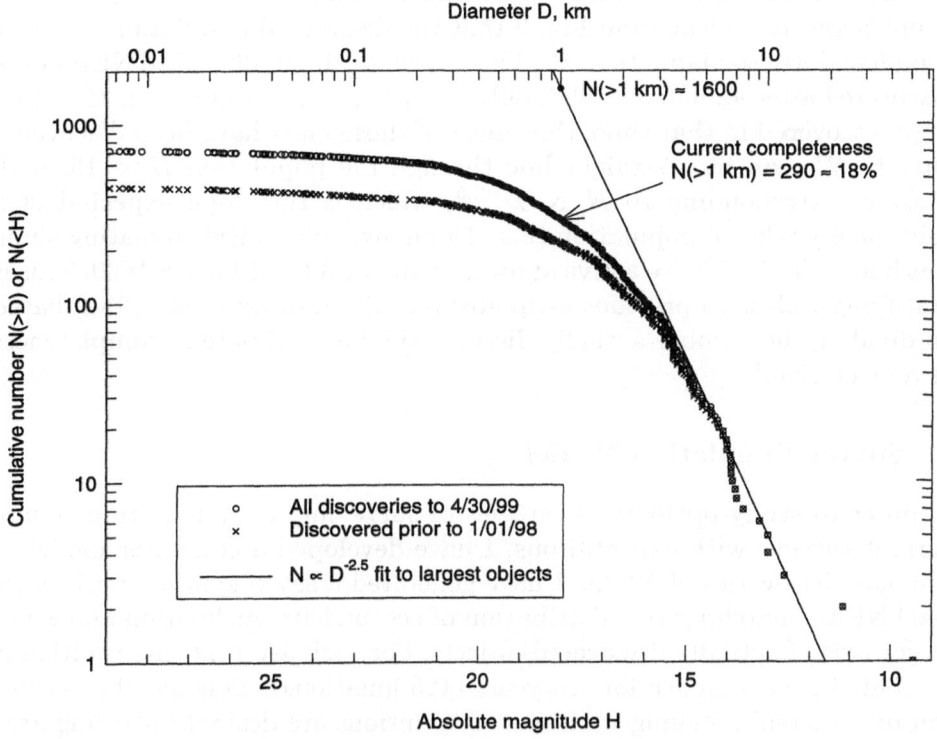

Figure 2.   Cumulative number vs. size of discovered NEAs.

largest ones, and we can extrapolate this population to make a reasonable estimate of the numbers down to ~1 km size. Figure 2 is a plot of the cumulative number of NEAs vs. size, discovered prior to the beginning of 1998, and as of the end of April, 1999. The actual observed measure relating to size is the absolute magnitude, H. We infer physical size, diameter D, by assuming an albedo. There are very few albedo measurements of NEAs, but from those we have, plus mean values of albedo in the main belt, we can infer a likely mean of around 10%. For the sake of simplicity, we take an absolute magnitude $H = 18.0$ to correspond to a diameter $D = 1.0$ km, which implies a mean albedo of 0.11. Diameter and albedo are thus related by $H = 18.0 - 5 \log D$. The top and bottom scales of Fig. 2 are adjusted to agree with this relationship. From this figure, we see that the largest NEA, (1036) Ganymed, would be nearly 50 km in diameter according to this albedo equivalence. It is actually known to have a higher albedo, so the best estimate of its diameter is 32 km. The largest asteroid known with an orbit that actually crosses the Earth's is (1866) Sisyphus, with $H = 13.0$. This corresponds to a diameter of 10 km, although as with Ganymed, it is known to have a somewhat higher albedo and an estimated diameter of about 8 km. It is clear from Fig. 2 that the discovered population is nearly complete down to about $H = 15$. That is, while about 40% of all NEAs were discovered after January 1 1998, only a few percent of those with $H < 15.0$ were discovered in that time, thus most of those must have been discovered already. We can fit a straight line through the population $H < 15$, with a slope corresponding to $N \propto D^{-2.5}$, which is the slope expected of a collisionally relaxed population (*e.g.*, Dohnanyi, 1969). Extrapolating along this line to $H = 18.0$, we arrive at an estimate of $N(> 1 \text{ km}) \approx 1600$. This is gratifyingly close to previous estimates (cf., Rabinowitz *et al.*, 1994) based on dividing the number actually discovered by an estimate of completeness at a given size.[1]

3. Survey Simulation Model

In order to study optimum strategies of surveying, as well as to compare current surveys with expectations, I have developed a computer model to simulate discoveries of NEAs. I have generated a set of synthetic orbits for 1000 NEAs, matching the distribution of eccentricity, inclination and semi-major axis of actually discovered objects. For each asteroid, the position is computed once a month for ten years (125 lunations). This file of positions can be "filtered", making whatever assumptions are desired regarding area of sky searched, horizon and other limitations, rate of motion, or limiting magnitude/absolute magnitude. These last two parameters, the limiting magnitude of the putative survey system and the absolute magnitude (and hence size) of the asteroid are very nearly 100% correlated, so it is possible to generate a "universal completeness curve" that describes completeness (fraction of objects discovered) as a function of the difference of these two numbers, and simply scale the curve appropriately to give results in terms

---

[1] *Note added in proof:* As of 13 March 2000, the total number of NEAs of all sizes discovered was 947. Of these, 362 have cataloged values of $H \leq 18.0$, corresponding to $D \geq 1$ km. Applying the linear extrapolation method used in Fig. 2, these latest discovery numbers yield about the same estimate of population, *i.e.*, $N(H < 18) \approx 1500$. However, Rabinowitz *et al.* (2000) have recently published an estimate of $N(H < 18) \approx 700 \pm 230$, and Bottke *et al.* (2000) found $N(H < 18) \approx 900$. I and other groups have more recently analyzed the statistics of new vs. re-detections of NEAs by the LINEAR project for the calendar year 1999, to estimate the number of NEAs vs. H magnitude. I obtain a preliminary result of $N(H < 18) \approx 837$, but known selection effects are one-directional leading to the conclusion that this number is a lower limit. My best current estimate of the total population is $N(H < 18) \approx 1000$, with an error bar probably about the same as the Rabinowitz *et al.* estimate. These lower population estimates suggest that current survey efforts may be only a factor of 3 or so below that needed to accomplish the *Spaceguard Goal*. However, if this is the case, then the current discovery rate is inconsistent with that which I would predict for a magnitude 19.0 all-sky survey and a total population $N(H < 18) < 1000$.

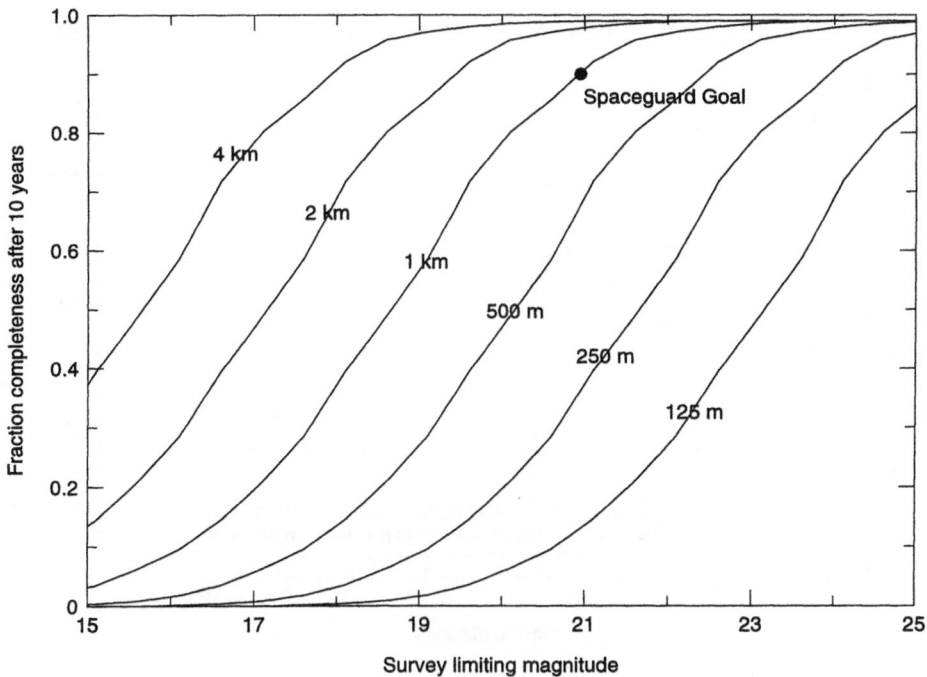

Figure 3. Model completeness of a ten-year, all-sky survey.

of various limiting magnitudes and/or sizes of asteroids. This computer simulation is described in greater detail in Harris (1998).

The first result that was immediately apparent from the simulations is that it is more effective to survey as much sky area as possible, rather than surveying a smaller area to fainter magnitude. This can be simply understood by noting that the discovery rate is to first order proportional to area of sky observed, but is proportional to only the square root of exposure time for a given sky area. Thus by covering twice as much sky area, one gains a factor of two by area but loses only a factor of the square root of two by depth of coverage. With this result in hand, I considered only all-sky surveys, where it was assumed that all available sky (with reasonable limits for horizon, darkness, Moon position, Galactic plane interference, etc.) was surveyed each month. The resulting completeness curve, scaled for specific asteroid sizes over a wide range of survey limiting magnitudes, is presented in Fig. 3. The goal of surveys which NASA and other government agencies have declared for NEA surveys is to catalog > 90% of all NEAs larger than 1 km in diameter ($H < 18.0$) in ten years. I have called this the *Spaceguard Goal*. That goal is marked on Fig. 3. Note, however, that the curves in the figure are differential completeness. That is, the point

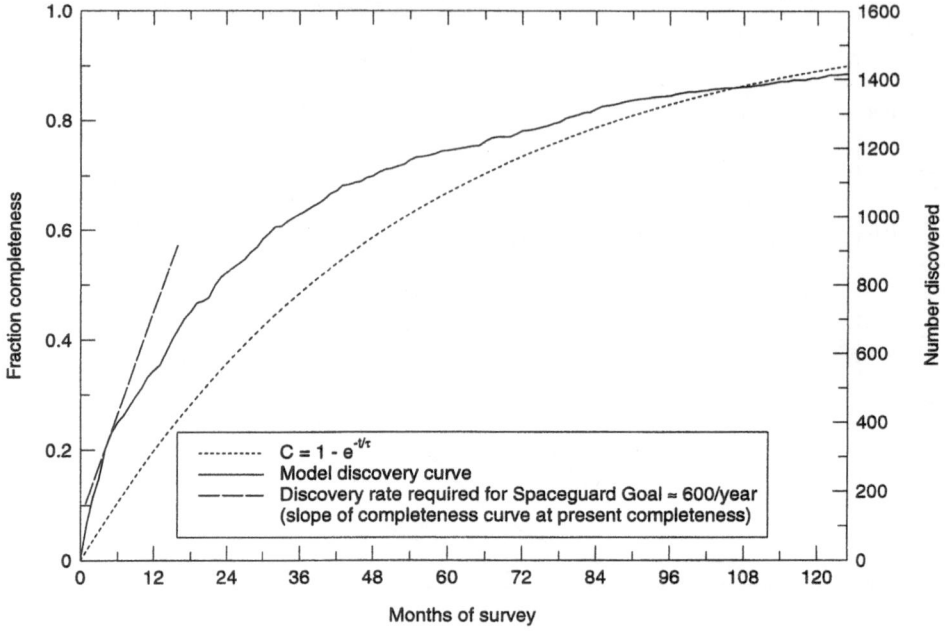

Figure 4.   Completeness vs. time for a survey to 90% completeness in ten years.

marked corresponds to 90% completeness of asteroids $D = 1$ km ($H = 18.0$). Because the population index (slope in Fig. 2) is so steep, there is not much difference between integral vs. differential completeness, but it does amount to about 0.3 in survey limiting magnitude (or H value at a given level of completeness). Thus for an integral completeness of 90% for $H < 18.0$, the required limiting magnitude is closer to 20.5 than to 21. Thus the *Spaceguard Goal*, translated into survey requirements, is to cover the entire visible sky to magnitude 20.5 each month.

I have also investigated the completeness vs. time of discoveries to estimate the rate of discoveries needed to achieve the *Spaceguard Goal*. Figure 4 is a plot of the modeled cumulative discoveries vs. time for the differential size/limiting magnitude at which completeness reaches ~ 90% in ten years. For comparison, an exponential curve is shown, which would be the completion curve if NEAs behaved like "particles in a box", so that the chance of discovery of a given object was random in time. Instead real NEAs follow repetitive orbits, so some are easier than average, and others harder, to discover. Thus the real curve rises faster than exponential at first as the easy objects are discovered, but slower near the end since the last few asteroids are much harder than average to find (for example, those in

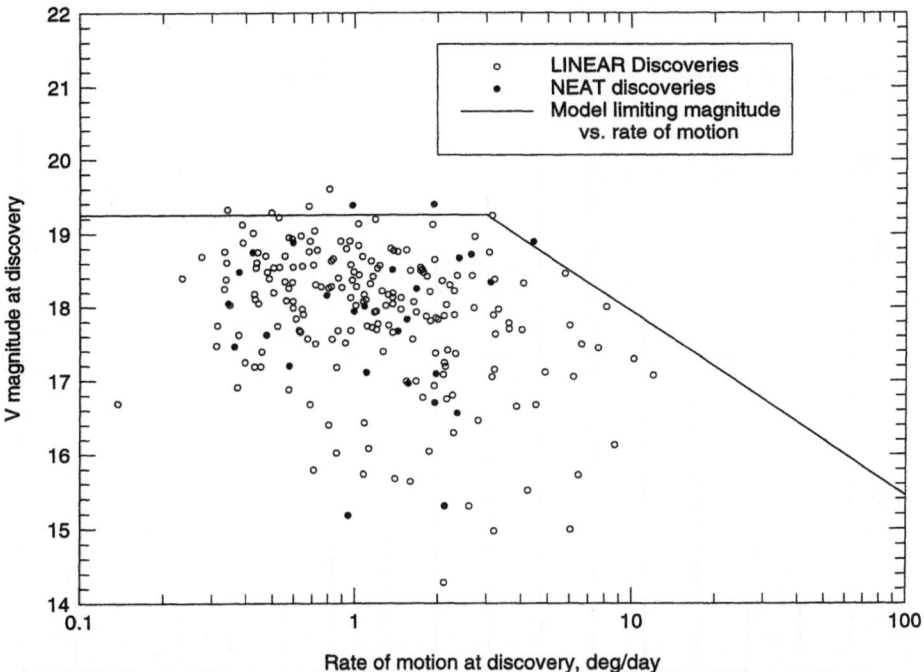

Figure 5. NEA discoveries from GEODSS-based systems through April 30, 1999.

orbits with periods nearly commensurate with the Earth's may not come into good apparitions very often). As with Fig. 3, the model curve applies to the differential completeness (single size objects) for which completeness in ten years equals 90%, not the integral completeness for all sizes, but as before, the integral curve should not be much different.

4. Current Status of Surveys

I have been tabulating the discoveries made by all survey systems since the beginning of the GEODSS-based surveys in 1996. As the systems have improved and the operators have gained experience, the rate of discoveries has steadily increased. I noticed early on that the reported discovery magnitudes were often unreliable. Thus I have used a data file that includes magnitude measurements from all subsequent observations to obtain absolute magnitudes (Bowell, ftp: //ftp.lowell.edu/pub/elgb/astorb.html), and from that back-compute an estimated sky magnitude at the time of discovery. In Fig. 5, I have plotted each of the discoveries made by GEODSS-based systems, sky magnitude vs. rate of motion. The reason for the rate of motion coordinate is that beyond a rate of $\sim 3°/$day, images start to trail and limiting magnitude diminishes. The broken line in the plot is a model of the

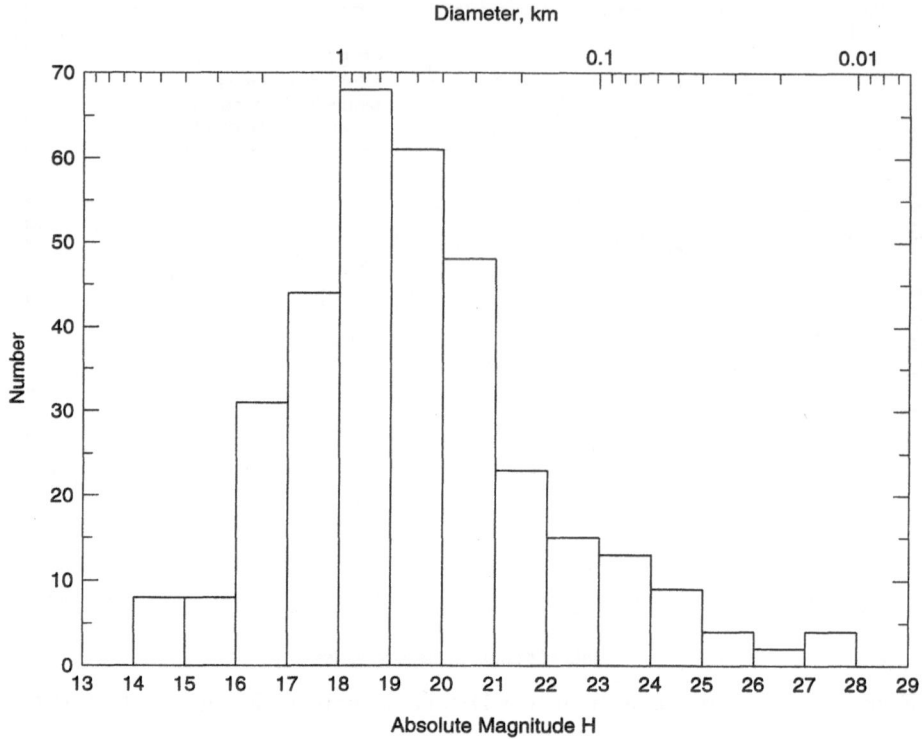

Figure 6.   Histogram of number of discovered NEAs vs. size (H) from September 15, 1996 to April 30, 1999.

expected limiting magnitude, adjusted to the level for slow-moving objects. It is clear from this plot that the current GEODSS-based systems (both LINEAR and NEAT) have a limiting magnitude near 19.0. Even though a few detections were made at fainter magnitudes, one can tell from the fall-off in numbers even somewhat brighter than 19.0 that detection efficiency is already down by a factor of about two by 19.0.

It is interesting also to note the statistics of the numbers of discoveries as a function of absolute magnitude (size) of asteroid. Figure 6 is a histogram of this distribution. Note that most of the discoveries are of objects smaller than 1 km ($H > 18.0$). At the very largest sizes, the paucity of objects is in part due to prior completeness. This plot includes discoveries from all programs, although down to $H \approx 22$ it is dominated by the GEODSS-based systems. Below that, the Spacewatch Camera is responsible for a substantial fraction of the discoveries.

In order to compare the current rate of discoveries *vis a vis* the *Space-guard Goal*, we consider only the first four bins of the histogram, $H < 18.0$,

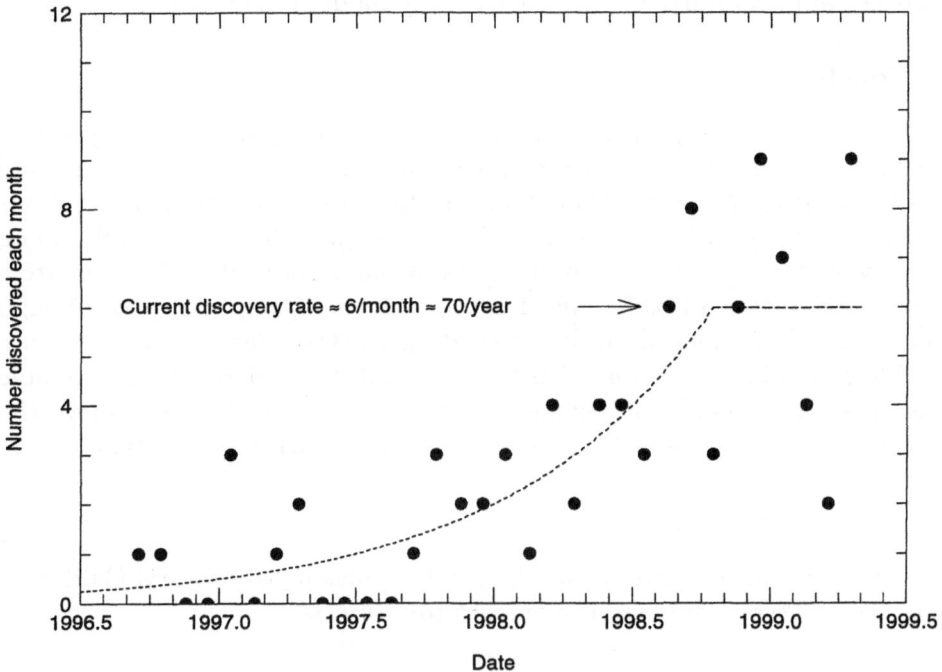

Figure 7.   Rate of discoveries of NEAs, $H < 18.0$.

which is a sample of 91 NEAs. Figure 7 is a plot of the number of discoveries vs. date for this sample. The dashed line is a rough fit to the numbers, showing an encouraging doubling of rate each six months, up through about the third quarter of 1998, after which the rate has been more or less constant at about 6/month: 37 in the last six months, or ~ 74/year. This leveling off can be understood in that the survey systems operating now are covering about half of the available sky each month, so in order to significantly further increase the rate of discovery it will be necessary to reach a fainter limiting magnitude. The rate of discovery required to achieve the *Spaceguard Goal* can be estimated from the slope of the line in Fig. 4, drawn tangent to the model discovery curve at about the present level of completeness, $\sim 18\%$. The slope at that point corresponds to a rate of ~ 600/year, or ~ 50/month. Thus the present surveys are falling about a factor of eight short of achieving the *Spaceguard Goal*. The present limiting magnitude of 19.0 is about a factor of 4 brighter in intensity units than the predicted requirement of limiting magnitude 20.5. The model rate of discovery is approximately proportional to limiting brightness in intensity units, so this accounts for about a factor of 4. The other factor of two is roughly the shortfall in current sky coverage, thus the present rate of discovery appears fairly consistent with the model prediction.

5. Conclusion

Current all-sky surveys are yielding discoveries of NEAs at an impressive rate, although still a factor of ~8 under that required to complete a survey of 90% of all NEAs larger than 1 km in diameter (estimated to be about 1600 in number), in ten years. The 1 m telescopes which are mainly doing the survey at present are reaching to about magnitude 19.0. We estimate a limiting magnitude of 20.5 is needed to reach the *Spaceguard Goal*. A high priority should be placed on demonstrating whether these instruments are capable at all of reaching magnitude 20.5. Until that is done, it is impossible to estimate how many such camera systems will be needed to achieve the *Spaceguard Goal*, or indeed if it is possible at all with these systems.

6. Acknowledgment

This research was carried out at the Jet Propulsion Laboratory, California Institute of Technology, under contract from NASA.

References

Bottke, W.F., Jedicke, R., Morbidelli, A., Petit, J.–M., Gladman B. (2000) Understanding the distribution of near-Earth asteroids, *Science* **288**, 2190-2194.

Dohnanyi, J.S. (1969) Collisional model of asteroids and their debris, *J. Geophys. Res.* **74**, 2531-2554.

Harris, A.W. (1998) Evaluation of ground-based optical surveys for near-Earth asteroids, *Planet. Space Sci.* **46**, 283-290.

Rabinowitz, D., Bowell, E., Shoemaker, E. and Muinonen, K. (1994) The population of Earth-crossing asteroids, in *Hazards due to Comets and Asteroids* (T. Gehrels, ed.), U. Arizona Press, Tucson, pp. 285-312.

Rabinowitz, D., Helin, E., Lawrence, K., Pravdo, S. (2000) A reduced estimate of the number of kilometre-sized near-EArth asteroids, *Nature* **403**, 165-166.

NEO, THE SPACEGUARD SYSTEM AND THE SPACEGUARD FOUNDATION

A. CARUSI

The Spaceguard Foundation
IAS-CNR, Area Tor Vergata, V. Fosso del Cavaliere 100,
I–00133 Roma, Italy

Abstract. The increased awareness of the threat posed by Near-Earth Objects (NEOs) to humankind has already produced a moderate enhancement of researches, from both a theoretical and an observational viewpoint, and has fostered our knowledge on the phenomena associated with impacts. However, the amount of efforts currently devoted to this important problem is far from being sufficient to significantly reduce the associated risk on a short time scale. This paper will present an overview of the initiatives that have been started in the past ten years, together with an assessment of the current situation. The set-up of The Spaceguard Foundation has provided scientists, in particular observers, with a potentially powerful tool to increase the co-ordination and efficiency of investigations.

1. Introduction

The Earth is subject to continuing bombardment by celestial objects (asteroids, comets and their fragments). In the majority of cases these events do no harm to living beings and to the terrestrial ecosystem but, in some cases, they may cause serious destructions on a local or regional scale. Even more rarely, cosmic impacts are able to trigger climatic perturbations that may lead to the disappearance of groups of species. It is now acknowledged that similar events may have been the cause of dramatic episodes in the Earth's past, the proofs of which have been found only recently.

Following these discoveries, a vast movement has grown worldwide which asks for a continuing and thorough activity of discovery and tracking, with the purpose to identify the next relevant impact before it might take place. This activity is a prerequisite for the planning of possible countermeasures

M. Ya. Marov and H. Rickman (eds.), Collisional Processes in the Solar System, 333–349.

to be applied if an object on a collision course with the Earth would be identified.

It has been evaluated that the probability of a major event taking place in the next century – an event capable to seriously endanger the survival of human civilisation – is of the order of 1/10,000 (Chapman and Morrison, 1994), an estimate that shows how the threat posed by Near-Earth Objects (NEOs) is comparable to, and sometimes higher than, the hazard posed by major natural catastrophes such as eruptions, earthquakes, floodings.

NEOs are usually subdivided into five broad classes: asteroids (of Aten, Apollo and Amor types) and comets (short- and long-period, see Table 1). A description of their characteristics and of their probable physical properties, as well as of their dynamical histories, is contained elsewhere in this book (see papers by Jewitt and Fernandez, Zappalà and Cellino, and Morbidelli); here, I want only to remark that the common character that makes objects in any class dangerous is the dynamical possibility to cross the Earth's orbit; only in that case could an impact take place, provided that the object and the Earth be located in the common point at the same time. Therefore, no Amor is dangerous at the moment, because Amors practically cannot collide with Earth. (In fact, there would be a slight possibility of an impact for marginal cases with orbits exactly tangent to the Earth's orbit in its aphelion.) However, their orbital evolution over rather long periods of time (of the order of centuries or millennia) can transform them into objects of the Apollo type, which can encounter the Earth closely all along the orbit (Carusi and Dotto, 1997).

In addition, the real quantity that matters is the actual minimum distance, at a given epoch, between the osculating orbits of the Earth and the object. This parameter is usually called MOID (Minimum Orbit Intersection Distance) and is a very good indicator of the level of risk presented by a specific NEO.

In fact, the vast majority of NEOs does not represent a real threat at the moment, because their MOID values forbid a very close approach, even taking into account the Earth's gravitational pull. This situation, however, may change in the future, over timescales ranging from centuries to millions of years, as perturbations by the major planets will change smoothly the orbits of NEOs.

For the purpose of discussing the necessity to set up a monitoring system, it is therefore important to keep in mind this distinction between objects which may impact Earth *now* (say, within the next 100 years) and objects which will possibly have a chance of hitting our planet in a more or less distant future. The important point here is that we have discovered only a fraction of the objects that are thought to belong to the afore-mentioned classes, and for long-period comets there is no hope to discover and cata-

TABLE 1. Dynamical classification of NEOs

Object class	Semimajor axis (AU)	Perihelion distance (AU)	Aphelion distance (AU)
Asteroids			
Aten	$a < 1.000$	$q < 1.000$	$Q > 0.983$
Apollo	$a > 1.000$	$q < 1.017$	any
Amor	$a > 1.000$	$1.017 < q < 1.300$	any
Comets			
Long-period	$a > 35.000$	$q < 1.017$	$Q > 33.983$
Short-period	$a < 35.000$	$q < 1.017$	$Q > 0.983$

logue even a majority of them, because it is impossible to observe them, with the current technology, at large distances and because their passages through the inner regions of the Solar System are effectively randomly distributed in time.

NEOs that may pass very close to Earth (that is, with values of MOID sufficiently small, conventionally smaller than 0.05 AU) are usually called *Potentially Hazardous Objects* (PHOs). The purpose of any system devoted to reduce the risk associated to impacts should therefore be to discover as soon as possible the majority, if not the totality, of PHOs larger than a given dimension (with the exception of long-period comets) and to track their motion in the immediate future in order to ascertain if an impact is going to take place in the next 5 − 10 decades.

In the next Section I will briefly outline the development of this field in the past twenty years or so, from the point of view of the increased effort being devoted to discovery and organization of research; the process is, however, under way and we are still far from a satisfactory situation. The subsequent Section will present the status of the current situation in terms of observing centers, and the last Section will be devoted to an overview of the possible role that national agencies and international organizations may play for the successful completion of the survey. As the reader will see, I will try to demonstrate that the only practicable way to substantially reduce the risk associated to NEOs is to put into continuous operation a well co-ordinated, international system, flexible enough to react promptly to emergencies. A certain degree of competition among the teams participating in the system is inevitable and welcome; however, it will never be stressed enough that the real key to success will not be competition, but rather an extended co-operation at an international level.

2. Growing awareness

2.1. FROM THE K/T EVENT TO THE MORRISON REPORT

The possibility that impacts might produce extended devastations over the whole Earth, possibly leading to severe damages to the living species, has been suggested several times in the past. As an example, Ralph Baldwin wrote, in 1949:

> ...since the Moon has always been the companion of the Earth, the history of the former is only a paradigm of the history of the latter... [Its mirror on Earth] contains a disturbing factor. There is no assurance that these meteoritic impacts have all been restricted to the past. Indeed we have positive evidence that [sizeable] meteorites and asteroids abound in space and occasionally come close to the Earth. The explosion that formed the crater Tycho...would, anywhere on Earth, be a horrifying thing, almost inconceivable in its monstruosity. (Baldwin, 1949; citation contained in Morrison, 1992.)

However, it was not until the publication of a famous paper by L. Alvarez, W. Alvarez, Asaro and Michel (Alvarez et al., 1980) that the scientific community started to consider very seriously this possibility. As a matter of fact, the pioneering observational work conducted by E.F. Helin since the early 70's, and later by E.M. and C.S. Shoemaker, although demonstrating that the number of near-Earth asteroids was relevant, did not propagate very far beyond the boundaries of the astronomical community. On the contrary, within that community scientists have always been aware of a possible danger, although the resources available for this type of work have never been sufficient to make extended and thorough surveys.

During the 80's and early 90's, several groups and individuals have started to dedicate a substantial fraction of their observing time to the discovery of asteroids and comets in general, with particular emphasis on NEOs. It will be sufficient to mention, among others, the PCAS and PACS programs at Palomar, under the leadership of E.F. Helin and E.M. Shoemaker respectively, the AANEAS program in Australia, led by D.I. Steel, and the project pursued at the University of Arizona (Tucson) by the team led by T. Gehrels. The set-up of the *Spacewatch Telescope* at Kitt Peak, the first system of this kind equipped with state-of-the-art CCD technology, has represented a dramatic improvement in the discovery rate of NEOs and has pushed other teams and individuals, both professional and amateur, to follow this line of research. (For a comprehensive review of these programs, their *modi operandi*, and their results see Carusi et al., 1994.)

Following a specific request of the US Congress, in 1990 NASA appointed two Working Groups with the purpose to examine the NEO problem in detail. The objective of the first of these WGs (the "discovery" WG, led by D. Morrison) was to make a scientific assessment of the problem and to design a system able to identify and catalogue all NEOs. The charter of the second WG (the "interception" WG, led by J. Rahe and J. Rather) was to examine the possibility to modify the path of an incoming object in order to eliminate, or at least substantially diminish, the likelihood of an impact. Both WGs produced a report (Morrison, 1992; Rahe and Rather, 1992) with the results of their work.

The report of the "discovery" WG (that will be called the Morrison Report in the following) had a very significant title: *The Spaceguard Survey*. The final recommendation of the WG was to build a system (the Spaceguard System) able to make an extended and complete survey of the sky, in order to discover and compute the orbits of the majority of objects larger than 1 km (the ones whose impacts could have global catastrophic effects). The name Spaceguard was suggested by Duncan I. Steel and is taken from the novel *Rendezvous with Rama* by Arthur C. Clarke (1973), in which an international organization with this name is created after a devastating impact.

The proposed Spaceguard System would have been formed by six 2-meter telescopes, evenly distributed in latitude and longitude, scanning continuously the sky in search for NEOs. It was calculated that such a system would have discovered about 95% of NEOs larger than 1 km in about 20 − 25 years.

2.2. THE WGNEO, THE SHOEMAKER REPORT AND THE COUNCIL OF EUROPE

While the NASA Working Groups were at work, the problem of NEO discovery and tracking was put to the attention of the International Astronomical Union (IAU), at its XXIst General Assembly in Buenos Aires (1991). The General Assembly approved a motion, presented jointly by six Commissions, which asked for the formation of an inter-commission working group with the purpose to bring the problem into an international context. As a matter of fact, it was felt by the proponents that NEO studies were underdeveloped in many countries and that an international organization such as the IAU could have helped in supporting these activities worldwide. The *Working Group on Near-Earth Objects* (WGNEO) was composed by members of the six proposing Commissions, and chaired by the president of Commission 20. The mandate of the working group was to examine the problem, especially its international connotations, and to report back to the General Assembly at the next opportunity.

Between 1991 and 1995 the number of meetings, workshops and confer-
ences dealing with NEOs increased continuously, thus demonstrating the
growing interest in the scientific (and not only astronomical) community.
However, the number of observing programs did not follow the same trend;
on the contrary, one of the most efficient programs, AANEAS, was terminated because of lack of funding by the Australian government, notwith-
standing the protests of the WGNEO and of the international community.
The closing down of AANEAS has left the observers without a continuous
monitoring program in the southern hemisphere.

In the same period an increased awareness of the threat presented by
NEOs was growing in the public, especially in conjunction with the exceptional event of the collision with Jupiter of the fragments of comet
D/Shoemaker-Levy 9 (July 1994). It was also thanks to this event, followed live by many media in many countries, that NASA was requested
by the US Congress to form another committee, under the chairmanship
of Eugene M. Shoemaker, in 1995. The purpose of the new committee was
to devise a system able to discover 90% of the NEOs larger than 1 km
in ten years. It was a good opportunity to revise the conclusions of the
Morrison Report, also because it was evident to all scientists that a system like the one proposed there would never have been funded. The revised
scheme, contained in the Report by the Shoemaker Committee (Shoemaker
et al., 1995), was less demanding than the previous one, taking also into
account that the anticipated duration of the discovery campaign was now
much less than before. However, NASA did not support the conclusions of
the Shoemaker committee before the Congress to the level needed to get a
substantial financial support to proceed with the survey.

A new event, during 1995, was represented by the discussion of the NEO
problem by the Committee on Science and Technology of the Council of
Europe. The original motion, presented to the Committee by a group of national representatives, was extensively discussed within the WGNEO, that
decided to support the initiative. After a presentation to the Committee by
the Chairman of the WGNEO, a Resolution was drafted and then approved
by the Council (Council of Europe, 1996).

2.3. THE SPACEGUARD FOUNDATION

The Resolution of the Council of Europe mentioned the existence of an
international organization called *The Spaceguard Foundation*. The idea to
set up an international organization capable to ensure the co-ordination of
observational campaigns and of studies of NEOs was put forward several
times in the past years, mainly because it was felt that the intrinsically international nature of the problem would have required ample concurrence

from different nations. Because of this peculiar character of the astronomical research on NEOs, the WGNEO presented at the XXIInd General Assembly of the IAU in The Hague (1994) its report (Carusi *et al.*, 1996) and recommended to prolong its activity for another triennium. The purpose of this additional mandate was to examine the possibility to found an international organization that would take care of the co-ordination of all activities in the world, so that studies and initiatives dealing with NEOs be put under the auspices of an international authority.

The WGNEO then organized a Workshop in the island of Vulcano (Italy, September 1995), entitled "Beginning the Spaceguard Survey". The aim of this Workshop was to underline the need for a co-ordinated effort and to lay the foundations for an effective international cooperation on the subject. During a long and stimulating discussion on the situation, the participants in the Workshop decided to set up a "spaceguard foundation", an organization which would contribute to supporting and co-ordinating the current researches worldwide.

On the last day of the Workshop the WGNEO decided to form a restricted committee (E.M. Shoemaker, D.I. Steel and A. Carusi) to explore the possible ways to establish such an organization. In a few months it was decided that the first step would be the creation of an Italian Association, called *The Spaceguard Foundation*, with ample participation of members and consultants of the WGNEO. The Spaceguard Foundation has been officially set up in Rome, on March 26, 1996.

The Spaceguard Foundation (SGF) is a private, international organization among professional and amateur astronomers. It has three main objectives:

- to promote and co-ordinate activities for the discovery, pursuit (follow-up) and orbital calculation of the NEO at an international level;
- to promote study activities – at theoretical, observational and experimental levels – of the physical-mineralogical characteristics of the minor bodies of the Solar System, with particular attention to the NEO;
- to promote and co-ordinate a ground network (the "Spaceguard System"), backed up by possible satellite network, for the discovery observations and for astrometric and physical follow-up.

It is evident from the above objectives that the principal goal of the SGF is "co-ordination": but, why is it reputed so important? The answer is in the peculiar character of the kind of astronomical research needed to discover and, more important, track NEOs. Usually, the chain of events following a new discovery is the following:

1. the astrometric positions of the newly discovered object (on two nights) are reported to the IAU collecting center, the *Minor Planet Center*;

2. the measured positions are used to compute a preliminary orbit and to check whether the object has already been observed in the past, using the previous observations of all minor bodies contained in the MPC databases;

3. the preliminary orbit is used to compute ephemerides for the following days, in order to help observers to make further observations;

4. if the object is a NEO (and especially if it is thought to be a PHO), the MPC puts it in a special listing, the "NEO Confirmation Page": the purpose of this list is to call for further observations (follow-up observations) needed to refine the preliminary orbit;

5. only when the preliminary orbit is good enough to ensure the recovery of the object at subsequent opportunities the new object is given a permanent designation.

There are two points in the above list which deserve consideration from our point of view: i) the fast transmission of the measurements to a collecting center and ii) the organization of follow-up observations. The first necessity is very well assured by the MPC, which collects observations 24 hours a day, seven days a week, all around the year. There is a weak aspect of this item, however; the MPC cannot stop working, or a redundant center must be put into operation, able to back-up the MPC in case of necessity. There has never been an event of this kind, but this does not imply that there *will* never be such a need.

The second point reported above, the organization of follow-up observations, is more crucial and concerns the SGF directly. As a matter of fact follow-up observations, although very well made by a number of dedicated, professional or amateur observers, need a high degree of co-ordination in order to be performed efficiently. For example, it is not necessary that many centers observe the same object on the same night (except if one wants to make use of parallax to get directly the distance of the object, something that could be useful in case of very close passages); it is, instead, more important to dilute the observations over a time span as long as possible, in order to lengthen the observational arc, and therefore improve the reliability of the computed orbit. In other words, it is better to have fewer observations, and even less precise, over an extended period of time, instead of many very precise observations at the same epoch. But to obtain this effect somebody should ask, or at least suggest, to the observers to make their observations at specific times, possibly making use of the knowledge of their observing capabilities, of the anticipated weather conditions, and of the geographical position of the observatories.

This first level co-ordination is exactly what the SGF plans to offer to the community with its *Spaceguard Central Node* (SCN), which is being set up in Rome with the financial support of the European Space Agency.

The support by ESA, in its turn, is at least partially a consequence of the invitation made by the Council of Europe with its Resolution 1080.

Higher level co-ordination is being studied by the SGF at this moment. It involves a specific agreement with the MPC on the share of specific responsibilities, the set-up of new observatories in locations where there are none or only a few, an inquiry on the content and location of image archives that may reveal a "gold mine" for past, undetected observations of newly discovered objects (usually called prediscoveries, or "precoveries"). This last possibility is of great interest, as has been recently demonstrated by the precovery of the NEO 1997 XF_{11} in films taken at Palomar in 1990 by the teams of E.F. Helin and of E.M. and C.S. Shoemaker: these precoveries have transformed a few-months arc into an 8-years arc, allowing a far better determination of the orbit of this rather intriguing PHO.

However, co-ordination of centers assumes that there are centers making observations. It is therefore useful now to overview what is the current situation in terms of observing centers, that is: what is the current status of the *Spaceguard System*?

3. Status of the *Spaceguard System*

3.1. OBSERVATIONAL CENTERS

An inquiry has been made by the SGF in connection with the establishment of the SCN in Rome. The purpose of the inquiry was to build a database containing all potentially useful data about observing centers, whatever their nature: from the technical details about the telescope(s) and associated sensing instrumentation, to the availability of computers and network connections, to the seeing and average weather conditions of the site, the sky visibility in various directions, manpower and facilities of any kind.

It has not been easy to collect these data. Nonetheless, 70 answers were received, and general statistics of the results obtained so far are illustrated in Fig. 1, where the numbers under the histogram colums represent the upper limit of the corresponding bin. (Note that the total number of items in the figure is sometimes larger than 70, because in some observatories there are more than one instrument.) In Fig. 1a the aperture of the instrument used for NEO observations is given, in cm. It is evident that the majority of telescopes used to this purpose is of very small aperture, smaller than 40 cm. This result should not be terribly biased by the moderate size of the sample, and reflects the fact that medium- to large-size telescopes are not available for this kind of research, mainly because of the little time allocated, for planetary observations in general, at the larger telescopes.

Even by using small instruments, however, the possibility to carry out a satisfactory activity is shown in Fig. 1b, where the distribution of limiting

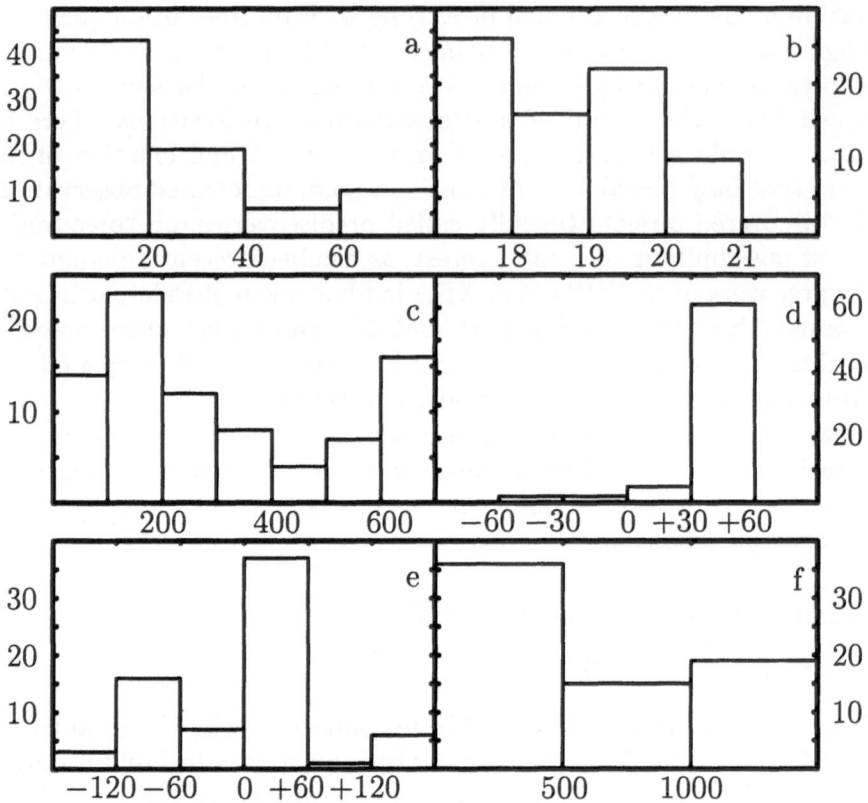

Figure 1. Various distributions related to the observational centres. **a**: telescope apertures (cm); **b**: Limiting visual magnitudes; **c**: Fields of view (square arcmin); **d**: Latitudes
of the sites (deg); **e**: Longitudes of the sites (deg); **f**: Altitudes above sea level (m). The
horizontal scales refer to the upper limit of the corresponding bin, while the vertical
scales indicate the number of centres in each bin.

magnitude M_{lim} among the available telescopes is reported. All the instruments (with only two exceptions) are coupled to CCD arrays, and this is
the reason why M_{lim} is never smaller than 17 (with one exception). A M_{lim}
of $18 - 19$ is sufficient for most cases of follow-up of larger objects, while
for discovery and tracking of smaller ones it would be necessary to reach
at least $M_{lim} = 21$. Again, the distribution in Fig. 1b seems to be quite
representative of the global situation.

Coming to the field of view (FOV) of the instruments (Fig. 1c), we see
that the distribution of this quantity is markedly bimodal. The majority of
sensors have a very small FOV, of the order of $100 - 200$ square arcmin,
due to the small size of the used CCD chips. A few instruments, among the
ones for which we got specifications, use sizeable CCDs, allowing FOVs of

the order of several square degrees. The upper limit of the last bin in Fig. 1c has been put to 100,000 in order to take into account the large FOVs obtained using photographic plates. Once again, small FOVs are sufficient for many cases of follow-up, when the orbit of the object is rather well known (in particular close to the discovery epoch). However, they are not very suitable for extended searches, either for new objects or intended to recover lost objects.

The geographical distribution of observatories, as previously mentioned, is very non-homogeneous (see Figs. 1d,e). Only few answers have been received from southern observatories and the large majority of the others are located (in the North) at mid-latitudes, reflecting the abundance of European, US and Japanese observatories in our sample. When a more complete dataset will be available we should have a less sharp distribution, while the situation for the South should be essentially the same.

The distribution in longitude (Fig. 1e) is again quite non-homogeneous; the three peaks correspond to the US, to Europe and to Japan. The first and last of these peaks should increase when more data will be available.

The last distribution (Fig. 1f) refers to the altitude above sea level. Only 13 centres are located at more than 1000 meters, and only 5 above 2000. The small average elevation makes observations less efficient, basically because of air and light pollution, and must be related to the limiting magnitude. A more detailed analysis on this point, essential for a good co-ordination of observations, will be possible when the sample will be complete.

3.2. ARCHIVING AND DATA COLLECTION

Once taken, data must be archived in some manner. This issue is not of minor importance for a number of reasons.

- The first necessity (*i.e.*, to have good astrometric measurements of the positions of objects on the sky at a given epoch) might be satisfied by the originators of data; however, the data reduction methods and – more important – the reference star catalogues, change and improve with time. It is therefore important to store in some way the original image, or at least the photometric centers in an image, for possible subsequent re-processing.
- Astronomical images constitute a "treasure chest" for many astronomical investigations (for example the search of optical counterparts of γ-ray bursts; Costa, 1999, personal communication). It is wise to keep memory of what has been observed, as it has been done since the inception of astronomical photography. A project like the Spaceguard Survey, aimed at scanning continuously the sky at limiting magnitude above 21 may represent an invaluable source of data for many other

astronomical and astrophysical researches, and every effort should be made in order to ensure a secure storage of all data.

- As mentioned before, even past images certainly contain useful information on NEOs that still need to be extracted from the archives. From this viewpoint it would be of extreme importance that not a single bit of an image be lost.

It is necessary to underline that storage of astronomical images represents a difficult task. As a matter of fact, a continuous scanning of the sky, even if performed with instruments with a limited field of view, produce an enormous amount of data. Furthermore, these data, to be really useful, must be accessible in a short time and readable by everybody.

There are two orders of difficulties related to the on-line availability of astronomical data for NEO research. The first is represented by the ownership of such data, which belongs to the observers; some of the teams have no objection to put their data immediately at disposal of a larger community, but in other cases the time lapse between the collection of an image and its availabilty for consultation may become large, and in other cases the originators of the data would not make their images available anyway. Obviously it is not fair, nor correct, to ask that data resulting from searches that have required a lot of efforts be freely distributed but, on the other hand, it has already been demonstrated that in critical cases it would be of extreme importance to find a newly discovered object on a previous image in a short time. As will be suggested in the next Section, a possible – or at least partial – solution to this problem is represented by international collaboration.

The second order of problems related to data storage is represented by their nature. Not only CCD images may differ in size and coding technique, but also a great portion of past image archives is not yet in a digital format. While some of the most important archives are being digitized, for the majority of smaller scale surveys this task is very heavy, and perhaps impossible for the individual centers. Nevertheless, it is important to remark again that the availability of images taken in the past 60 − 70 years would be extremely beneficial for NEO researches. In addition, sometimes one does not even know what are the existing archives and where they are located. In other words, an "archive of archives" is missing.

3.3. DO WE MEET THE REQUIREMENTS?

The Spaceguard Survey, as depicted in the original Morrison Report, was intended to discover 95% of NEOs larger than 1 km in 25 years. The requirement contained in the charter of the Shoemaker Committee was to discover 90% of such objects in ten years. At seven years of distance from the Morrison Report, and four from the Shoemaker Committee, it is therefore useful to see if we are moving in the right direction or, at least, if we are approaching a satisfactory discovery rate.

Figure 2 reports the discovery rate of Near-Earth Asteroids, of the Aten, Apollo and Amor classes, in time. Comets are not included in the picture. 651 NEAs have been discovered up to the end of 1998, 204 of which ($\sim$ 31%) in the last year; in comparison, the number of objects discovered before 1971 is 28. This trend will probably continue in the near future, as new discovery programs will become effective; it is probably possible to state that we are not terribly far from the requirements, in terms of discovery.

It may be interesting to further analyze the composition of the sample of NEOs at our disposal. First, it is composed by 45 Atens, 309 Apollos, and 297 Amors. We don't know for certainty their dimensions and can give only a rough estimate based on the values of the absolute visual magnitude. Taking $H = 18$ as the value of this quantity indicating dimensions of the order of 1 km (something really questionable for a number of reasons), we see that the percentage of discovered objects larger than 1 km is the following:

- **Aten:** 10/45 (=22%)
- **Apollo:** 132/309 (=43%)
- **Amor:** 140/297 (=47%)

The different percentages of larger objects in the three classes of asteroids may be indicative of an observational bias; as a matter of fact, many NEAs are discovered at opposition, at low phase angles and therefore under favourable viewing conditions. Atens, whose aphelion distance is not so far from the Earth's orbit as for the other objects, tend to be closer to our planet when discovered, thus allowing the identification of smaller objects (for equal magnitude limit at discovery). On the contrary, both Apollos and Amors, when first observed, may be located at larger distances; it is therefore understandable that a larger fraction of larger objects has been detected for these two classes.

However, it has to be noted that, if we examine only the Apollos and Amors which were closer to Earth at discovery, the percentage of absolutely brighter objects diminishes – as one should expect – but does not match that of Atens, anyway. All 45 Atens have been discovered at a distance from Earth less than 0.688 AU; there are 244 (out of 297) Amors discovered within that distance, and 255 (out of 309) Apollos. The previous percentages, with these limitations, amount to 39% for Amors and 27% for Apollos. Other selection effects could therefore be at work, possibly related to the different dynamical lifetime of objects of different dimensions. (83% of Amors and 89% of Apollos discovered at larger distances have $H \leq 18$.)

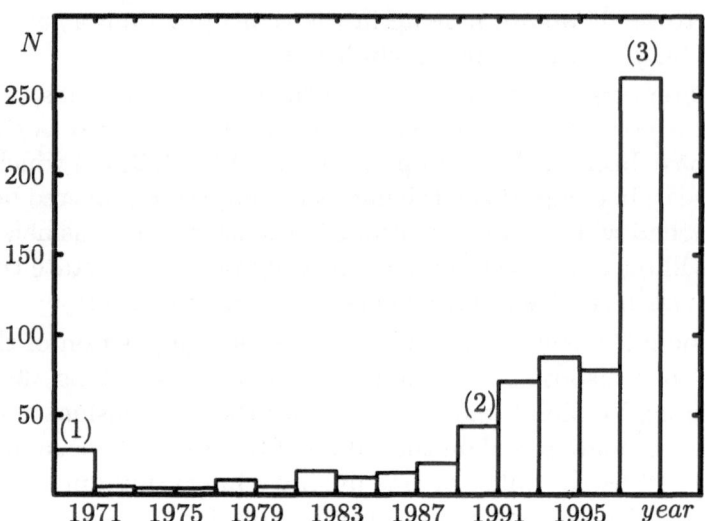

Figure 2. Discovery rate of NEAs: (1) all discoveries before 1971, (2) Spacewatch starts operation in 1989, (3) LINEAR and LONEOS start operations in 1997-1998

This brief, and certainly incomplete, analysis of the composition of the sample of discovered NEAs was intended to underline that it may be possible to optimize the search for new objects, taking into account the discovery capabilities of the observing centers. As an example, Spacewatch tends to discover fainter objects, while LINEAR and NEAT in general find the brighter ones. A co-ordination of the discovery programs would therefore be desirable, although not fundamental: it would principally shorten the time needed to meet the requirements of the Spaceguard Survey.

On the contrary, I think that a co-ordination of follow-up observations is really important, and it is lacking at the moment.

As previously mentioned, there are at least two types of searches that can be done in this context: the "real" follow-up, consisting in repeated observations of known objects and aiming at an improved knowledge of their orbits, and the archival searches intended to identify precoveries. In both cases a better co-ordination would be extremely helpful, and could be obtained by providing support to the interested people and teams. This support could include:

— suggestions about the best search strategies for recovering lost objects, taking into account the characteristics of the used instruments;
— suggestions for specific observations of specific objects, with the purpose to maximize the effectiveness of astrometric determinations;

- provision of prioritized lists of objects in need of further observations, including an analysis of the "best places" and "best times" to observe them;
- computation of past opportunities in which the objects could have been visible, with an inquiry about the existence, in some archive, of plates or frames containing the related portion of the sky at the right epoch.

Several centers in various countries are now designing services of this type; their contribution, coupled with the activity of the Minor Planet Center and of the future Spaceguard Central Node, should help improving the efficiency of follow-up to the same level as the discovery programs. It is clear that the two kinds of activities must proceed at the same rate, or many discoveries will simply become useless.

4. An international problem

It has been remarked several times that impacts do not pay attention to political frontiers, and therefore that the impact threat is intrinsically international. It has also been underlined that the very nature of a system devoted to discovery and monitoring of NEOs will require the concurrence of centers spread all over the surface of the planet, the existence of international data-collecting centers, and possibly the contribution of at least one co-ordinating center. The problem now is: What kind of international organization is best suited to manage such a system? Does such an organization already exist?

4.1. THE ROLE OF SPACE AGENCIES

Some space agencies are contributing to the development of the Spaceguard System. However, one could ask why space agencies should take a leading role in support of ground-based observations; as a matter of fact, this role would better suit other scientific organizations, like national or international academies, or international astronomical centers like ESO. A possible answer to this question is that the two political bodies which have so far initiated an overview of this problem (the US Congress and the Council of Europe) have directly involved NASA and ESA instead of, for example, NSF and ESO.

Nonetheless, the contribution of space agencies to NEO researches may be very important, for several reasons. First of all, space agencies already take care of most of the space-based monitoring systems devoted to Earth observations and protection from natural catastrophes. In the long run, it seems natural to me that also this topic, which certainly deals with the protection of the Earth's ecosystem, should be included in the same category. In addition, there are studies regarding NEOs that cannot be done from the ground, or are better conducted from space. These studies include, for example, the physical and mineralogical characterization of objects, which typically require observations in the infrared domain of the spectrum. Such observations are difficult from the ground and require very powerful telescopes, not easily available for these researches. Furthermore, even the discovery of specific classes of objects, such as Atens or objects with aphelion just inside the Earth's orbit, may be greatly improved using space-borne instrumentation that can image the sky at very low elongations from the Sun.

But the most important contribution of space agencies to the Space-guard System may be of a different nature. One of the problems of a ground system is represented by the necessity to set up a complex network involving many centers and working continuously for some decades. Although projects of this nature are not uncommon in science, it is typical of space agencies to start and manage programs with the breath of a decade or two: These organizations have the structure, the manpower and the programmatic ability to pursue investigations like the ones needed here. Moreover, a high-level degree of co-ordination among space agencies has already been put in place for big programs, like the exploration of Mars or the building of the International Space Station. It would therefore be desirable and possible that a pool of space agencies might work together also for a program like Spaceguard.

4.2. INTERNATIONAL ORGANIZATIONS AT WORK

Even in this case, however, I think that a more "political" framework would be necessary. The impact problem has implications that go well beyond the limits of scientific and technical work; they touch fields like civil protection, international treaties, information to the public, not to speak about the possibility, although remote, that a defense system might become necessary sooner or later. It is out of question that a purely scientific organization may have the know-how to deal with all these aspects of the problem.

The Spaceguard Foundation was set up by a group of scientists as a possible answer to the more urgent needs of international co-ordination of researches and of support to the emplacement of observing projects, not for running continuously such a complex system. As a matter of fact, as soon as the Spaceguard System will be better defined in its components, and the search and follow-up activities will become more a routine than a pioneering work, The Spaceguard Foundation should dissolve, to leave place to an organization able to operate the System continuously. In other words, international organizations like the SGF and the IAU may help in getting the work organized in a scientific and technically correct way, but do not match the requirements of an operational agency.

It is my opinion that ultimately an international treaty on the matter should be prepared, like the ones already in place for the exploitation of the Moon and for the researches in Antarctica. Such a treaty could provide a framework for the entire program, defining roles and competences at an international level and, at the same time, assign the duty of running the Spaceguard System to a new international agency, or to a pool of existing ones.

References

Alvarez L.W., Alvarez, W., Asaro, F. and Michel, H.V. (1980) Extraterrestrial cause for the Cretaceous-Tertiary extinction, *Science* **208**, 1095–1108.

Baldwin, R.B. (1949) *The Face of the Moon.* University of Chicago Press, Chicago.

Carusi, A. and Dotto, E. (1996) Close Encounters of Minor Bodies with the Earth, *Icarus* **124**, 392–398.

Carusi, A., Gehrels, T., Helin, E.F., Marsden, B.G., Russell, K.R., Shoemaker, C.S., Shoemaker, E.M., Steel, D.I. (1994) Near-Earth Objects: Present Search Programs, in *Hazards due to Comets and Asteroids* (T. Gehrels ed.), University of Arizona Press, Tucson, pp. 127–148.

Carusi, A., Yeomans, D.K., Bowell, E. (1996) Commission 20: Positions and Motions of Minor Planets, Comets and Satellites, *Transactions of the IAU* (I. Appenzeller ed.) **Vol. XXIIB**, pp. 163–169.

Chapman, C.R., and Morrison, D. (1994) Impacts on the Earth by asteroids and comets: assessing the hazard, *Nature* **367**, 33–40.

Clarke, A.C. (1973) *Rendezvous with Rama.* Ballantine Books, New York.

Council of Europe, Parliamentary Assembly (1996) *Resolution 1080 (1996) on the detection of asteroids and comets potentially dangerous to humankind*, Strasbourg.

Morrison D. (1992) The Spaceguard Survey, *Report of the NASA International Near-Earth-Object Detection Workshop*, NASA Office of Space Science and Application, Washington.

Rather, J.D.G., and Rahe, J.H. (1992) *Report of the NASA Near-Earth-Object Interception Workshop*, NASA Office of Space Science and Application, Washington.

Shoemaker, E.M., Boyarchuk, A.A., Canavan, G., Carusi, A., Coradini, M., Darrah, J., Harris, A.J., Morrison, D., Mumma, M.J., Rabinowitz, D.L., Rikhova, R. (1995) *Report of the Near-Earth Objects Survey Working Group*, NASA Office of Space Science and Application, Washington.

Index